P9-CFE-203

Right Prism

h = height; P = perimeter of base;
B = area of base
Lateral area: $LA = Ph$
Surface area: $SA = 2B + LA$
Volume: $V = Bh$

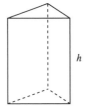

Rectangular Solid (Prism)

ℓ, w, h = edges
Surface area: $SA = 2\ell h + 2wh + 2\ell w$
Volume: $V = \ell wh$
If $\ell = w = h$, the solid is a *cube*.

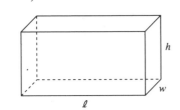

Cube

e = edge
Surface area: $SA = 6e^2$
Volume: $V = e^3$

Right Circular Cylinder

r = radius; h = height
Surface area: $SA = 2\pi rh + 2\pi r^2$
Volume: $V = \pi r^2 h$

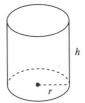

Sphere

r = radius
Surface area: $SA = 4\pi r^2$
Volume: $V = \dfrac{4}{3}\pi r^3$

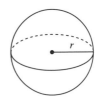

Regular Pyramid

h = height; p = perimeter of base;
ℓ = slant height; B = area of base

Lateral area: $LA = \dfrac{1}{2}p\ell$

Surface area: $SA = B + \dfrac{1}{2}p\ell$

Volume: $V = \dfrac{1}{3}Bh$

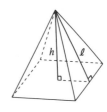

Right Circular Cone

r = radius; h = height
Lateral area: $LA = \pi r\sqrt{r^2 + h^2}$
Surface area: $SA = \pi r\sqrt{r^2 + h^2} + \pi r^2$
Volume: $V = \dfrac{1}{3}\pi r^2 h$

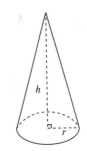

ESSENTIALS
OF GEOMETRY
for College Students

ESSENTIALS
OF GEOMETRY
for College Students

MARGARET L. LIAL
American River College

ARNOLD R. STEFFENSEN
Northern Arizona University

L. MURPHY JOHNSON
Northern Arizona University

HarperCollins*Publishers*

FOR THE STUDENT

If you need further help as you study geometry, you may want to get a copy of the *Student Solutions Manual* that accompanies this textbook from your college bookstore. This manual provides step-by-step solutions to the odd-numbered section exercises and to all review exercises in the text.

Unless otherwise acknowledged, all photographs are the property of Scott, Foresman and Company.

Cover photo: David Holt

Interior art: *p. 21,* NASA; *p. 31,* © Don & Pat Valenti 1989; *p. 38,* Camerique/H. Armstrong Roberts; *p. 49,* Scala/ Art Resource, New York; *p. 54,* © Don & Pat Valenti 1989; *p. 61,* © Don & Pat Valenti 1987; *p. 65,* © Craig Aurness/West Light; *p. 72,* Historical Pictures Service, Chicago; *p. 78,* Copyright © 1967 David Plowden; *p. 108,* Courtesy United Airlines; *p. 119,* Taken from "Rhythm/Color: Morris Men," an art quilt by Michael James, Somerset Village, Mass. Photo courtesy of The Quilt Digest Press, San Francisco. Photo by Sharon Risedorph; *p. 123,* Grant Heilman Photography, Inc.; *p. 124,* U.S. Air Force photo by MSgt Ken Hammond, released by the Department of Defense; *p. 135,* Courtesy United Airlines; *p. 142,* Camerique/ H. Armstrong Roberts; *p. 153,* © Dave Black—Sportschrome East/West; *p. 170,* © Everett C. Johnson; *p. 175,* © Don & Pat Valenti 1988; *p. 178,* H. Armstrong Roberts; *p. 180,* © Liane Enkelis; *p. 186,* © Everett C. Johnson; *p. 192,* Courtesy of the University Library, Gottingen, West Germany; *p. 201,* Syd Greenberg/FPG; *p. 211,* Camerique/H. Armstrong Roberts; *p. 219,* North Wind Picture Archives; *p. 220,* © Chicago Tribune, 1989, World Rights Reserved; *p. 221,* Syd Greenberg/FPG; *p. 230,* NASA; *p. 234,* U.S. Department of Agriculture; *p. 254,* NASA; *p. 265,* North Wind Picture Archives; *p. 286,* © Chuck O'Rear/West Light; *p. 292,* © Chuck O'Rear/West Light; *p. 306,* © Image Finders Photo Agency/West Light; *p. 316,* © Larry Lee/West Light; *p. 319,* University Library, Gottingen, West Germany; *p. 320,* © Don & Pat Valenti 1989; *p. 325,* NASA; *p. 326,* © Image Finders Photo Agency/West Light; *p. 327,* Camerique/H. Armstrong Roberts; *p. 334,* © Mary E. Goljenboom/Ferret; *p. 343,* © Mary E. Goljenboom/Ferret; *p. 350,* Copyright 1989/Comstock; *p. 351,* Camerique/H. Armstrong Roberts; *p. 352,* Photographie Giraudon; *p. 358,* Superstock; *p. 359,* © 1988 David Loxey/Comstock; *p. 360,* Copyright 1989/Comstock; *p. 367,* National Oceanic & Atmospheric Administration/National Severe Storms Laboratory; *p. 376,* FPG; *p. 379,* © Brian Drake—Sportschrome East/West; *p. 385,* from PORTRAITS AND LIVES OF ILLUSTRIOUS MEN by Andre Thevet, Keruert et Chaudiere, Paris, 1584. Courtesy of Burndy Library, Norwalk, CT; *p. 386,* FPG.

Library of Congress Cataloging-in-Publication Data

Lial, Margaret L.
  Essentials of geometry for college students / Margaret L. Lial,
Arnold R. Steffensen, L. Murphy Johnson.
    p.   cm.
  Includes index.
  1. Geometry.   I. Steffensen, Arnold R.   II. Johnson,
L. Murphy (Lee Murphy).   III. Title.
  QA453.L5   1990   89-10741
  516.2—dc20   CIP

ISBN 0-673-38419-5

Copyright © 1990 by Margaret L. Lial, Arnold R. Steffensen, and
L. Murphy Johnson.

Artwork, illustrations, and other materials supplied by the
publisher. Copyright © 1990 HarperCollins*Publishers*

All Rights Reserved.

Printed in the United States of America.

 2 3 4 5 6-WAK-94 93 92 91

PREFACE

Essentials of Geometry for College Students is an applications-oriented text designed to provide today's students with the sound geometric foundation necessary to pursue further courses in college mathematics. Some exposure to beginning algebra is the only prerequisite. Informal yet carefully worded explanations, motivating marginal information, pedagogical second color, ample examples and exercises, and comprehensive chapter reviews are hallmarks of the book. The text has been written for maximum flexibility of instructor use. Both core and peripheral topics can be selected to fit individual course needs. A test manual, an answer and solutions manual, and overhead transparencies are provided for the instructor. A solutions manual is also available for student purchase.

Key Features

Proofs and Constructions   While including the necessary degree of rigor, proofs and constructions have been incorporated in a manner that allows instructors flexibility in choosing those most appropriate for their particular course. Some statements normally proved as theorems have been postulated to avoid slowing the presentation and to keep students from losing sight of the essential application of the material. Most proofs are presented using the statement-reason format.

Applications   To demonstrate the usefulness and practicality of geometry, applications are integrated throughout the examples and exercises in the text. Each chapter introduction also features a relevant applied problem that is then solved later in the chapter.

Pedagogical Use of Second Color   Second color not found in many college geometry texts is used pedagogically throughout the book to highlight important information. Definitions, postulates, theorems, and corollaries are set off in colored boxes for increased emphasis. Text figures and constructions utilize color to clarify concepts presented. Examples present important steps and helpful side comments in color.

Examples About 175 carefully selected examples include detailed, step-by-step solutions and side annotations in color.

Exercises As a key feature of the text, over 2000 exercises, including over 500 review exercises, are provided. Each section ends with a comprehensive set of exercises ranging from routine to more challenging problems, including proofs and applications. Answers to the odd-numbered problems are included at the back of the book.

Practice Exercises Throughout the section discussions practice exercises with their answers are to be found. This integral element keeps students involved with the presentation by allowing them to immediately check their understanding of the material.

Review Exercises To provide ample opportunities for review, a comprehensive set of **Chapter Review Exercises** and a **Practice Test** are included after each chapter. Answers to *all* of these problems appear at the back of the book.

Figures The text includes approximately 750 figures, many of which utilize second color. In particular, the chapter introductions and marginal information feature photographs, sketches of famous mathematicians, or other art to provide interest and further motivate students.

Chapter Reviews In addition to the Chapter Review Exercises and the Practice Tests, comprehensive chapter reviews include **Key Terms and Symbols** and a helpful list of **Proof Techniques,** all page referenced.

Supplements

The supplemental package available for use with *Essentials of Geometry for College Students* includes the following:

For the Instructor

The Instructor's Test Manual contains two ready-to-duplicate tests for each chapter, two final examinations, and a set of easy-to-use answer keys for grading purposes. Additional problems, organized by section, are also included for use as in-class examples or on quizzes and tests. Answers are provided for all of these problems.

The Instructor's Answer and Solutions Manual contains answers and step-by-step solutions to the even-numbered text exercises.

Five-Color Overhead Transparency Overlays that illustrate 25 step-by-step constructions, and an additional 22 two-color transparencies are provided for classroom lectures and presentations.

For the Student

The Student Solutions Manual contains complete, step-by-step solutions to the odd-numbered text exercises and to all chapter review and practice test problems.

Acknowledgments

This project was initiated during the summer of 1987. At that time, Scott, Foresman and Company sent a geometry questionnaire to schools across the country; over 100 schools completed questionnaires and sent helpful syllabi. Thanks must be given to all those who took the time to respond.

Since then, many other people have been involved in this project. We extend our sincere gratitude to the following instructors who helped us determine the organization and features of the book: Jack Hennington, Johnson County Community College; Irma Holm, Long Beach City College; Leon F. Marzillier, Los Angeles Valley College; Madeline Masterson, Lansing Community College.

We are also indebted to the following reviewers who provided countless beneficial suggestions and criticisms during the writing of the text:

Mary Kay Beavers, City College of San Francisco
Jane Edgar, Brevard Community College
Charles S. Johnson, Los Angeles Valley College
Charles S.Kolsrud, Clinton Community College
Maurice Ngo, Chabot College
Judith F. Ross, San Diego Mesa College
Dorothy Schwellenbach, Hartnell College
Wesley W. Tom, Chaffey College

We express special appreciation to Joseph Mutter and to the students at Northern Arizona University who helped class test the material and to Diana Denlinger for typing the manuscript and supplements.

Thanks must go to our editors, Bill Poole and Terry McGinnis, whose support is most appreciated. We are also indebted to many others at Scott, Foresman/Little, Brown including Nancy Siadek, Elizabeth Stout, and Ellen Pettengell.

Margaret L. Lial
Arnold R. Steffensen
L. Murphy Johnson

TO THE STUDENT

The word *geometry* means "earth measure" and stems from the Greek words *geo,* meaning earth, and *metry,* meaning measure. Geometry originated in Egypt where periodic flooding of the Nile River made it necessary to resurvey the flood plains. As time passed, "earth measure" took on a broader meaning and involved measuring many things related to the earth. Eventually it came to be realized that geometry is more than just a collection of facts used for measurement; it is a system in which these facts are related in a precise and logical manner.

There are many answers to the question "Why study geometry?" Unfortunately, some students will reply "Because it's a required course." If we take this one step further and ask "Why is geometry required?" we discover the real answers to the original question.

First, *we use geometry as a tool in many areas of mathematics* as well as in practical situations. For example, you may have already used some of the simple geometry formulas such as those for finding areas, perimeters, and volumes.

A second reason for studying geometry is *to gain knowledge of symmetry and proportion* that will enable us to appreciate beauty in manmade art and architecture as well as in nature. For example, the lasting beauty of a building such as the Parthenon on the Acropolis of Athens results from the notion of the *golden ratio.* (See Chapter 5.) The golden ratio, which was known by the early Greeks, gives the most visually appealing dimensions for a rectangle. Or what could be more impressive geometrically than the honeycomb of a bee? (See Chapter 4.)

Perhaps a third reason for studying geometry is that *geometric figures are used in most areas of mathematics to help us "picture" complex concepts.* For example, in beginning algebra you were encouraged to draw sketches showing the pertinent information given in many "word problems."

On another less important but interesting level, a fourth reason for knowing geometry involves *solving puzzles.* Most of us are intrigued by simple puzzles and enjoy the challenge of trying to solve them. Actually the thought processes we follow when attempting to find a solution are more valuable than the solution itself. This fact will be illustrated with some classic puzzles in the exercises following Section 1.1.

The fifth and perhaps most important reason for studying geometry involves *recognizing geometry as an axiomatic system*. Every branch of mathematics is an axiomatic system, but studying geometry gives us one of the best ways to learn more about these systems and the thought processes used in them. In fact, we will start our study of geometry with logical systems in Section 1.1.

We hope that you will see the value of geometry more clearly as you proceed through the course. To help you, we begin each chapter with an example of a practical application that is solved later in the chapter. We also include a variety of other examples and exercises of a practical nature throughout the text.

Margaret L. Lial
Arnold R. Steffensen
L. Murphy Johnson

Contents

1

FOUNDATIONS OF GEOMETRY

Because the study of geometry requires an understanding of the way we think, we begin this chapter by discussing logical systems. This discussion leads us to consider some basic geometric terms such as *point, line, plane,* and *angle.*

Throughout our presentation, we emphasize ways that you can use geometry to solve applied problems. One application of the concepts presented in this chapter follows and is discussed in Example 4 of Section 1.1.

The following warranty is given with a new radio: "This radio is warranted for one year from the date of purchase against defects in materials or workmanship. During this period, any such defects will be repaired, or the radio will be replaced at the company's option without charge. This warranty is void in the case of misuse or negligence."

Assume the premises stated in this warranty and answer the following.

(a) Bill purchased a radio and when he got home and opened the box, he discovered that the case was cracked. What could Bill conclude?

(b) Beth purchased a radio and fourteen months later the digital display burned out. What could Beth conclude?

(c) Jamie purchased a radio and two months later left it outside after listening to it while sunbathing. That night it rained, and the next day the radio would not play. What could Jamie conclude?

1.1 LOGICAL SYSTEMS

To determine the preferred candidate in an upcoming student-body presidential election, a student interviews the first 50 students who enter the student-union building. He discovers that 42 people prefer Abby Bayona and 8 people prefer Shawn Herman. Based on these 50 specific observations, he concludes that Abby Bayona will win the election. Is this conclusion necessarily correct? Can you think of situations in which the results of his poll might inaccurately reflect the feelings of the total student body?

To understand the proofs used in geometry, we begin by considering the two basic ways that we reason, or think. The first can be illustrated by studying the following list of numbers:

$$4, \ 11, \ 18, \ 25.$$

What is the next number in this list? Most of us will try to discover what 4, 11, 18, and 25 have in common, and shortly realize that $4 + 7 = 11$, $11 + 7 = 18$, $18 + 7 = 25$, so it would seem that each number after 4 is obtained from the one in front of it by adding 7. As a result, we would probably conclude that the next number should be 32 because $25 + 7 = 32$. Our reasoning process involved considering several *specific* observations, and based on these, we formulated the *general* conclusion that the list will continue in the same pattern if we always add 7 to one number to obtain the next.

DEFINITION: INDUCTIVE REASONING

We use **inductive reasoning** when we reach a general conclusion based on a limited collection of specific observations.

Natural and social scientists frequently use inductive reasoning. When a laboratory experiment is performed several times with the same result, the physicist might form a general conclusion based on this experimentation. Or a sociologist might collect information from a limited number of people and attempt to draw a general conclusion about the total population. Also, when a new medicine is tested on a sample of several hundred people, the test results might lead a scientist to draw conclusions about the drug's effectiveness. In all these cases, a general conclusion is drawn from specific observations.

Does inductive reasoning always lead to the same conclusion? The answer is no, not always. If you return to our earlier example of four numbers, in which we concluded that the next number must be 32, you will see that another conclusion is also possible. The next number in the series can be 1, followed by 8,

15, 22, etc. This sequence gives the dates of the Mondays in the year 1988, starting with Monday, January 4, 1988!

As you can see, although inductive reasoning is widely used, there are no guarantees the conclusion drawn is always correct or that it is the only possible conclusion. The primary flaw with inductive reasoning is that we cannot be sure of what will generally be true based on a limited number of cases.

EXAMPLE 1 Use inductive reasoning to determine the next element in each list. Remember, there might be more than one answer.

(a) 2, 4, 8, 16, 32
We might conclude that the next number is 64 because each number after the first is twice the one before it.

(b) o, t, t, f, f, s, s, e
This one is a bit more difficult. This is a list of the first letters in the words *one, two, three, four,* etc. Thus, the next letter would be *n,* the first letter in the word *nine.* ◪

PRACTICE EXERCISE 1

Use inductive reasoning to determine the next element in each list.

(a) 1, 4, 9, 16, 25

(b) 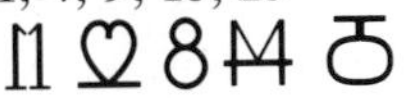

Although inductive reasoning might not always lead to the same conclusion, it is still an important process, one that is used even by mathematicians. For example, consider the following observations:

$$1 + 1 = 2, \quad 1 + 3 = 4, \quad 3 + 3 = 6, \quad 3 + 5 = 8,$$
$$5 + 1 = 6, \quad 7 + 3 = 10.$$

After examining these equations, a mathematician might conclude that the sum of two odd integers is always an even integer. Certainly these six observations do not prove that this statement is true. What if we listed several hundred such observations with the same results? Would that have proved the statement? The mathematician would say you cannot prove a general statement by giving any number of specific cases, unless, of course, the number of specific cases is limited. The mathematician usually requires a different kind of proof, which is the basis of an *axiomatic system.*

DEFINITION: AXIOMATIC SYSTEM

An **axiomatic system** consists of four parts: undefined terms, definitions, axioms or postulates, and theorems.

George Cantor (1845–1918)

German mathematician George Cantor is credited with creating a new area of mathematics, set theory, in about 1875. Some of his ideas were viewed as radical by other mathematicians of the day, and a tremendous controversy developed. Cantor was so unfairly treated that he eventually lost his mind and later died in a mental hospital. Today, however, his work with sets serves as a foundation for many areas in mathematics, and his efforts are so appreciated that he is referred to as the "Father of Set Theory."

Undefined terms are the starting points in a system. Every statement we make is composed of words that have meaning to us. It is impossible to truly define every term because definitions are also formed with words that have meaning. Some terms must be assumed in order to go forward. For example, in geometry the word *set* is an undefined term. Intuitively, we recognize a set to be a collection, or group, or bunch of objects, but this is a definition of *set* only if the terms *collection, group,* or *bunch* are also defined. We could continue to find synonyms for the word *set,* but in doing so we would not be expanding the system. Because we have an intuitive understanding of the word *set,* we simply accept it as an undefined term and use it as a building block for the system.

Definitions are statements that give meaning to new terms that will be used in a system. The words used to form a definition are either undefined terms or previously defined terms. **Postulates,** or **axioms,** are statements about undefined terms and definitions that are accepted as true without verification or proof; they also serve as starting points in a system. In geometry the word *postulate* is used most often.

Finally, when we have undefined terms, definitions, and postulates, we are ready to begin building the system by deducing results called *theorems.* A **theorem** is a statement that we can prove by using definitions, postulates, and the rules of deduction and logic. Most theorems can be expressed as *"If . . . then"* statements. The phrase that follows the word *if* includes the given information, or the **hypothesis** (plural—**hypotheses**) of the theorem. The phrase following the word *then* includes the statement we are to verify, or the **conclusion** of the theorem. We often begin with a hypothesis, and in a step-by-step manner, obtain other statements by using undefined terms, definitions, postulates, or previously proved theorems, until we reach a conclusion. This process illustrates *deductive reasoning.*

> ### DEFINITION: DEDUCTIVE REASONING
> We use **deductive reasoning** when we reach a specific conclusion based on a collection of generally accepted assumptions.

Let's look at an example of a simple axiomatic system.

EXAMPLE 2   Consider the following axiomatic system.

Undefined terms: happy, pleasant, person

Definition: Terri is a happy person.

Postulate: Every happy person is pleasant.

Using this information we can state a theorem.

Theorem: Terri is pleasant.

Although we have not yet discussed the procedures for using deductive reasoning to write a proof of a theorem, informally, we might present the following "proof."

PROOF: Because Terri is a happy person, and every happy person is pleasant, it follows that Terri, as one of the happy persons, must also be pleasant.   ◪

The next example involves an axiomatic system that is more abstract. Don't worry if the terms appear unfamiliar at first, simply concentrate on recognizing the parts of the axiomatic system.

Euclid (about 300 B.C.)

After the death of Alexander the Great (356 B.C.–323 B.C.), the city that bore his name, Alexandria, Egypt, became the center of world knowledge. Euclid, a teacher at the University of Alexandria, collected all known geometry facts into a text called *Elements*. *Elements* contained a systematic and logical arrangement of geometry and was divided into thirteen chapters called *books*. The text was unique in that it began with a few basic assumptions and logically derived everything else from them.

EXAMPLE 3   Consider the following axiomatic system.

Undefined terms: SQA, PAG, RAL, contains

Definition: A REC is a PAG that contains a RAL.

Postulate: Every SQA is a REC.

Using this given information we could state two theorems.

Theorem 1: Every SQA is a PAG.

Theorem 2: Every SQA contains a RAL.

We might present the following "proof" of Theorem 1.

PROOF OF THEOREM 1:   Because every SQA is a REC by the postulate given, and every REC is a PAG by the definition given, we can conclude that every SQA is a PAG.

Can you see that Theorem 2 can be "proved" in much the same way? On first glance, you probably feel uncomfortable with these undefined terms, the definition, and the postulate, but what happens if we think of SQA as *square*, PAG as *parallelogram*, and RAL as *right angle*? You will now recognize that this system is not as abstract as first imagined, and that the two theorems simply state that "Every square is a parallelogram" and "Every square contains a right angle." We will work more with squares, parallelograms, rectangles, and right angles in later chapters.   ◪

Using a general formula to find the area of a specific geometric figure is a good example of deductive reasoning. Remember, when we reason deductively, we start with one or more **premises** (undefined terms, definitions, axioms or postulates, or previously proved theorems) and attempt to arrive at a conclusion that logically follows if the premises are accepted. The next example uses deductive reasoning to solve the application given in the chapter introduction.

The following warranty is given with a new radio: "This radio is warranted for one year from the date of purchase against defects in materials or workmanship. During this period, any such defects will be repaired, or the radio will be replaced at the company's option without charge. This warranty is void in the case of misuse or negligence."

EXAMPLE 4 Assume the premises stated in the warranty in the margin and answer the following.

(a) Bill purchased a radio and when he got home and opened the box, he discovered that the case was cracked. What could Bill conclude?

   Because Bill had just purchased the radio, one year had not gone by. He had not been negligent, nor had he misused the radio. He concluded that the company would either repair or replace the radio.

(b) Beth purchased a radio and fourteen months later the digital display burned out. What could Beth conclude?

   Because fourteen months exceeds the time period of the warranty, Beth concluded that the company would not be required by the warranty to repair or replace the radio.

(c) Jamie purchased a radio and two months later left it outside after listening to it while sunbathing. That night it rained, and the next day the radio would not play. What could Jamie conclude?

Although the radio became defective during the warranty period, the defect was due to her negligence. The company would not be required by the warranty to repair or to replace it. ▨

The next example gives practice in recognizing the two types of reasoning—inductive and deductive.

EXAMPLE 5 Determine if each conclusion follows logically from the premises, and state whether the reasoning is inductive or deductive.

(a) *Premise:* My coat is tan.
   Premise: Bob's coat is tan.
   Premise: Di's coat is tan.
   Conclusion: All coats are tan.

   Because we are reasoning from three specific examples and drawing a general conclusion, the process involves inductive reasoning. It is obvious that the conclusion does not logically follow from the premises.

(b) *Premise:* All athletes are in good physical condition.
   Premise: Shelly is an athlete.
   Conclusion: Shelly is in good physical condition.

   In this case, if we accept the premises, then the conclusion logically follows. This is an example of deductive reasoning. Notice that the conclusion may be true or false; we don't actually know if Shelly is or is not in good condition. However, this is beside the point because the conclusion logically follows from the premises. Certainly, if the premises are true and a conclusion follows, then the conclusion is also true. In mathematics we assume premises are true so that deduced conclusions are also true statements. ▨

> ### PRACTICE EXERCISE 2
>
> Determine if each conclusion follows logically from the premises, and state whether the reasoning is inductive or deductive.
>
> **(a)** *Premise:* This year is leap year.
> *Conclusion:* Next year will not be leap year.
>
> **(b)** *Premise:* The Boston Celtics won the NBA Championship in the years 1960–66.
> *Premise:* The Boston Celtics won the NBA Championship in the years 1968–69
> *Conclusion:* The Celtics won the NBA Championship every year during the 1960's.

A **fallacy** occurs when we reach a conclusion that does not necessarily follow from the premises.

EXAMPLE 6 Consider the following premises.

Premise: A coach tells his team, "If we are to win tonight, then we will have to play very hard."

Premise: The team played extremely hard.

What can we conclude? Some of you might conclude that "the team won," but this is *not* correct based on the premises. The team was told that *If* they win *then* they will have played hard, but nothing was said about what would happen *If* they played hard. In fact, you can see that even if the team members played their hearts out, the other team might have been far superior and whipped them. In this case, a conclusion does not logically follow from the premises. However, if the second premise were replaced with the following:

Premise: The team won the game.

We could then logically conclude that the team played hard. ◼

Aristotle (384–322 B.C.)

Aristotle, a Greek logician and philosopher, is credited with being the first to systematically study the logic and reasoning used in everyday life. He was a student of Plato and eventually became a tutor of Alexander the Great. The logic of Aristotle forms the basis for how we learn to reason deductively. It also serves as a foundation for the more formalized, symbolic logic studied and used by mathematicians today.

ANSWERS TO PRACTICE EXERCISES: **1. (a)** One answer is 36 since each number is the square of the numbers 1, 2, 3, 4, and 5. That is, $1 = 1^2$, $4 = 2^2$, $9 = 3^2$, $16 = 4^2$, $25 = 5^2$, and $36 = 6^2$. **(b)** The next element is ⌀⌀ since each element is one of the numbers 1, 2, 3, 4, and 5 placed next to its mirror image. **2. (a)** Using deductive reasoning, the conclusion follows logically from the given premise and the accepted but unstated premise that leap year occurs every four years. **(b)** The conclusion does not necessarily follow from the premises. It was reached by inductive reasoning. It is a false statement because in 1967 the Philadelphia 76ers won the NBA Championship.

1.1 EXERCISES

Use inductive reasoning in Exercises 1–10 to determine the next element in each list.

1. 3, 8, 13, 18, 23

2. 12, 7, 2, -3, -8

3. 1, 3, 9, 27, 81

4. 1, 5, 25, 125

5. 1, -2, 4, -8, 16

6. 40, -20, 10, -5

7. 1, 1, 2, 3, 5, 8, 13, 21

8. 1, 3, 4, 7, 11, 18, 29

9. S, M, T, W, T, F

10. J, F, M, A, M

11. What is the purpose of having undefined terms in an axiomatic system?

12. What is the difference between a postulate and a definition?

13. What is the difference between a postulate and an axiom?

14. What is the difference between a postulate and a theorem?

15. Give a definition of the word *happy*. What synonyms did you use in your definition? If you used a dictionary of synonyms to find those for *happy*, look up the meaning of each synonym. Continuing in this manner, what will eventually happen?

16. Most theorems are *"If . . . then . . ."* statements. What is the information that follows *If* called? What is the information that follows *then* called?

17. Give an informal "proof" of Theorem 2 in Example 3.

18. In the Declaration of Independence, the statement is made that "All men are created equal." Is this statement a postulate or a theorem?

19. What do you suppose would happen if two axiomatic systems had the same undefined terms and definitions but different postulates?

20. In a democracy such as that of the United States, we assume that the government exists to serve the people. In a dictatorship, it is assumed that the people exist to serve the government. What happens sometimes when two countries that support different structures of government try to negotiate?

In Exercises 21–32, determine if each conclusion follows logically from the premises, and state whether the reasoning is inductive or deductive.

21. *Premise:* If you are a mathematics major, then you can compute discounts on sale items.
Premise: Becky is a mathematics major.
Conclusion: Becky can compute discounts on sale items.

22. *Premise:* If you are a mathematics major, then you can compute discounts on sale items.
Premise: Becky can compute discounts on sale items.
Conclusion: Becky is a mathematics major.

23. *Premise:* If you are a home buyer, then you make payments.
Premise: Doug makes payments.
Conclusion: Doug is a home buyer.

24. *Premise:* If you are a home buyer, then you make payments.
Premise: Doug is a home buyer.
Conclusion: Doug makes payments.

25. *Premise:* It rained on Tuesday.
Premise: It rained on Wednesday.
Conclusion: It will rain on Thursday.

26. *Premise:* Last year I won money in Las Vegas.
Premise: The year before last I won money in Las Vegas.
Conclusion: I will win money in Las Vegas this year.

27. *Premise:* If I buy a car, then it will be a Buick.
Premise: If I receive a check from my father, then I will buy a car.
Premise: I received a check from my father.
Conclusion: I will buy a Buick.

28. *Premise:* If you are going to be an engineer, then you will study mathematics.
Premise: If you study mathematics, then you will get a good job.
Premise: Roy is going to be an engineer.
Conclusion: Roy will get a good job.

29. *Premise:* If you are an ogg, then you are an arg.
Premise: If you are a pon, then you are an ogg.
Conclusion: If you are a pon, then you are an arg.

30. *Premise:* If it is a frog, then it is green.
Premise: If it hops, then it is a frog.
Conclusion: If it hops, then it is green.

31. Explain why
Conclusion: You are an arg.
does not logically follow from the premises in Exercise 29.

32. Explain why
Conclusion: It is green.
does not logically follow from the premises in Exercise 30.

The puzzles in Exercises 33–44 are classic examples and a certain amount of deductive reasoning is required to solve them. Some of these puzzles are quite challenging, so don't be discouraged if you have trouble finding the solution immediately. Ideally they will make you think a bit, and along the way provide a bit of entertainment.

33. Mary has two U.S. coins in her purse. Together they total 55¢. One is not a nickel. What are the two coins? [Hint: We did *not* say that neither is a nickel.]

34. A young dog and an older dog are in the backyard. The young dog is the older dog's daughter but the older dog is not the young dog's mother. Explain.

35. If you take 5 apples from 8 apples, what do you have?

36. A rancher had 20 cattle. All but 12 died. How many did he have left?

37. We know there are 12 one-cent stamps in a dozen, but how many two-cent stamps are in a dozen?

38. A museum fired an archaeologist who claimed she found a coin dated 300 B.C. Why?

39. What arithmetic symbol can be placed between 2 and 3 to form a number greater than 2 and less than 3?

40. The number of marbles in a jar doubles every minute and is full in 10 minutes. When was the jar half full?

41. A boat will carry at maximum 200 lb. How can a man weighing 200 lb and his two daughters, each of whom weighs 100 lb, use the boat to reach an island?

42. How many times can you subtract 5 from 25?

43. A judge wishing to convict a defendant puts two pieces of paper in a hat. He tells the jury that if the defendant draws the piece marked "guilty" he will be convicted, but if he draws the piece marked "innocent" he will be set free. The hitch is that the judge wrote "guilty" on both pieces of paper. But when the crafty defendant showed the jury one piece of paper, the judge was forced to let him go free. How did the defendant outwit the judge?

44. You have 3 sacks, each containing 3 coins. Two of the sacks contain real coins and each coin weighs 1 lb. The third contains counterfeit coins, and each weighs 1 lb 1 oz. A scale is available, but it can be used one time and one time only to obtain a particular measure of weight. How might you use the scale to determine which sack contains the counterfeit coins? [Note: You cannot add or subtract coins to a total because any change of reading up or down on the scale will cause it to zero out.]

The puzzles in Exercises 45–52 have all been attributed to Englishman Henry Ernest Dudeney (1857–1930), called by some the greatest puzzle writer of all time.

45. A block of wood in the shape of a cube, shown in the accompanying figure, is to be sawed into two pieces with one cut so that each resulting piece has a surface in the shape of a regular hexagon (a six-sided figure with sides of equal length). How should the cut be made? [Hint: A cube has six surfaces, and all must be cut to create the two pieces.]

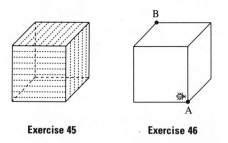

Exercise 45   Exercise 46

46. A bug is sitting on the surface of a solid cube of wood at point *A* as shown in the figure below. If it wants to get to the opposite corner at point *B* by the shortest possible route, show the path it should take along the surface of the cube.

47. A square piece of cardboard, 8 inches on a side and marked like a checkerboard, is cut into four pieces as shown below on the left. The pieces are reassembled to form a rectangle like the one shown on the right. The original square contains 64 little squares whereas the rectangle contains 65 little squares. Where does the extra square come from?

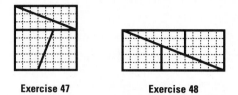

| Exercise 47 | Exercise 48 |

48. Rearrange the four pieces of the square in Exercise 47 to form a new figure that contains only 63 little squares. In this case we seem to have lost a little square, whereas in the preceding exercise we seem to have gained a square. Explain what happened to this square.

49. Consider the six matches arranged to form a regular hexagon as shown below. Take three more matches and arrange the nine to show another regular six-sided figure. The matches cannot be placed on top of one another, cannot be broken, and there should be no loose ends when you are finished.

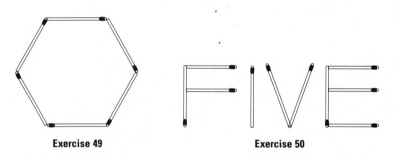

Exercise 49   Exercise 50

50. Consider ten matches arranged to form FIVE as shown above. Remove seven of the matches so that what is left is four.

51. Consider the following eight "postulates."
   (a) Smith, Jones, and Rodriquez are the engineer, brakeman, and fireman on a train, not necessarily in that order.
   (b) Riding on the train are three passengers with the same last names as the crewmembers, identified as Passenger Smith, Passenger Jones, and Passenger Rodriquez in the following statements.
   (c) The brakeman lives in Denver.
   (d) Passenger Rodriquez lives in San Francisco.
   (e) Passenger Jones long ago forgot all the algebra that he learned in high school.
   (f) The passenger with the same name as the brakeman lives in New York.

(g) The brakeman and one of the passengers, a professor of mathematical physics, attend the same church.

(h) Smith beat the fireman in a game of tennis at a court near their homes.

Can you discover a theorem that tells the name of the engineer? The brakeman? The fireman?

52. Consider the following six "postulates."

(a) Art, Bev, Cheryl, Dot, and Ed all attend the same college where one is a freshman, one is a sophomore, one is a junior, one is a senior, and one is a graduate student.

(b) Art, Bev, and Cheryl have not yet completed their undergraduate work.

(c) Bev is one year ahead of Ed.

(d) Art is not a freshman.

(e) Ed is not a freshman.

(f) Art is in a higher class than Ed.

Can you discover theorems that give each student's class in college?

1.2 POINTS, LINES, AND PLANES

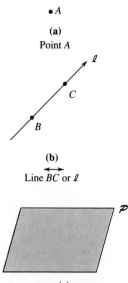

(a)
Point A

(b)
Line $\overleftrightarrow{BC}$ or ℓ

(c)
Plane $\mathcal{P}$

Figure 1.1

In Section 1.1 we discussed axiomatic systems and discovered that such systems consist of undefined terms, definitions, postulates (or axioms), and theorems. The undefined terms provide a starting point for developing an axiomatic system. In geometry, we begin with the following undefined terms:

set,   point,   line,   plane.

Any attempt to define these terms would require more words that are undefined. Intuitively, we have some idea of their meaning. A **set** is a collection or group of objects. A **point** can be thought of as an object that determines a position but that has no dimension (length, width, or height). We can symbolize a point with a dot and label it with a capital letter, such as the point A shown in Figure 1.1(a).

A **line** can be thought of as a set of points in a one-dimensional straight figure that extends in opposite directions without ending. Figure 1.1(b) shows a representation of a line passing through the two points B and C. The arrowheads indicate that the line continues in that direction without ending. We symbolize a line such as this using $\overleftrightarrow{BC}$ or $\overleftrightarrow{CB}$, or when appropriate, by using a lowercase letter such as ℓ.

Finally, a **plane** is a set of points in a flat surface, such as the face of a blackboard, having two dimensions and extending without boundary. We often represent a plane as shown in Figure 1.1(c) and symbolize planes using a script letter such as $\mathcal{P}$.

Have you ever looked closely at the screen on a television set? The surface is composed of thousands of small dots that glow when hit by an electron beam. When viewed from a distance, the individual dots blend together forming the picture. Each minute dot serves as a model for a point, one of the simplest geometric figures that we study.

DEFINITION: SPACE AND GEOMETRIC FIGURES

The set of all points is called **space.** Any set of points, lines, or planes in space is called a **geometric figure.**

We can think of geometry as the study of properties of geometric figures. Some of these properties must be assumed in the form of postulates.

POSTULATE 1.1

Given any two distinct points in space, there is exactly one line that passes through them.

Intuitively, Postulate 1.1 tells us that we can draw one and only one straight line through two different points.

POSTULATE 1.2

Given any three distinct points in space not on the same line, there is exactly one plane that passes through them.

Figure 1.2(a) shows the plane containing the three points *A*, *B*, and *C*. Notice that *A*, *B*, and *C* are not on the same line. If three points such as *D*, *E*, and *F* are on the same line as in Figure 1.2(b), we can see that more than one plane can contain them.

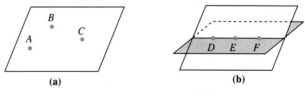

Figure 1.2   Points in Planes

POSTULATE 1.3

The line determined by any two distinct points in a plane is also contained in the plane.

From Postulate 1.3 we see that when we draw lines between points in a given plane, we always remain in the plane. The next postulate guarantees that there are more points in space than those found in any given plane.

POSTULATE 1.4

No plane contains all points in space.

The next postulate has numerous applications in algebra.

POSTULATE 1.5   RULER POSTULATE

There is a one-to-one correspondence between the set of all points on a line and the set of all real numbers.

If we draw a line, select a point on it (called the **origin**), mark off equal units in both directions, and label these points with integers, the result is called a **number line.** See Figure 1.3. The number corresponding to a given point on the line is called the **coordinate** of the point, and when we identify a point with a given real number, we are **plotting** the point associated with the number.

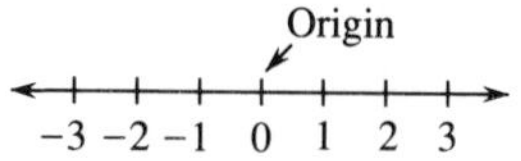

Origin

$$\xleftarrow{\hspace{0.5em}}\!\!\!\!\!\! \underset{-3\;-2\;-1\;\;\;0\;\;\;1\;\;\;2\;\;\;3}{+\;\;+\;\;+\;\;+\;\;+\;\;+\;\;+}\!\!\!\!\!\!\xrightarrow{\hspace{0.5em}}$$

Figure 1.3   Number Line

EXAMPLE 1   Plot the points associated with the real numbers $\frac{1}{2}$, $-\frac{3}{4}$, $\sqrt{2}$, and 2.5 on a number line.

Start with a number line like the one in Figure 1.3. Because $\frac{1}{2}$ is halfway between 0 and 1, we locate the point midway between them. The points corresponding to $-\frac{3}{4}$ and 2.5 are determined similarly. We can find the approximate location of the point corresponding to $\sqrt{2}$ by recognizing that $\sqrt{2}$ is approximately 1.4. The four points are plotted in Figure 1.4. ◪

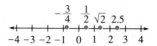

Figure 1.4   Points on a Number Line

PRACTICE EXERCISE 1

Approximate the coordinates of points *A, B, C,* and *D* on the number line.

Because there are infinitely many real numbers, there are infinitely many points on any given line. This conclusion is actually a theorem, a statement that can be proved. When proving certain theorems in geometry, we will use some of the following postulates from algebra.

POSTULATE 1.6   THE REFLEXIVE LAW

Any quantity is equal to itself. ($x = x$)

POSTULATE 1.7   THE SYMMETRIC LAW

If x and y are any two quantities and $x = y$, then $y = x$.

POSTULATE 1.8   THE TRANSITIVE LAW

If x, y, and z are any three quantities with $x = y$ and $y = z$, then $x = z$.

POSTULATE 1.9   THE ADDITION-SUBTRACTION LAW

If w, x, y, and z are any four quantities with $w = x$ and $y = z$, then $w + y = x + z$ and $w - y = x - z$.

The addition-subtraction law states that if equal quantities are added to or subtracted from equal quantities, the results are also equal.

POSTULATE 1.10   THE MULTIPLICATION-DIVISION LAW

If w, x, y, and z are any four quantities with $w = x$ and $y = z$, then $wy = xz$ and $\dfrac{w}{y} = \dfrac{x}{z}$ (provided $y \neq 0$ and $z \neq 0$).

The multiplication-division law states that if equal quantities are multiplied or divided by equal quantities (division by zero excluded), the results are also equal.

POSTULATE 1.11   *THE SUBSTITUTION LAW*

If x and y are any two quantities with $x = y$, then x can be substituted for y in any expression containing y.

POSTULATE 1.12   *THE DISTRIBUTIVE LAW*

If x, y, and z are any three quantities, then $x(y + z) = xy + xz$.

Postulates 1.6–1.12 are used extensively when solving algebraic equations. We illustrate this in the following exercises.

ANSWER TO PRACTICE EXERCISE: **1.** A: 3; B: -2; C: $-\dfrac{1}{2}$; D: $\dfrac{7}{4}$

1.2 EXERCISES

1. How many lines can be drawn between two distinct points?
2. How many planes are determined by three distinct points that are not on the same line?
3. If distinct points A and B are in plane $\mathcal{P}$, and point C is on the line determined by A and B, what can be said about C relative to $\mathcal{P}$?
4. If $\mathcal{P}$ is a plane and A is a point in space, must A be on $\mathcal{P}$?

In Exercises 5–8, A, B, and C are distinct points, ℓ and m are lines, and $\mathcal{P}$ is a plane. Name the postulate illustrated by each statement.

5. If A and B are on ℓ, and A and B are on m, then $m = \ell$.
6. If A and B are on ℓ, ℓ is in $\mathcal{P}$, and C is on ℓ, then C is in $\mathcal{P}$.
7. There is a point A such that A is not in $\mathcal{P}$.
8. If A and B are on ℓ, A is on m, and $\ell \neq m$, then B is not on m.
9. Construct a number line and plot the points associated with the real numbers $3, \dfrac{7}{8}, -\dfrac{1}{3}, -1.5$, and $\sqrt{3}$.
10. Construct a number line and plot the points associated with the real numbers $-3, \dfrac{1}{4}, -\dfrac{4}{3}, 3.2$, and $-\sqrt{5}$.

In Exercises 11–12, approximate the coordinates of points *A, B, C,* and *D* on the given number lines.

11.

```
        D            C
 +--+--+--•--+--+--•--+•+--•+•+•+-->
 -4  B -2 -1  0  1  A  3   4
```

12.

```
    D                      C
 +--•+•--+--•--+--•--+--+-•+•--+-->
 -4 -3  A -1  B  1  2  3  4
```

In Exercises 13–18, complete each statement using the specified postulate.

13. Reflexive law: $5 = \underline{\quad?\quad}$.

14. Transitive law: If $a = b$ and $b = 3$, then $a = \underline{\quad?\quad}$.

15. Addition-subtraction law: If $a = b$, then $a + 7 = b + \underline{\quad?\quad}$.

16. Multiplication-division law: If $2x = 6$, then $x = \underline{\quad?\quad}$.

17. Symmetric law: If $a = -3$, then $-3 = \underline{\quad?\quad}$.

Answer *true* or *false* in Exercises 18–22.

18. By the reflexive law, $-2 = -2$.

19. If $w = 7$ and $7 = x$, then $w = x$ by the symmetric law.

20. If $x + 1 = 8$, then $x = 7$ by the addition-subtraction law.

21. If $\frac{1}{3}y = 2$, then $y = 6$ by the addition-subtraction law.

22. If $8 = x$ then $x = 8$ by the reflexive law.

In Exercises 23–24, give the postulate that supports each indicated step in the solution of the equation.

23. Solve $3x + 2 = 4 + 5x$.

STATEMENTS	REASONS
1. $3x + 2 = 4 + 5x$	1. Given
2. $3x + 2 - 4 = 4 - 4 + 5x$	2. _____
3. $3x - 2 = 5x$	3. Simplify
4. $3x - 3x - 2 = 5x - 3x$	4. _____
5. $-2 = 2x$	5. Simplify
6. $\frac{1}{2}(-2x) = \frac{1}{2}(2x)$	6. _____
7. $-1 = x$	7. Simplify
8. $x = -1$	8. _____
Check: -1 in $3x + 2 = 4 + 5x$	
9. $3(-1) + 2 = 4 + 5(-1)$	9. _____
10. $-1 = -1$	10. Simplify

24. Solve $\frac{2}{3}x + 1 = x$.

STATEMENTS	REASONS
1. $\frac{2}{3}x + 1 = x$	1. Given
2. $3\left[\frac{2}{3}x + 1\right] = 3x$	2. _____
3. $2x + 3 = 3x$	3. _____
4. $2x - 2x + 3 = 3x - 2x$	4. _____
5. $3 = x$	5. Simplify
6. $x = 3$	6. _____

Check: 3 in $\frac{2}{3}x + 1 = x$

7. $\frac{2}{3}(3) + 1 = 3$	7. _____
8. $3 = 3$	8. Simplify

25. Can a given point exist on two distinct lines? On five distinct lines?

26. Can two distinct points both exist on two distinct lines?

27. Can a given line exist in two distinct planes? In five distinct planes?

28. Can two given lines with no point in common exist in the same plane?

29. Explain why stools are often made with three legs rather than four, to provide greater stability.

1.3 SEGMENTS, RAYS, AND ANGLES

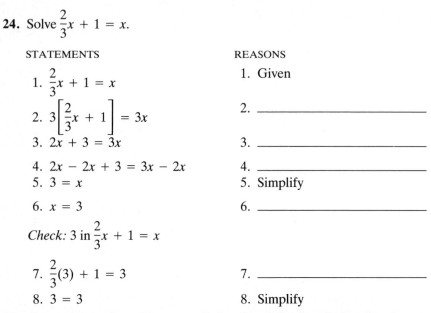

Figure 1.5   Points Between A and B

A line can be thought of as a geometric figure consisting of points that extend infinitely far in opposite directions. We now consider two new figures that are parts of a line. Another undefined term, *between*, describes the position of points on a line in relation to two given points on that line. Intuitively, points on the colored portion of the line in Figure 1.5 are said to be *between* points A and B.

> **DEFINITION: LINE SEGMENT**
>
> Let A and B be two distinct points on a line. The geometric figure consisting of all points between A and B, including A and B, is called a **line segment** or **segment,** denoted by $\overline{AB}$ or $\overline{BA}$. The points A and B are called **endpoints** of $\overline{AB}$.

Recall that the line determined by distinct points A and B is denoted by $\overleftrightarrow{AB}$ or $\overleftrightarrow{BA}$, which distinguishes it from the segment $\overline{AB}$ or $\overline{BA}$. The **length** of segment $\overline{AB}$ is the distance between the endpoints A and B and is often denoted by AB.

The next postulate provides a way to find the length of a segment that is made up of other segments. Refer to Figure 1.6.

POSTULATE 1.13   SEGMENT ADDITION POSTULATE

Let A, B, and C be three points on the same line with B between A and C. Then $AC = AB + BC$, $BC = AC - AB$, and $AB = AC - BC$.

The segment addition postulate can be applied to many problems whose solutions also require basic algebra.

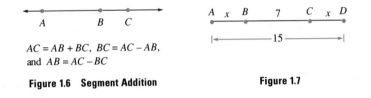

$AC = AB + BC,\ BC = AC - AB,$
and $AB = AC - BC$

Figure 1.6   Segment Addition   **Figure 1.7**

E X A M P L E 1   In Figure 1.7, $AD = 15$, $BC = 7$, and $AB = CD$. Find AB.

Let $AB = x$. Then because $AB = CD$, $CD = x$. By extending the segment addition postulate we have

$$AD = AB + BC + CD.$$

Substitute 15 for AD, 7 for BC, and x for both AB and CD to obtain

$$15 = x + 7 + x.$$
$$15 = 2x + 7 \qquad \text{Collect like terms}$$
$$8 = 2x \qquad \text{Subtract 7 from both sides}$$
$$4 = x \qquad \text{Divide both sides by 2}$$

Thus, $AB = 4$. ◾

PRACTICE EXERCISE 1

Use the figure below to find the value of x.

$$x + 2 \qquad 2x + 1 \qquad x$$

$$|\!\longleftarrow\!\!\!\longrightarrow\!| \quad 23$$

DEFINITION: RAY

Let A and B be two distinct points on a line. The geometric figure consisting of the point A together with all points on $\overleftrightarrow{AB}$ on the same side of A as B is called a **ray,** denoted by $\overrightarrow{AB}$. The point A is called the **endpoint** of $\overrightarrow{AB}$.

In Figure 1.8, the ray $\overrightarrow{AB}$ is shown in color. Unlike the notations for segments and lines, the rays $\overrightarrow{AB}$ and $\overrightarrow{BA}$ are different. A ray has only one endpoint and it is always written first. Taken together, rays $\overrightarrow{AB}$ and $\overrightarrow{BA}$ make up the line $\overleftrightarrow{AB}$.

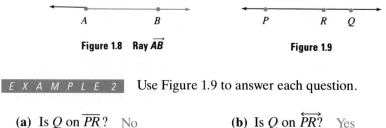

Figure 1.8   Ray $\overrightarrow{AB}$

Figure 1.9

E X A M P L E 2   Use Figure 1.9 to answer each question.

(a) Is Q on $\overleftrightarrow{PR}$?   No
(b) Is Q on $\overleftrightarrow{PR}$?   Yes
(c) Is Q on $\overrightarrow{PR}$?   Yes
(d) Is Q on $\overrightarrow{RP}$?   No
(e) What are the endpoints of $\overline{PR}$? P and R
(f) What are the endpoints of $\overrightarrow{PR}$?   Only P
(g) What are the endpoints of $\overleftrightarrow{PR}$?   There are no endpoints.
(h) Are $\overline{PQ}$ and $\overline{QP}$ the same?   Yes
(i) Are $\overrightarrow{PQ}$ and $\overrightarrow{QP}$ the same?   No
(j) If $PQ = 5$ cm and $RQ = 2$ cm, what is the length of $\overline{PR}$?   3 cm   ◩

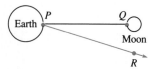

Scientists have measured the distance between the earth and the moon to within a few centimeters. A laser beam is directed from point P and reflected back to the earth from a mirror that was left by the Apollo astronauts at point Q. The time it takes for the beam to return to earth can be measured and used to calculate the distance from the earth to the moon. The two points P and Q along with the laser beam provide a model of a segment. A laser beam directed into space from P through R serves as a model for a ray.

PRACTICE EXERCISE 2

Draw two points X and Y and place point Z on $\overleftrightarrow{XY}$ but not on $\overline{XY}$. Use your figure to answer the following questions.

(a) Is Z on $\overrightarrow{YX}$?
(b) Is Z on $\overline{YX}$?
(c) Is Y on $\overrightarrow{ZX}$?
(d) What are the endpoints of $\overleftrightarrow{XY}$?
(e) What are the endpoints of $\overline{XY}$?
(f) Are $\overrightarrow{XY}$ and $\overrightarrow{YX}$ the same?
(g) If $ZX = 3$ cm and $XY = 5$ cm, what is the length of $\overline{ZY}$?

The *angle* is one of the most important figures studied in geometry.

> **DEFINITION: ANGLE**
>
> An **angle** is a geometric figure consisting of two rays that share a common endpoint, called the **vertex** of the angle. The rays are called **sides** of the angle.

The degree as a unit of angular measure originated with the ancient Sumerians. The Sumerians thought that it took the earth 360 days to revolve around the sun. They assumed the earth's orbit was a circle and that the earth traveled at a constant speed. Thus, to traverse completely around a circle required 360 units of time (days) so that in 1 day, the earth would travel $\frac{1}{360}$ of a circle. The Sumerians defined the measure of the angle formed by $\frac{1}{360}$ of a circle as a degree.

Angles are named in three ways. Consider the angle formed by rays $\overrightarrow{AC}$ and $\overrightarrow{AB}$ in Figure 1.10. We use the three points on the angle, A, B, and C, and call the angle $\angle CAB$ or $\angle BAC$ (the symbol $\angle$ is read "angle"). Notice in both cases the vertex is written between the other two points. When no confusion can arise, we simply name the angle $\angle A$, using only the vertex point. The third possibility is to write a small letter or number such as 1 in the position shown in Figure 1.10 and call the angle $\angle 1$.

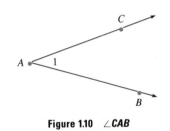

Figure 1.10 $\angle CAB$

We are familiar with measuring the length of a segment using a suitable unit of measure such as inch, centimeter, foot, or meter. In order to measure an angle, we need a measuring unit. The most common unit is the **degree** (°). An angle with measure 0° is formed by two coinciding rays such as $\overrightarrow{AB}$ and $\overrightarrow{AC}$ in Figure 1.11.

Figure 1.11 $\angle BAC$ Has Measure 0°

If we rotate ray $\overrightarrow{AB}$ in a counterclockwise direction from ray $\overrightarrow{AC}$ in Figure 1.11, the two rays form larger and larger angles as shown in Figure 1.12.

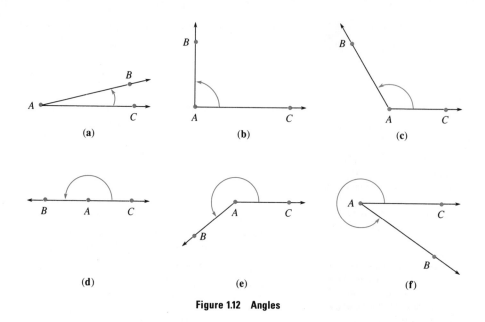

Figure 1.12   Angles

If $\overrightarrow{AB}$ in Figure 1.11 is allowed to rotate completely around until it coincides with ray $\overrightarrow{AC}$ again, the resulting angle is said to measure 360°. Thus, an angle of measure 1° is formed by making $\dfrac{1}{360}$ of a complete rotation. An angle of measure 1° is shown in Figure 1.13.

Figure 1.13   Angle Measuring 1°

The approximate measure of an angle in degrees can be found by using a protractor shown in Figure 1.14.

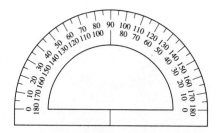

Figure 1.14   Protractor

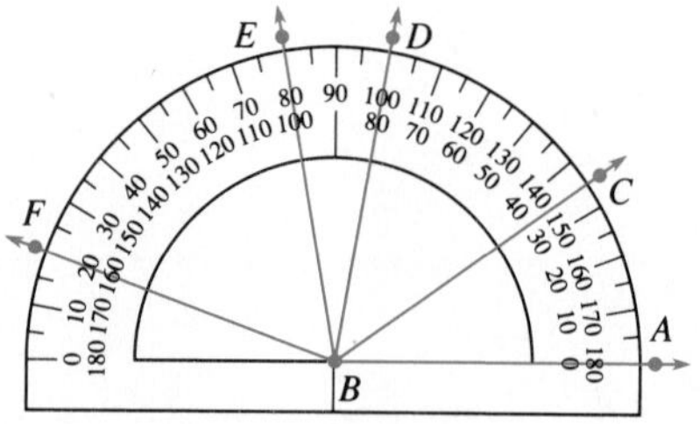

Figure 1.15   Measuring Angles

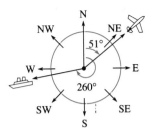

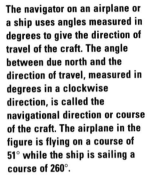

The navigator on an airplane or a ship uses angles measured in degrees to give the direction of travel of the craft. The angle between due north and the direction of travel, measured in degrees in a clockwise direction, is called the navigational direction or course of the craft. The airplane in the figure is flying on a course of 51° while the ship is sailing a course of 260°.

Figure 1.15 shows how a protractor is used to measure various angles. ∠ABC has measure 35°, ∠ABD has measure 80°, ∠ABE has measure 100°, and ∠ABF has measure 160°. Rather than say "∠ABC has measure 35°," we simply write ∠ABC = 35°. Also, when two angles have the same measure, we simply say that the angles are **equal.**

In some practical applications, angles must be measured with greater precision. One degree is divided into 60 equal parts called **minutes** (′), and one minute is divided into 60 equal parts called **seconds** (″). Thus, the measure of an angle might be given as 55°28′48″.

Certain angles are given special names.

DEFINITION: SPECIAL ANGLES

An angle whose sides form a line is called a **straight angle** and has measure 180°. An angle with measure 90° is called a **right angle.**
An angle with measure between 0° and 90° is called an **acute angle.**
An angle with measure between 90° and 180° is called an **obtuse angle.**

EXAMPLE 3   Use Figure 1.16 to answer each question.

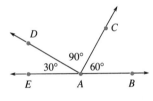

Figure 1.16

(a) Is ∠BAE a straight angle?   Yes

(b) Is ∠EAD an acute angle?   Yes

(c) Is ∠CAE a right angle?   No

(d) Is ∠DAB an obtuse angle?   Yes ◼

Two acute angles or two obtuse angles need not be equal. However, right angles are equal.

POSTULATE 1.14   RIGHT-ANGLE POSTULATE
All right angles are equal.

Some pairs of angles have properties that are useful in many applications.

> ### DEFINITION: COMPLEMENTARY AND SUPPLEMENTARY ANGLES
>
> Two angles whose measures total 90° are called **complementary angles** and each is called the **complement** of the other. Two angles whose measures total 180° are called **supplementary angles,** and each is called the **supplement** of the other.

EXAMPLE 4

(a) If $\angle A = 35°$ and $\angle B = 55°$, then $\angle A$ and $\angle B$ are complementary because $35° + 55° = 90°$.

(b) If $\angle C = 120°15'45''$ and $\angle D = 59°44'15''$, then because
$120°15'45'' + 59°44'15''$

$= 120°15' + 59°44' + 60''$	$45'' + 15'' = 60''$
$= 120°15' + 59°44' + 1'$	$60'' = 1'$
$= 120°15' + 59°45'$	$44' + 1' = 45'$
$= 120° + 59° + 15' + 45'$	
$= 120° + 59° + 60'$	$15' + 45' = 60'$
$= 120° + 59° + 1°$	$60' = 1°$
$= 180°,$	

$\angle C$ and $\angle D$ are supplementary angles.

(c) If $\angle P = 20°$, $\angle Q = 30°$, and $\angle R = 40°$, although $20° + 30° + 40° = 90°$, we do not call the angles complementary. The definition of complementary angles involves only two angles (not three or more) that sum to 90°. ▨

PRACTICE EXERCISE 3

(a) If $\angle A = 27°$ and $\angle A$ and $\angle B$ are complementary, find the measure of $\angle B$.

(b) If $\angle P = 74°26'52''$ and $\angle P$ and $\angle Q$ are supplementary, find the measure of $\angle Q$.

EXAMPLE 5

If $\angle A = (2x)°$, $\angle B = (x - 6)°$, and $\angle A$ and $\angle B$ are complementary, find x.

Because $\angle A$ and $\angle B$ are complementary, $\angle A + \angle B = 90°$. Substituting, we have

$$2x + (x - 6) = 90.$$
$$2x + x - 6 = 90$$
$$3x - 6 = 90$$
$$3x = 96$$
$$x = 32 \quad ▨$$

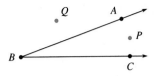

Figure 1.17   Points Interior and Exterior to ∠ABC

PRACTICE EXERCISE 4

If $\angle P = (2y - 9)°$, $\angle Q = (7y)°$, and $\angle P$ and $\angle Q$ are supplementary, find y.

Two undefined terms that are useful when working with angles are *interior* and *exterior*. Consider $\angle ABC$ and points P and Q as shown in Figure 1.17. We say that P is in the **interior** of $\angle ABC$ and Q is **exterior** to $\angle ABC$.

> **DEFINITION: ADJACENT ANGLES**
>
> Two angles are called **adjacent angles** if they have a common vertex and share a common side.

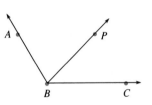

Figure 1.18   Adjacent Angles

For example, in Figure 1.18, $\angle ABP$ and $\angle PBC$ are adjacent, but $\angle ABC$ and $\angle ABP$ and $\angle ABC$ and $\angle PBC$ are not. Notice that intuitively, $\angle ABP$ and $\angle CBP$ are adjacent angles if P is in the interior of $\angle ABC$.

Figure 1.18 can also be used to clarify the next postulate.

> **POSTULATE 1.15   ANGLE ADDITION POSTULATE**
>
> Let A, B, and C be points that determine $\angle ABC$ with P a point in the interior of the angle. Then $\angle ABC = \angle ABP + \angle PBC$, $\angle PBC = \angle ABC - \angle ABP$, and $\angle ABP = \angle ABC - \angle PBC$.

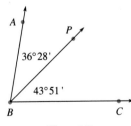

Figure 1.19

E X A M P L E 6   Suppose that $\angle ABP$ and $\angle PBC$ are adjacent angles and that $\angle ABP = 36°28'$ and $\angle PBC = 43°51'$. Find the measure of $\angle ABC$. Figure 1.19 shows a sketch of the given angles from which we can see that because P is in the interior of $\angle ABC$, by Postulate 1.15,

$$\angle ABC = \angle ABP + \angle PBC$$
$$= 36°28' + 43°51'$$
$$= 79°79'$$
$$= 80°19' \qquad 79' = 60' + 19' = 1°19' \quad \blacksquare$$

ANSWERS TO PRACTICE EXERCISES: **1.** $x = 5$ **2. (a)** yes **(b)** no **(c)** yes **(d)** only X **(e)** X and Y **(f)** no **(g)** 8 cm **3. (a)** 63° **(b)** 105°33'8" **4.** $y = 21$

1.3 EXERCISES

Consider the line $\overleftrightarrow{AB}$ with point C between A and B and ray $\overrightarrow{CD}$ as shown below. Use this figure and answer *true* or *false* in Exercises 1–20.

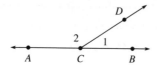

1. Point B is on $\overrightarrow{AC}$.

2. Point A is on $\overrightarrow{CB}$.

3. Point C is on $\overline{AB}$.

4. Point B is on $\overrightarrow{AC}$.

5. If $AC = 10$ cm and $CB = 13$ cm, then $AB = 23$ cm.

6. If $AB = 30$ cm and $AC = 12$ cm then $CB = 18$ cm.

7. $\overrightarrow{CA}$ and $\overrightarrow{AC}$ are the same.

8. $\angle 1$ is another name for $\angle DCB$.

9. $\angle ACD$ is another name for $\angle 2$.

10. C is the endpoint of $\overrightarrow{BC}$.

11. A and B are endpoints of $\overleftrightarrow{AB}$.

12. A and C are endpoints of $\overline{AC}$.

13. The vertex of $\angle 1$ is C.

14. The vertex of $\angle DCA$ is A.

15. $\angle ACB$ is a right angle.

16. $\angle BCA$ is a straight angle.

17. $\angle 1$ and $\angle 2$ are adjacent angles.

18. $\angle 1$ and $\angle 2$ are complementary angles.

19. $\angle DCB$ is an acute angle.

20. $\angle 1$ is the supplement of $\angle DCA$.

Exercises 21–26 refer to the figure below. Give the measure of each angle.

21. $\angle ABC$

22. $\angle EBF$

23. $\angle ABD$

24. $\angle ABE$

25. $\angle FBC$

26. $\angle CBE$

Find the complement of each angle in Exercises 27–30.

27. $18°$

28. $64°$

29. $36°40'$

30. $71°45'20''$

Find the supplement of each angle in Exercises 31–34.

31. $74°$

32. $136°$

33. $57°35'$

34. $110°35'40''$

35. What angle has the same measure as its complement?

36. What angle has the same measure as its supplement?

37. What is the complement of the supplement of an angle measuring $130°$?

38. What is the supplement of the complement of an angle measuring $50°$?

39. What is the complement of the complement of an angle measuring $25°$?

40. What is the supplement of the supplement of an angle measuring $160°$?

41. What is the measure of an angle whose supplement is four times its complement?

42. What is the measure of an angle whose supplement is three times its complement?

State whether each angle given in Exercises 43–51 is straight, right, acute, or obtuse.

43. 65°   **44.** 115°   **45.** 180°   **46.** 90°

47. The complement of an angle measuring 42°.

48. The supplement of an angle measuring 42°.

49. The complement of any acute angle.

50. The supplement of any obtuse angle.

51. The supplement of a right angle.

In Exercises 52–55, $\angle ABP$ and $\angle PBC$ are adjacent angles. Find the measure of $\angle ABC$.

52. $\angle ABP = 62°20'$ and $\angle PBC = 31°50'$

53. $\angle ABP = 49°55'$ and $\angle PBC = 57°15'$

54. $\angle ABP = 27°25'41''$ and $\angle PBC = 52°51'35''$

55. $\angle ABP = 120°38'22''$ and $\angle PBC = 18°41'54''$

56. On $\overleftrightarrow{AB}$, how many points are located 5 cm from point A?

57. On $\overleftrightarrow{AB}$, how many points are located 5 cm from point B?

58. On $\overrightarrow{AB}$, how many points are located 5 cm from point A?

59. On $\overrightarrow{AB}$, how many points are located 5 cm from point B if $AB = 10$ cm?

60. On $\overline{AB}$, how many points are located 5 cm from point A if $AB = 10$ cm?

61. On $\overrightarrow{AB}$, how many points are located 5 cm from point B?

62. Give the number of acute or obtuse angles formed by each of the following. Assume that no two rays are on the same line. Make a sketch in each case.
   (a) Two distinct rays with the same endpoint.
   (b) Three distinct rays with the same endpoint.
   (c) Four distinct rays with the same endpoint.
   (d) Five distinct rays with the same endpoint.

63. After solving Exercise 62, can you determine a formula (in terms of n) that gives the number of acute or obtuse angles formed by n distinct rays with the same endpoint, no two of which are on the same line?

In Exercises 64–65, find the value of x in each figure.

64.
$x+5$	$3x$	x

⊢────── 30 ──────⊣

65.
$25-x$	x	$3x$

⊢────── 52 ──────⊣

66. If $\angle A = (5y)°$, $\angle B = (y + 6)°$, and $\angle A$ and $\angle B$ are complementary, find y.

67. If $\angle R = (30 - y)°$, $\angle S = (9y - 10)°$, and $\angle R$ and $\angle S$ are supplementary, find y.

KEY TERMS AND SYMBOLS

1.1 inductive reasoning, p. 2
   axiomatic system, p. 3
   undefined terms, p. 4
   definitions, p. 4
   axioms, p. 4
   postulates, p. 4
   theorems, p. 4
   hypothesis, p. 4
   conclusion, p. 4
   deductive reasoning, p. 4
   premises, p. 5
   fallacy, p. 7
1.2 set, p. 12
   point, p. 12
   line, p. 12

plane, p. 12
space, p. 13
geometric figure, p. 13
origin, p. 14
number line, p. 14
coordinate, p. 14
plot, p. 14
1.3 line segment, p. 18
   endpoints, p. 18
   length, (of a segment),
    p. 19
ray, p. 20
angle ($\angle$), p. 21
vertex, p. 21
sides (of an angle), p. 21

degree (°), p. 21
equal angles ($=$), p. 23
minutes ($'$), p. 23
seconds ($''$), p. 23
straight angle, p. 23
right angle, p. 23
acute angle, p. 23
obtuse angle, p. 23
complementary angles,
  p. 24
supplementary angles,
  p. 24
interior (of an angle), p. 25
exterior (of an angle), p. 25
adjacent angles, p. 25

REVIEW EXERCISES

Section 1.1

1. What are the four parts to any axiomatic system?

2. What role do undefined terms play in an axiomatic system?

Use inductive reasoning in Exercises 3–5 to determine the next element in each list.

3. $-8, -3, 2, 7, 12$

4. A, C, E, G, I

5. $1, \dfrac{1}{2}, \dfrac{1}{4}, \dfrac{1}{8}, \dfrac{1}{16}$

In Exercises 6–9, determine if each conclusion follows logically from the premises, and state whether the reasoning is inductive or deductive.

6. *Premise:* If I wash my car, then it will rain.
   Premise: I washed my car.
   Conclusion: It will rain.

7. *Premise:* If I wash my car, then it will rain.
   Premise: It is raining.
   Conclusion: I washed my car.

8. *Premise:* Bob Begay owns a VCR.
   Premise: Bob Jones owns a VCR.
   Premise: Bob Santos owns a VCR.
   Conclusion: Everyone named Bob owns a
    VCR.

9. *Premise:* If it is Sunday, then tomorrow is
    Monday.
   Premise: If tomorrow is Monday, then it is a
    holiday.
   Conclusion: If it is Sunday, tomorrow is a
    holiday.

10. Explain why
   Conclusion: Tomorrow is a holiday.
   does not follow logically from the premises in Exercise 9.

11. If 3 cats kill 3 mice in 3 minutes, how long will it take 100 cats to kill 100 mice?

12. There are 5 white socks and 5 black socks in a drawer. How many socks must be removed without looking to be assured of having a pair that match?

Section 1.2

Answer *true* or *false* in Exercises 13–16.

13. Exactly one line can be drawn between two distinct points.

14. Any three distinct points determine a unique plane.

15. All points on a line determined by two distinct points in a plane are also in the plane.

16. No plane contains all points in space.

17. Plot the points associated with 4, $-\dfrac{3}{4}$, and $\sqrt{6}$ on a number line.

In Exercises 18–22, give the name of the postulate illustrated by the given statement.

18. $3 = 3$

19. If $x = -2$, then $-2 = x$.

20. If $x = 3$ and $3 = w$, then $x = w$.

21. If $\dfrac{1}{2}x = 6$, then $x = 12$.

22. If $x = 3$ and $x + 2 = 5$, then $3 + 2 = 5$.

Section 1.3

Use the figure below to answer *true* or *false* in Exercises 23–34.

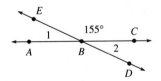

23. C is on $\overrightarrow{BA}$.

24. B is on $\overline{ED}$.

25. $\angle 1$ is another name for $\angle EBA$.

26. B is the vertex of $\angle 2$.

27. $\angle EBD$ is a right angle.

28. $\angle ABD$ and $\angle CBD$ are adjacent angles.

29. $\angle 1$ and $\angle 2$ are complementary.

30. E is an endpoint of $\overleftrightarrow{ED}$.

31. $\angle 2$ and $\angle EBC$ are supplementary.

32. $\overrightarrow{BE}$ and $\overrightarrow{EB}$ are the same.

33. $\angle 1 = 25°$

34. $\angle 2$ is an obtuse angle.

35. Find the complement of 24°30′45″.

36. Find the supplement of 136°42′51″.

37. What is the complement of the supplement of an angle measuring 110°?

38. Is an angle of measure 145° acute?

39. How many points on $\overrightarrow{AB}$ are 3 cm from A?

40. How many points on $\overleftrightarrow{AB}$ are 3 cm from A?

PRACTICE TEST

1. What is the difference between a postulate and a theorem?

2. Discuss the difference between inductive and deductive reasoning?

3. Which type of reasoning is used in the proof of a theorem in mathematics?

4. Does inductive reasoning always lead to the same conclusion? Explain.

5. Use inductive reasoning to give the next element in the list: $125, -25, 5, -1$

6. Does the conclusion below follow logically from the premises? What type of reasoning is being used?
Premise: Michael Jordan eats oatmeal.
Premise: Larry Bird eats oatmeal.
Premise: Magic Johnson eats oatmeal.
Conclusion: All great basketball players eat oatmeal.

7. How many lines can be drawn through one point?

8. Use the addition-subtraction law to complete the following:
If $x = y$, then $x + 2 = \underline{\quad?\quad}$.

Use the figure below to answer *true* or *false* in Exercises 9–16.

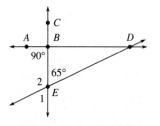

9. $\angle ABC$ is adjacent to $\angle ABE$.   **10.** $\angle DBE = 90°$ **11.** $\angle 2 = 65°$

12. $\angle 1 = 65°$ **13.** A is on $\overrightarrow{BD}$ **14.** A is on $\overrightarrow{BD}$

15. If $\angle DBE + \angle BED + \angle EDB = 180°$, then $\angle EDB = 25°$.

16. $\angle BED$ and $\angle EDB$ are supplementary.

17. Find the complement of the supplement of an angle measuring 146°20′.

30

2

*I*NTRODUCTION TO *P*ROOFS

In this chapter we begin our discussion of geometric proofs and work with problems that involve lines and angles. We conclude the chapter with a discussion of geometric constructions.

The following application depends on an understanding of topics discussed in this chapter, and its solution appears in Section 2.3, Example 1.

A family is building a recreational cabin at the edge of a wide mountain valley containing a stream that flows north to south. To supply the cabin with water, a pipe must be laid from the cabin to the stream. To minimize construction costs, the family plans to use the least amount of pipe possible. How can they determine the point on the stream bank that is closest to the cabin?

2.1 DIRECT PROOFS

Theorems are important parts of an axiomatic system. We said in Section 1.1 that a theorem is a statement that requires proof. What is a proof? Although mathematicians might disagree on the best proof for a statement, they will agree that a proof involves deductive reasoning rather than inductive reasoning. That is, a theorem is not proved by showing that it is true in a specific number of examples. We must use deductive reasoning to show that a theorem is true based on accepted assumptions and previously proved theorems.

Most theorems we prove in geometry can be stated in the form "If . . . , then . . . ." Such statements are called **conditional statements,** and are often symbolized by using $\longrightarrow$. For example, if we let P represent the statement "the sun is shining" and Q represent the statement "I can see my shadow," then the conditional statement

"If the sun is shining, then I can see my shadow,"

has the form "If P, then Q," and can be symbolized by

$P \longrightarrow Q$. Read "If P then Q."

The following example offers practice in recognizing and writing conditional statements.

EXAMPLE 1 Let P represent the statement "an animal is a dog," Q represent the statement "it has four legs," and R represent the statement "it barks."

(a) The symbolic form $P \longrightarrow R$ represents the statement "If an animal is a dog, then it barks."

(b) The symbolic form $P \longrightarrow Q$ represents the statement "If an animal is a dog, then it has four legs."

(c) The symbolic form $R \longrightarrow P$ represents the statement "If an animal barks, then it is a dog."

(d) The symbolic form $Q \longrightarrow P$ represents the statement "If an animal has four legs, then it is a dog." ☑

A variation of a conditional statement $P \longrightarrow Q$ that is often confused with the given statement is its **converse,** $Q \longrightarrow P$. A common error in reasoning is to assume that the converse of a conditional statement is true. We see from Example 1 that if we accept the statement $P \longrightarrow Q$, "If an animal is a dog, then it has four legs," it would not be correct to assume that $Q \longrightarrow P$, "If an animal has four legs, then it is a dog," follows from it. Clearly $Q \longrightarrow P$ is not true. Thus, a conditional statement and its converse are not the same.

Suppose P represents the statement "we are to win tonight" and Q represents the statement "we must play very hard." Then the conditional statement

"If we are to win tonight, then we must play very hard,"

can be symbolized by $P \longrightarrow Q$. In Example 6 of Section 1.1, we used deduc-

A politician once stated "All politicians are liars." This is a classic example of a *paradox*— a statement that is seemingly absurd or self-contradictory. If the politician is telling the truth, then by making this statement he is lying, and if the politician is lying, then by making this statement he is telling the truth.

tive reasoning to conclude that Q is true; that is, "we played hard" is true, when we assumed $P \longrightarrow Q$ and P to be true. We can symbolize this form of deductive reasoning as follows:

Premise: If we are to win tonight, then we must play very hard.

$$P \longrightarrow Q$$

Premise: We won the game tonight.

$$P$$

Conclusion: We played very hard.

$$\therefore Q$$

The horizontal line is used to separate the premises $P \longrightarrow Q$ and P from the conclusion Q, and the three dots symbolize the word *therefore*. This type of reasoning, recognized and used informally in the preceding section, serves as the basis for writing proofs, often called *direct proofs*. The classic format of a direct proof of a theorem $P \longrightarrow Q$ shows a series of statements, starting with the hypothesis P. Each statement follows from the preceding one using the reasoning of the preceding statement, and the final statement is the conclusion Q. The reasons given for the truth of each statement, written to the right of the statement, must be accepted or previously proved statements.

DIRECT PROOF OF $P \longrightarrow Q$

Suppose we already have $P \longrightarrow Q_1$, $Q_1 \longrightarrow Q_2$, $Q_2 \longrightarrow Q_3$, and $Q_3 \longrightarrow Q$ as accepted or previously proved statements. The format used to write a **direct proof** of $P \longrightarrow Q$ is:

Given: P

Prove: Q

Proof:

STATEMENTS	REASONS
1. P	1. Given
2. Q_1	2. $P \longrightarrow Q_1$
3. Q_2	3. $Q_1 \longrightarrow Q_2$
4. Q_3	4. $Q_2 \longrightarrow Q_3$
5. Q	5. $Q_3 \longrightarrow Q$
$\therefore P \longrightarrow Q.$	

The format of this direct proof shows five statements, but another proof might consist of any number of steps. Notice that P (given in statement 1) together with $P \longrightarrow Q_1$ (assumed true) gives Q_1 (statement 2). Then Q_1 with $Q_1 \longrightarrow Q_2$ gives Q_2 (statement 3), and so on.

EXAMPLE 2 Assume that the following statements are true.

Premise 1: If you are a politician, then you are elected by the people.

Premise 2: If you serve the people, then you are trustworthy.

Premise 3: If you are elected by the people, then you serve the people.

Give a direct proof of the following "theorem."

Theorem: If you are a politician, then you are trustworthy.

Given: You are a politician.

Prove: You are trustworthy.

Proof:

STATEMENTS	REASONS
1. You are a politician.	1. Given
2. You are elected by the people.	2. Premise 1
3. You serve the people.	3. Premise 3
4. You are trustworthy.	4. Premise 2

∴   If you are a politician, then you are trustworthy.   ▰

NOTE: In Example 2 we have not proved "You are trustworthy." We have shown that *if* you are a politician, *then* you are trustworthy. That is, being trustworthy depends on the condition assumed, namely that "you are a politician."   □

PRACTICE EXERCISE 1

Assume that the following statements are true.

Premise 1: If I live in the dorm, then I will have a roommate.

Premise 2: If I go to college, then I will live in the dorm.

Premise 3: If I have a roommate, then I will not be able to study in my room.

Premise 4: If I can't study in my room, then I will fail all my courses.

Give a direct proof of the following "theorem."

Theorem: If I go to college, then I will fail all my courses.

Given: I am going to college.

Prove: I will fail all my courses.

Proof:

STATEMENTS	REASONS
1. I am going to college.	1. _____
2. I will live in the dorm.	2. _____
3. I will have a roommate.	3. _____
4. I will not be able to study in my room.	4. _____
5. I will fail all my courses	5. _____

∴ If I go to college, then I will fail all my courses.

We said earlier that the converse of a conditional $P \longrightarrow Q$ is the statement $Q \longrightarrow P$. The **negation** of P is the statement "not P" and is denoted $\sim P$. For example, if P is the statement

"My car is white,"

then the negation of P ($\sim P$) is

"My car is not white."

If a statement P is true, then $\sim P$ is false; and if P is false, then $\sim P$ is true.

For the statement $P \longrightarrow Q$, we define its **inverse** as $\sim P \longrightarrow \sim Q$ and its **contrapositive** as $\sim Q \longrightarrow \sim P$. Suppose P is the statement "it is a wheel" and Q is the statement "it is round." Then

$P \longrightarrow Q$ "If it is a wheel, then it is round" has the following three variations:

$Q \longrightarrow P$	"If it is round, then it is a wheel"	Converse
$\sim P \longrightarrow \sim Q$	"If it is not a wheel, then it is not round"	Inverse
$\sim Q \longrightarrow \sim P$	"If it is not round, then it is not a wheel"	Contrapositive

In general, the inverse (and converse) of a given conditional need not be true when the conditional is true. However, the contrapositive of a given conditional is always true when the conditional is true. Notice how these facts are supported by the above example. A conditional statement can always be replaced with its contrapositive.

ANSWER TO PRACTICE EXERCISE: **1.** 1. Given, 2. Premise 2, 3. Premise 1, 4. Premise 3, 5. Premise 4

2.1 EXERCISES

Complete the direct proof of each "theorem" in Exercises 1–2.

1. *Premise 1:* If I have enough money, then I will take a trip.
   Premise 2: If I lose my job, I will be unhappy.
   Premise 3: If I take a trip, then I will lose my job.
   Theorem: If I have enough money, then I will be unhappy.

   Given: I have enough money.

   Prove: I will be unhappy.

   Proof:

STATEMENTS	REASONS
1. I have enough money.	1. _____
2. I will take a trip.	2. _____
3. I will lose my job.	3. _____
4. I will be unhappy.	4. _____

   ∴ If I have enough money, then I will be unhappy.

2. *Premise 1:* If taxes rise, then the people will be unhappy.
Premise 2: If people are unhappy, then they will go to the polls.
Premise 3: If the president gets his budget passed, then taxes will rise.
Premise 4: If people go to the polls, the president will be voted out of office.
Theorem: If the president gets his budget passed, then he will be voted out of office.

Given: _____

Prove: _____

Proof:

STATEMENTS	REASONS
1. The president gets his budget passed.	1. _____
2. _____	2. Premise 3
3. The people will be unhappy.	3. _____
4. _____	4. Premise 2
5. _____	5. Premise 4
∴ _____	

Give a direct proof of each "theorem" in Exercises 3–4.

3. *Premise 1:* If the weather report is accurate, then we will get 12 inches of snow.
Premise 2: If we get 12 inches of snow, then the streets will be treacherous.
Premise 3: If the streets are treacherous, then school will be canceled.
Theorem: If the weather report is accurate, then school will be canceled.

4. *Premise 1:* If I watch TV, then I will not do my homework.
Premise 2: If I fail geology, then I will lose my scholarship.
Premise 3: If I do not do my homework, then I will fail geology.
Premise 4: If I lose my scholarship, then my parents will be upset.
Theorem: If I watch TV, then my parents will be upset.

5. In Exercise 3, do we prove that "School will be canceled"?

6. In Exercise 4, do we prove that "My parents will be upset"?

7. If Premise 2 is left out in Exercise 3, can the same theorem be proved?

8. If Premise 3 is left out in Exercise 4, can the same theorem be proved?

Give the converse of each statement in Exercises 9–12, and note that each converse makes a different statement.

9. If a person is president, then he/she must be a United States citizen.

10. If it is Sunday, then I will watch football.

11. If a figure is a square, then it is a rectangle.

12. If you fly in an airplane, then you will go to the airport.

Give the negation of each statement.

13. The tree is a pine.

14. The city is large.

15. I received an A in the course.

16. Joe did not run in the race.

Give the inverse of each statement in Exercises 17–20. Note that each inverse makes a different statement.

17. If it rains, then I will stay indoors.

18. If this animal is a bird, then it has two legs.

19. If I have the flu, then I will run a fever.

20. If it's gold, then it glitters.

Give the contrapositive of each statement in Exercises 21–24.

21. If I take a shower, then I will get wet.

22. If a figure is a rectangle, then it is a parallelogram.

23. If I drink orange juice, then I am healthy.

24. If we beat Central State, then we are the conference champions.

Charles Dodgson (1832–1898), who wrote *Alice's Adventures in Wonderland* under the pseudonym Lewis Carroll, was also a mathematics lecturer at Oxford University in England. He invented a logic puzzle that gives a series of premises from which the reader draws a conclusion. The conclusion is actually a theorem that can be proved using a direct proof. Instead of using "If . . . then. . . ." statements, however, Lewis Carroll used statements in the form "All . . . are. . . ." Assume that "All P are Q" translates to $P \longrightarrow Q$, and remember that the contrapositive of $P \longrightarrow Q$, $\sim Q \longrightarrow \sim P$, can be used in place of $P \longrightarrow Q$. Use this assumption to rewrite each premise in Exercises 25–26 as a conditional statement and give a logical conclusion (theorem) in each Lewis Carroll puzzle.

25. *Premise 1:* All sane people can do logic.
Premise 2: All people serving on a jury are not lunatics.
Premise 3: All of your sons cannot do logic.

26. *Premise 1:* All promise-breakers are untrustworthy.
Premise 2: All wine-drinkers are very communicative.
Premise 3: All pawnbrokers are not teetotalers. (A teetotaler is not a wine-drinker.)
Premise 4: All people who keep promises are honest.
Premise 5: All communicative persons can be trusted.

2.2   PROOFS INVOLVING LINES AND ANGLES

A talented pool player knows a great deal about angles. When a ball hits the side cushion on a pool table, it will rebound along a line making the same angle with the cushion as the angle at which it struck. When a direct line between the cue ball and the intended target ball is obstructed by another ball, the player must play a "bank shot" off the cushion using a knowledge of angles.

In Section 2.1 we discussed direct proofs and introduced some of the techniques used to write them. Now we are ready to prove theorems in geometry. Recall that most theorems are "If . . . , then. . . ." statements. The word *if* directs us to the hypotheses of the theorem, the information that is given or assumed true, and the word *then* tells us the conclusion of the theorem, what must be shown to be true. We will follow a precise format in a proof by listing statements in one column and a reason for each statement in a second column. Remember that the reason for each statement can consist of given information (one of the hypotheses), a postulate, a definition, or a previously proved theorem.

The first step in writing a proof is to draw a figure showing the given information. Next, write *Given:* and list all assumed statements using notation from your drawn figure. Then write *Prove:* and state what is to be proved. Finally, write *Proof:* and head two columns with the words *STATEMENTS* and *REASONS*. The first statements written are usually taken from the *Given* statements, and the final statement will always be the *Prove* statement. The examples of proofs in this section will help you to see the format clearly.

THEOREM 2.1   ADDITION THEOREM FOR SEGMENTS

If B is a point between A and C on segment $\overline{AC}$, Q is a point between P and R on segment $\overline{PR}$, $AB = PQ$, and $BC = QR$, then $AC = PR$.

Given: B is between A and C on $\overline{AC}$
Q is between P and R on $\overline{PR}$
$AB = PQ$
$BC = QR$

Prove: $AC = PR$

Proof:

Figure 2.1   Segment Addition

STATEMENTS	REASONS
1. B is between A and C on $\overline{AC}$	1. Given
2. Q is between P and R on $\overline{PR}$	2. Given
3. $AB = PQ$	3. Given
4. $BC = QR$	4. Given
5. $AB + BC = PQ + QR$	5. Addition-subtraction law (Postulate 1.9)
6. $AC = AB + BC$ and $PR = PQ + QR$	6. Segment addition postulate (Postulate 1.13)
7. $AC = PR$	7. Substitution law (Postulate 1.11)

> **THEOREM 2.2   SUBTRACTION THEOREM FOR SEGMENTS**
>
> If B is a point between A and C on segment $\overline{AC}$, Q is a point between P and R on segment $\overline{PR}$, $AC = PR$, and $AB = PQ$, then $BC = QR$.

The proof of Theorem 2.2 is similar to that of Theorem 2.1. It is outlined in the exercises at the end of this section where you are asked to supply reasons for each statement.

> **THEOREM 2.3   ADDITION THEOREM FOR ANGLES**
>
> If D is a point in the interior of $\angle ABC$, S is a point in the interior of $\angle PQR$, $\angle ABD = \angle PQS$, and $\angle DBC = \angle SQR$, then $\angle ABC = \angle PQR$.

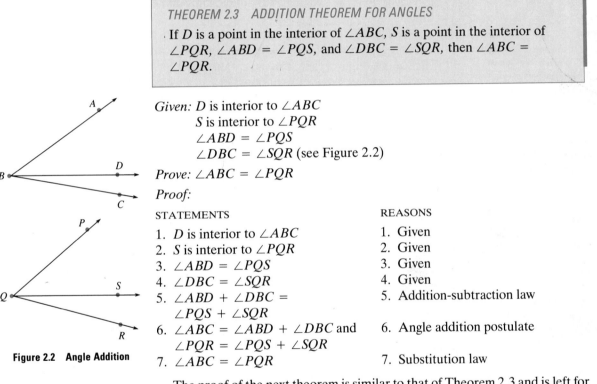

Figure 2.2   Angle Addition

Given: D is interior to $\angle ABC$
  S is interior to $\angle PQR$
  $\angle ABD = \angle PQS$
  $\angle DBC = \angle SQR$ (see Figure 2.2)

Prove: $\angle ABC = \angle PQR$

Proof:

STATEMENTS	REASONS
1. D is interior to $\angle ABC$	1. Given
2. S is interior to $\angle PQR$	2. Given
3. $\angle ABD = \angle PQS$	3. Given
4. $\angle DBC = \angle SQR$	4. Given
5. $\angle ABD + \angle DBC =$ $\angle PQS + \angle SQR$	5. Addition-subtraction law
6. $\angle ABC = \angle ABD + \angle DBC$ and $\angle PQR = \angle PQS + \angle SQR$	6. Angle addition postulate
7. $\angle ABC = \angle PQR$	7. Substitution law

The proof of the next theorem is similar to that of Theorem 2.3 and is left for you to complete as an exercise.

> **THEOREM 2.4   SUBTRACTION THEOREM FOR ANGLES**
>
> If D is a point in the interior of $\angle ABC$, S is a point in the interior of $\angle PQR$, $\angle ABC = \angle PQR$, and $\angle DBC = \angle SQR$, then $\angle ABD = \angle PQS$.

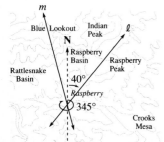

In recent years, the sport of *orienteering* has gained popularity among wilderness enthusiasts. Using a topographic map and a magnetic compass, you can find your way through a wilderness area by sighting specific landmarks on a map. For example, suppose you look at Raspberry Peak and determine its direction is 40° from your position. You then know that you are somewhere on line ℓ in the figure. If you look toward Blue Lookout and determine its direction is 345° you also know you are somewhere on line *m*. The point of intersection of ℓ and *m* indicates your approximate location in the wilderness area.

For later work, we will need several theorems that involve supplementary and complementary angles. For these theorems we will be required to illustrate the given information with a figure.

THEOREM 2.5

Two equal supplementary angles are right angles.

Notice that Theorem 2.5 is not in the form "If . . . , then. . . ." It does state, however, that "If two angles are equal and supplementary, then they are right angles."

First make a sketch of two equal supplementary angles as in Figure 2.3. The statements we make about what is given and what is to be proven refer to the figure. The figure, then, is a part of the proof.

Given: $\angle ABD = \angle DBC$
$\angle ABD$ and $\angle DBC$ are supplementary

Prove: $\angle ABD$ and $\angle DBC$ are right angles

Proof:

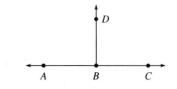

Figure 2.3   Equal Supplementary Angles

STATEMENTS	REASONS
1. $\angle ABD = \angle DBC$	1. Given
2. $\angle ABD$ and $\angle DBC$ are supplementary	2. Given
3. $\angle ABD + \angle DBC = 180°$	3. Definition of supplementary angles
4. $\angle ABD + \angle ABD = 180°$	4. Substitute $\angle ABD$ for $\angle DBC$ in Statement 3 using the substitution law
5. $2\angle ABD = 180°$	5. Distributive law
6. $\angle ABD = 90°$	6. Divide both sides of Statement 5 by 2 using the multiplication-division law
7. $\angle DBC = 90°$	7. Substitute $\angle DBC$ for $\angle ABD$ in Statement 6 using the substitution law
8. $\angle ABD$ and $\angle DBC$ are right angles	8. Definition of right angle

---

THEOREM 2.6

Complements of equal angles are equal.

Make a sketch as in Figure 2.4.

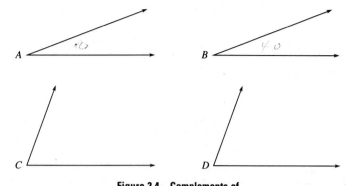

**Figure 2.4   Complements of
Equal Angles**

Given: ∠A and ∠C are
complementary
∠B and ∠D are complementary
∠A = ∠B

Prove: ∠C = ∠D

Proof:

STATEMENTS	REASONS
1. ∠A and ∠C are complementary	1. Given
2. ∠A + ∠C = 90°	2. Definition of complementary angles
3. ∠B and ∠D are complementary	3. Given
4. ∠B + ∠D = 90°	4. Definition of complementary angles
5. ∠A + ∠C = ∠B + ∠D	5. Transitive law and symmetric law
6. ∠A = ∠B	6. Given
7. ∠C = ∠D	7. Addition-subtraction law

A **corollary** is a theorem that is easy to prove as a direct result of a previously proved theorem. The next corollary follows immediately from Theorem 2.6.

COROLLARY 2.7

Complements of the same angle are equal.

A similar theorem and corollary exist for supplementary angles.

> **THEOREM 2.8**
>
> Supplements of equal angles are equal.

The proof of Theorem 2.8 is left for you to do as an exercise.

> **COROLLARY 2.9**
>
> Supplements of the same angle are equal.

Two adjacent angles whose noncommon sides lie on the same line are supplementary. For example, in Figure 2.5, $\angle 1$ and $\angle 2$ are supplementary adjacent angles. We prove this assertion in the next theorem.

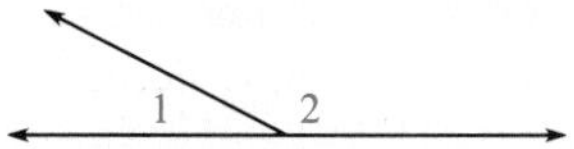

**Figure 2.5   Supplementary
Adjacent Angles**

> **THEOREM 2.10**
>
> If A, B, and C are three points on a line, with B between A and C, and $\angle ABD$ and $\angle DBC$ are adjacent angles, then $\angle ABD$ and $\angle DBC$ are supplementary.

Given: A, B, and C are on the same
line
B is between A and C
$\angle ABD$ and $\angle DBC$ are adjacent
angles (See Figure 2.6.)

Prove: $\angle ABD$ and $\angle DBC$ are
supplementary

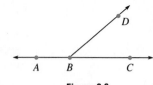

Figure 2.6

Proof:

STATEMENTS	REASONS
1. A, B, and C are on the same line	1. Given
2. B is between A and C	2. Given
3. $\angle ABC$ is a straight angle	3. Definition of straight angle
4. $\angle ABC = 180°$	4. Definition of straight angle
5. $\angle ABD + \angle DBC = \angle ABC$	5. Angle addition postulate

6. $\angle ABD + \angle DBC = 180°$ 6. Transitive law
7. $\angle ABD$ and $\angle DBC$ are   7. Definition of supplementary
   supplementary   angles

DEFINITION: VERTICAL ANGLES

Two nonadjacent angles formed by two intersecting lines are called
vertical angles.

In Figure 2.7, $\angle 1$ and $\angle 2$ are vertical angles as are $\angle 3$ and $\angle 4$. It appears
that $\angle 1 = \angle 2$ and $\angle 3 = \angle 4$ and the final theorem of this section affirms this.

THEOREM 2.11

Vertical angles are equal.

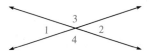

Figure 2.7   Vertical Angles

Given: $\angle 1$ and $\angle 2$ are vertical angles

Prove: $\angle 1 = \angle 2$

Proof:

STATEMENTS

1. $\angle 1$ and $\angle 2$ are vertical angles
2. $\angle 1$ and $\angle 3$ are adjacent angles
3. $\angle 1$ is supplementary to $\angle 3$

4. $\angle 2$ and $\angle 3$ are adjacent angles
5. $\angle 2$ is supplementary to $\angle 3$

6. $\angle 1 = \angle 2$

REASONS

1. Given
2. Def. of adj. $\angle$
3. Adj. $\angle$'s with 2 sides in a line are
   supp.
4. Def. of adj. $\angle$
5. Adj. $\angle$'s with 2 sides in a line are
   supp.
6. Supp. of the same $\angle$ are $=$

NOTE: When giving reasons in a proof, we often abbreviate words and use
symbols. In the proof of Theorem 2.11 we used *def.* to mean *definition, adj.* to
mean *adjacent,* and $\angle$ to mean *angle.* It is also appropriate to abbreviate the
statement of a theorem or corollary making the proof easier to read without reference to particular numbers. For example, for Reason 6 we gave "Supp. of the
same $\angle$ are $=$," instead of "Corollary 2.9" and "Reasons 3 and 5 paraphrase
Theorem 2.10."   ☐

2.2 EXERCISES

In Exercises 1–2, supply reasons for the statements in each proof.

1. Prove Theorem 2.2.

Given: B is between A and C on $\overline{AC}$
Q is between P and R on $\overline{PR}$
$AC = PR$
$AB = PQ$

Prove: $BC = QR$

Proof:

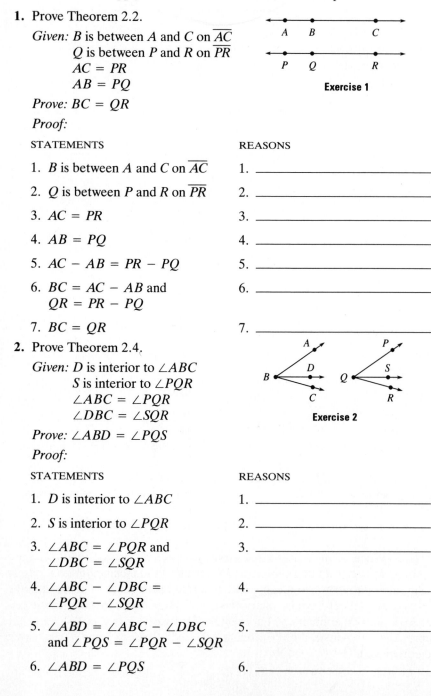

Exercise 1

STATEMENTS	REASONS
1. B is between A and C on $\overline{AC}$	1. _____
2. Q is between P and R on $\overline{PR}$	2. _____
3. $AC = PR$	3. _____
4. $AB = PQ$	4. _____
5. $AC - AB = PR - PQ$	5. _____
6. $BC = AC - AB$ and $QR = PR - PQ$	6. _____
7. $BC = QR$	7. _____

2. Prove Theorem 2.4.

Given: D is interior to $\angle ABC$
S is interior to $\angle PQR$
$\angle ABC = \angle PQR$
$\angle DBC = \angle SQR$

Prove: $\angle ABD = \angle PQS$

Proof:

Exercise 2

STATEMENTS	REASONS
1. D is interior to $\angle ABC$	1. _____
2. S is interior to $\angle PQR$	2. _____
3. $\angle ABC = \angle PQR$ and $\angle DBC = \angle SQR$	3. _____
4. $\angle ABC - \angle DBC = \angle PQR - \angle SQR$	4. _____
5. $\angle ABD = \angle ABC - \angle DBC$ and $\angle PQS = \angle PQR - \angle SQR$	5. _____
6. $\angle ABD = \angle PQS$	6. _____

Complete each proof in Exercises 3–8.

3. *Given:* B is the midpoint of $\overline{AC}$

Prove: $AB = \dfrac{AC}{2}$

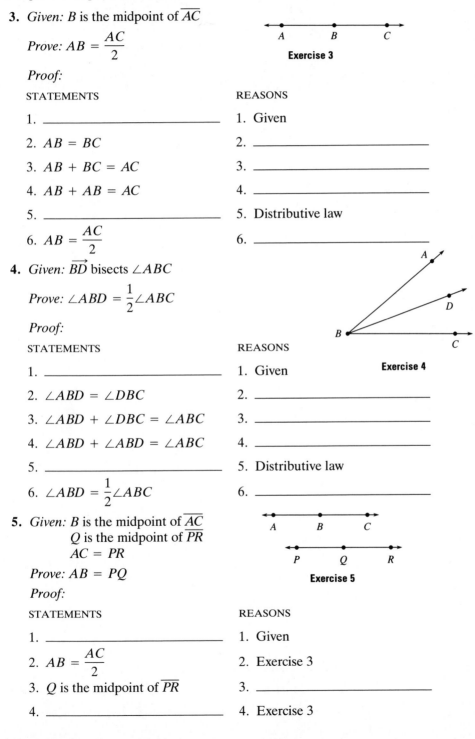

Exercise 3

Proof:

STATEMENTS	REASONS
1. _____	1. Given
2. $AB = BC$	2. _____
3. $AB + BC = AC$	3. _____
4. $AB + AB = AC$	4. _____
5. _____	5. Distributive law
6. $AB = \dfrac{AC}{2}$	6. _____

4. *Given:* $\overrightarrow{BD}$ bisects $\angle ABC$

Prove: $\angle ABD = \dfrac{1}{2}\angle ABC$

Proof:

Exercise 4

STATEMENTS	REASONS
1. _____	1. Given
2. $\angle ABD = \angle DBC$	2. _____
3. $\angle ABD + \angle DBC = \angle ABC$	3. _____
4. $\angle ABD + \angle ABD = \angle ABC$	4. _____
5. _____	5. Distributive law
6. $\angle ABD = \dfrac{1}{2}\angle ABC$	6. _____

5. *Given:* B is the midpoint of $\overline{AC}$
 Q is the midpoint of $\overline{PR}$
 $AC = PR$

Prove: $AB = PQ$

Exercise 5

Proof:

STATEMENTS	REASONS
1. _____	1. Given
2. $AB = \dfrac{AC}{2}$	2. Exercise 3
3. Q is the midpoint of $\overline{PR}$	3. _____
4. _____	4. Exercise 3

5. _____   5. Given

6. $\dfrac{AC}{2} = \dfrac{PR}{2}$ 6. _____

7. $AB = PQ$ 7. _____

6. *Given:* $\overrightarrow{BD}$ bisects $\angle ABC$
   $\overrightarrow{QS}$ bisects $\angle PQR$
   $\angle ABC = \angle PQR$

   Prove: $\angle ABD = \angle PQS$

   Proof:

Exercise 6

STATEMENTS   REASONS

1. _____   1. Given

2. $\angle ABD = \dfrac{1}{2}\angle ABC$ 2. Exercise 4

3. $\overrightarrow{QS}$ bisects $\angle PQR$ 3. _____

4. _____   4. Exercise 4

5. _____   5. Given

6. $\dfrac{1}{2}\angle ABC = \dfrac{1}{2}\angle PQR$ 6. _____

7. _____   7. Substitution law

7. *Given:* $\angle ABC$ is a right angle
   $\angle DBC$ and $\angle 1$ are
   complementary

   Prove: $\angle ABD = \angle 1$

   Proof:

Exercise 7

STATEMENTS   REASONS

1. _____   1. Given

2. $\angle ABC = 90°$ 2. _____

3. $\angle DBC + \angle ABD = \angle ABC$ 3. _____

4. $\angle DBC + \angle ABD = 90°$ 4. _____

5. $\angle DBC$ and $\angle 1$ are
   complementary   5. _____

6. $\angle DBC + \angle 1 = 90°$ 6. _____

7. $\angle DBC + \angle ABD =$
   $\angle DBC + \angle 1$ 7. _____

8. _____   8. Addition-subtraction law

8. *Given:* Three lines *m, n,* and *ℓ*
    as shown
        $\angle 1 + \angle 2 = 180°$
  Prove: $\angle 4 = \angle 3$

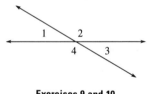

Exercise 8

Proof:

STATEMENTS	REASONS
1. _____	1. Given
2. $\angle 1$ and $\angle 3$ are supplementary	2. _____
3. $\angle 1 + \angle 3 = 180°$	3. _____
4. $\angle 1 + \angle 2 = \angle 1 + \angle 3$	4. _____
5. $\angle 2 = \angle 3$	5. _____
6. $\angle 4$ and $\angle 2$ are vertical angles	6. _____
7. _____	7. Vert. $\angle$'s are $=$
8. _____	8. Transitive law

9. Use the figure below and name all pairs of adjacent angles.

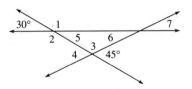

Exercises 9 and 10

10. Use the figure above and name all pairs of vertical angles.

In Exercises 11–16, use the figure below to answer each question.

11. What is the measure of $\angle 1$?   **12.** What is the measure of $\angle 3$?
13. What is the measure of $\angle 2$?   **14.** What is the measure of $\angle 4$?

15. If $\angle 3 + \angle 5 + \angle 6 = 180°$, what is the measure of $\angle 6$?

16. If $\angle 3 + \angle 5 + \angle 6 = 180°$, what is the measure of $\angle 7$?

17. If $\angle A$ and $\angle B$ are vertical angles, must $\angle A = \angle B$?

18. If $\angle A = \angle B$, must $\angle A$ and $\angle B$ be vertical angles?

19. If $\angle A$ and $\angle B$ are supplementary, can $\angle A$ and $\angle B$ be vertical angles too?

20. If $\angle A$ and $\angle B$ are complementary, can $\angle A$ and $\angle B$ be vertical angles too?

21. Can two supplementary angles both be obtuse?

22. Can two complementary angles both be acute?

23. Can two vertical angles both be obtuse?

24. Can two vertical angles both be acute?

25. Can two right angles be vertical angles?

26. Can two vertical angles both be right angles?

27. If two angles are vertical angles, can one be obtuse and the other acute?

28. If two angles are adjacent angles, can one be obtuse and the other acute?

29. If two angles are adjacent angles, can they be supplementary?

30. If two angles are adjacent angles, can they be complementary?

31. Prove Theorem 2.8.

32. In the figure below, $PQ = RS$. Prove $PR = QS$.

P   Q   R   S

33. In the figure above, $PR = QS$. Prove $PQ = RS$.

34. Prove that the sum of the measures of the complements of complementary angles is 90°.

35. Prove that the sum of the measures of the supplements of complementary angles is 270°.

2.3 CONSTRUCTIONS INVOLVING LINES AND ANGLES

Part of geometry involves making accurate drawings of geometric figures. Such drawings are called **geometric constructions.** Two instruments are used to make constructions, a **straightedge** and a **compass.** A straightedge, a ruler with no marks of scale, is used to draw a line between two points. (We often use a standard ruler but do not use it to measure lengths.) A compass is used to draw circles, or portions of circles, called **arcs.** (Circles and arcs will be studied in more detail in Chapter 7.)

The first construction we consider involves duplicating a given line segment.

The bronze sculpture, Geometria, by Renaissance sculptor Antonio del Pollaiolo (1433–98) depicts the study of geometry. It shows the construction of a geometric figure using a compass and a straightedge. The sculpture appears on the base of the tomb of Pope Sixtus IV and is located in St. Peter's Cathedral in Rome.

CONSTRUCTION 2.1

Construct a line segment with the same length as a given line segment.

Given: Line segment $\overline{AB}$ (See Figure 2.8.)

To Construct: Line segment $\overline{CD}$ with $CD = AB$.

Construction:

1. Draw a line $\overleftrightarrow{CE}$ containing point C.
2. Place the point of a compass at point A and the pencil point at point B.
3. Without changing the distance between these points on the compass, place the point of the compass at point C and with the pencil point draw an arc that intersects $\overleftrightarrow{CE}$. The point of intersection of the arc and $\overrightarrow{CE}$ determines the point D. $\overline{CD}$ is the desired line segment.

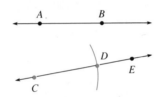

Figure 2.8   Reconstructing Segment Length

The next construction involves duplicating a given angle.

CONSTRUCTION 2.2

Construct an angle with the same measure as a given angle.

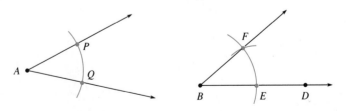

Figure 2.9   Reconstructing Angle Measure

Given: $\angle A$ (See Figure 2.9.)

To Construct: $\angle B$ such that $\angle A = \angle B$

Construction:

1. Draw ray $\overrightarrow{BD}$.
2. Choose a convenient setting of the compass, place the point at A and draw an arc that intersects the sides of $\angle A$ at two points, for instance P and Q.
3. Without changing the compass setting, place the point at B and draw a suitable arc that intersects $\overrightarrow{BD}$ at a point, for instance E.
4. Place the points of the compass so that one is at Q and the other is at P.
5. Without changing the compass setting, place the point at E and draw an arc that intercepts the arc drawn in Step 3. Label this point F.
6. Draw ray $\overrightarrow{BF}$ to form the second side of the desired angle, $\angle B$.

Next we divide a line segment into two equal parts.

DEFINITION: LINE SEGMENT BISECTION

Let $\overline{AB}$ be a line segment. To **bisect** $\overline{AB}$ is to identify a point C between A and B such that $AC = CB$. The point C is called the **midpoint** of $\overline{AB}$. Any line or line segment that contains the midpoint C but no other point of $\overline{AB}$ is called a **bisector** of $\overline{AB}$.

CONSTRUCTION 2.3

Construct a bisector of a given line segment.

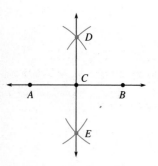

Figure 2.10   Segment Bisector

Given: Line segment $\overline{AB}$ (See Figure 2.10.)

To Construct: A bisector $\overleftrightarrow{CD}$ of $\overline{AB}$

Construction:

1. Set the compass so that the distance between the point and the pencil is greater than $\dfrac{AB}{2}$.
2. Set the point at A and draw two arcs, one above $\overline{AB}$ and the other below $\overline{AB}$. Repeat this procedure using the same compass setting with the point at B.
3. The arcs drawn in Step 2 should intersect at two points we label D and E.
4. Draw the line $\overleftrightarrow{DE}$. Then $\overleftrightarrow{DE}$ is a bisector of $\overline{AB}$ and determines midpoint C.

Construction 2.3 shows that a given line segment has a midpoint. The next postulate guarantees that this midpoint is unique.

POSTULATE 2.1 MIDPOINT POSTULATE

Each line segment has exactly one midpoint.

When two lines intersect and form right angles, we call the lines *perpendicular.*

DEFINITION: PERPENDICULAR LINES

Two lines are **perpendicular** if they intersect and form equal adjacent angles. If $\overleftrightarrow{AB}$ is perpendicular to $\overleftrightarrow{CD}$, we write $\overleftrightarrow{AB} \perp \overleftrightarrow{CD}$. Two line segments are perpendicular if they intersect and are contained in perpendicular lines.

In Figure 2.11, lines ℓ and m are perpendicular if and only if $\angle 1 = \angle 2$. When $\ell \perp m$, we represent this fact in figures using the symbol ⌐ as shown. Also, in Figure 2.12, $\overline{AB} \perp \overline{CD}$ provided $\overleftrightarrow{AB} \perp \overleftrightarrow{CD}$.

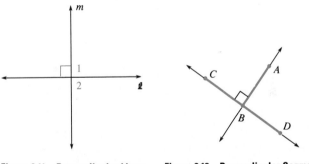

Figure 2.11   Perpendicular Lines   Figure 2.12   Perpendicular Segments

Although a line segment can have many bisectors, the most important one is the *perpendicular bisector.* Construction 2.3 shows this bisector.

DEFINITION: PERPENDICULAR BISECTOR OF A SEGMENT

A line that both bisects and is perpendicular to a given line segment is called a **perpendicular bisector** of the segment.

We assume that the perpendicular bisector of a segment is unique.

POSTULATE 2.2   PERPENDICULAR BISECTOR POSTULATE

Each given line segment has exactly one perpendicular bisector.

In the next construction we find a line perpendicular to a given line passing through a given point on the line.

CONSTRUCTION 2.4

Construct a line perpendicular to a given line passing through a given point on the line.

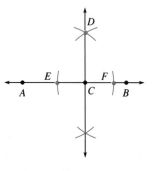

Figure 2.13

Given: Line $\overleftrightarrow{AB}$ with point C on $\overleftrightarrow{AB}$ (See Figure 2.13.)

To Construct: A line $\overleftrightarrow{DC}$ such that $\overleftrightarrow{DC} \perp \overleftrightarrow{AB}$

Construction:

1. Choose a convenient setting for the compass, place the point at C, and make two arcs that intersect line $\overleftrightarrow{AB}$. Label these points E and F.

2. Use Construction 2.3 to draw the perpendicular bisector of the segment $\overline{EF}$. The result, $\overleftrightarrow{DC}$ in Figure 2.13, is the desired line perpendicular to $\overleftrightarrow{AB}$ at the point C.

POSTULATE 2.3

There is exactly one line perpendicular to a given line passing through a given point on the line.

Next we consider finding a line perpendicular to a given line passing through a given point *not* on the line.

CONSTRUCTION 2.5

Construct a line perpendicular to a given line passing through a given point not on that line.

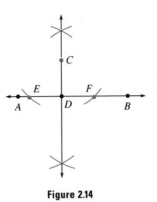

Figure 2.14

Given: Line $\overleftrightarrow{AB}$ with point C not on $\overleftrightarrow{AB}$ (See Figure 2.14.)

To Construct: A line $\overleftrightarrow{CD}$ such that $\overleftrightarrow{CD} \perp \overleftrightarrow{AB}$

Construction:

1. Choose a convenient setting for the compass, place the point at C, and make two arcs that intersect line $\overleftrightarrow{AB}$. Label these points E and F.

2. Use Construction 2.3 to draw the perpendicular bisector of the segment $\overline{EF}$. The result, $\overleftrightarrow{CD}$ in Figure 2.14, is the desired line perpendicular to $\overleftrightarrow{AB}$ and passing through C.

POSTULATE 2.4

There is exactly one line perpendicular to a given line passing through a given point not on that line.

We said that the **distance** between two points A and B is the length of the segment $\overline{AB}$, denoted by AB. Construction 2.5 and Postulate 2.4 give us a way to find the *distance* from a point to a line. Use Figure 2.14 as a reference for the following definition.

DEFINITION: DISTANCE FROM A POINT TO A LINE

Let $\overleftrightarrow{AB}$ be a line with C a point not on $\overleftrightarrow{AB}$. If D is the point on $\overleftrightarrow{AB}$ such that $\overleftrightarrow{CD} \perp \overleftrightarrow{AB}$, the **distance** from C to $\overleftrightarrow{AB}$ is CD, the length of $\overline{CD}$.

We now have a way to solve the applied problem given in the chapter introduction.

E X A M P L E 1   A family is building a recreational cabin at the edge of a wide mountain valley containing a stream that flows north to south. To supply the cabin with water, a pipe must be laid from the cabin to the stream. To minimize construction costs, the family plans to use the least amount of pipe possible. How can they determine the point on the stream bank that is closest to the cabin?

If we think of the stream as a straight line, and the cabin as a point not on the line, the desired point on the bank of the stream, labeled P in Figure 2.15 below, is found by constructing the line ℓ perpendicular to the stream passing through a point, the cabin, not on that line.   ◪

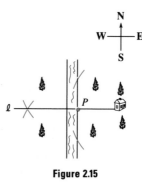

Figure 2.15

We have considered the problem of bisecting a line segment, and now we'll try to bisect an angle.

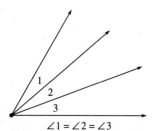

$\angle 1 = \angle 2 = \angle 3$

For years mathematicians tried to find a way to trisect (divide into three equal angles) an angle using only a straightedge and compass. Some special angles (such as a straight angle) can be trisected, but the methods used do not carry over to more general angles. In fact, it has now been proved that it is impossible to trisect an arbitrary angle by construction.

DEFINITION: ANGLE BISECTION

Let $\angle ABC$ be an angle. To **bisect** $\angle ABC$ is to identify a ray $\overrightarrow{BD}$, where D is in the interior of $\angle ABC$ and $\angle ABD = \angle DBC$. The ray $\overrightarrow{BD}$ is called the **bisector** of $\angle ABC$.

CONSTRUCTION 2.6

Construct a bisector of a given angle.

Given: $\angle ABC$ (See Figure 2.16 on the next page.)

To Construct: Ray $\overrightarrow{BD}$ that bisects $\angle ABC$

Construction:

1. Choose a convenient setting for the compass, place the point at B, the vertex of $\angle ABC$, and make two arcs that intersect $\overrightarrow{BA}$ and $\overrightarrow{BC}$. Label these points E and F.

2. Choose a convenient setting for the compass so the distance between the pencil and the point is greater than $\dfrac{EF}{2}$. Set the point at E and draw an arc interior to $\angle ABC$. Then set the point at F and draw a second arc with the same compass setting that intercepts the first arc at a point we label D.

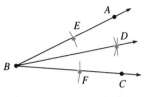

Figure 2.16   Angle Bisector

3. Draw the ray $\overrightarrow{BD}$. $\overrightarrow{BD}$ is the desired bisector of $\angle ABC$. We shall assume that the bisector of an angle is unique.

POSTULATE 2.5   ANGLE BISECTOR POSTULATE
Each angle has exactly one bisector.

2.3 EXERCISES

1. Draw a line segment $\overline{AB}$ approximately 3 inches in length, and a line ℓ. Construct $\overline{CD}$ on ℓ such that $AB = CD$.

2. Draw a line segment $\overline{AB}$ approximately 2 inches in length, and a line ℓ. Construct $\overline{CD}$ on ℓ such that CD is twice AB.

3. Draw an acute angle and then construct another acute angle with the same measure.

4. Draw an obtuse angle and then construct another obtuse angle with the same measure.

5. Draw an acute angle and then construct another angle whose measure is twice that of the first.

6. Draw an obtuse angle and then construct another angle whose measure is twice that of the first.

7. Draw a line segment $\overline{AB}$ approximately 2 inches in length and locate the midpoint of the segment by bisecting it.

8. Draw a line segment $\overline{AB}$ approximately 4 inches in length and divide the segment into four equal parts by first bisecting $\overline{AB}$ and then bisecting each resulting part.

9. Draw a line segment $\overline{AB}$ approximately 3 inches in length and construct the perpendicular bisector of $\overline{AB}$.

10. Draw a line ℓ and select two points P and Q on ℓ. Construct the perpendicular bisector of $\overline{PQ}$ and locate the midpoint of $\overline{PQ}$.

11. Draw a line ℓ and select a point P on ℓ. Construct the line through P and perpendicular to ℓ.

12. Draw a line ℓ and select a point P not on ℓ. Construct the line through P and perpendicular to ℓ.

13. Draw an acute angle and construct its bisector.

14. Draw an obtuse angle and construct its bisector.

15. Draw an acute angle and use construction techniques to divide the angle into four equal angles.

16. Draw an obtuse angle and use construction techniques to divide the angle into four equal angles.

17. What is the distinction between a ruler and a straightedge?

18. How many midpoints does a line segment have?

19. How many perpendicular bisectors does a line segment have?

20. How many bisectors does a line segment have?

21. Draw a line ℓ with a point P not on ℓ. Construct the line m through P and perpendicular to ℓ and label the point of intersection of ℓ and m as Q. The length of $\overline{PQ}$ is the distance from P to ℓ. Use a ruler to measure $\overline{PQ}$ and approximate this distance.

Assume that every point on the perpendicular bisector of a line segment with endpoints A and B is equidistant from A and B. Also, assume that every point on the bisector of an angle is equidistant from the sides of the angle. Use this information and the figure on the right in Exercises 22–24.

22. A ranger wishes to drill a well at the edge of the forest equidistant from the cabin and the ranger station. Explain how she should locate this point at the forest's edge.

23. A ranger wishes to drill a well at the edge of the forest equidistant from the stream and the road. Explain how he should locate this point at the forest's edge.

24. A meteorologist wishes to place a weather station in the meadow equidistant from the ranger station, the bridge, and the cabin. Explain how she should locate this point.

KEY TERMS AND SYMBOLS

2.1 conditional statement
($\longrightarrow$), p. 32
converse, p. 32
direct proof, p. 33
paradox, p. 32
negation, p. 35
inverse, p. 35
contrapositive, p. 35
2.2 corollary, p. 41

vertical angles, p. 43
2.3 geometric construction,
p. 48
straightedge, p. 48
compass, p. 48
arcs, p. 48
bisect (a line segment),
p. 50
midpoint, p. 50

perpendicular lines ($\perp$),
p. 51
perpendicular bisector,
p. 51
distance (from a point to a
line), p. 53
bisector (of an angle),
p. 54

PROOF TECHNIQUES

To Prove:

Two Angles Equal

1. Show they are both right angles. (Postulate 1.14, p. 23)
2. Show they are complements of the same or equal angles. (Theorem 2.6 or Corollary 2.7, pp. 40–41)
3. Show they are supplements of the same or equal angles. (Theorem 2.8 or Corollary 2.9, p. 42)
4. Show they are vertical angles. (Theorem 2.11, p. 43)
5. Show they can be formed as the sum or difference of equal corresponding angles. (Theorems 2.3 and 2.4, p. 39)

Two Angles Complementary

1. Show their sum is a right angle that measures 90°. (Definition of complementary angles, p. 24)

Two Angles Supplementary

1. Show their sum is a straight angle that measures 180°. (Definition of supplementary angles, p. 24)
2. Show they are adjacent angles whose noncommon sides lie on the same line. (Theorem 2.10, p. 42)

Two Segments Equal

1. Show that each can be formed by adding or subtracting equal corresponding segments. (Theorems 2.1 and 2.2, pp. 38–39)
2. Show that they are segments of a third segment formed by the midpoint of the third segment and its endpoints. (Definition of midpoint, p. 50)

Section 2.1

In Exercises 1–2 give a direct proof of the "theorems."

1. *Premise 1:* If it is Gleep, then it is Glop.
   Premise 2: If it is Gunk, then it is Grob.
   Premise 3: If it is Glop, then it is Gunk.
   Theorem: If it is Gleep, then it is Grob.

2. *Premise 1:* If it will burn, it can be destroyed by fire.
   Premise 2: If it is a tree, it is made of wood.
   Premise 3: If it is made of wood, it will burn.
   Theorem: If it is a tree, it can be destroyed by fire.

In Exercises 3–4 give the converse of each statement.

3. If it is ice, then it is cold.

4. If you love someone, then you will want the best for that person.

In Exercises 5–6 give the negation of each statement.

5. My car is red.

6. The moon is not made of green cheese.

In Exercises 7–8 give the inverse of each statement.

7. If the runner wins the race, then she must be in excellent condition.

8. If the painting is a Picasso, then it is valuable.

In Exercises 9–10 give the contrapositive of each statement.

9. If the weather is good, then I will climb the mountain.

10. If I drink, then I do not drive.

Section 2.2

Use the figure below in Exercises 11–14.

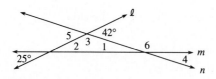

11. What is the measure of ∠5?

12. What is the measure of ∠3?

13. If ∠1 + ∠2 + ∠3 = 180°, what is the measure of ∠4?

14. If ∠1 + ∠2 + ∠3 = 180°, what is the measure of ∠6?

15. Complete the following proof.

Given: ∠1 and ∠2 are complementary
∠1 and ∠3 are vertical angles
∠2 and ∠4 are vertical angles

Prove: ∠3 and ∠4 are complementary

Proof:

STATEMENTS	REASONS
1. ∠1 and ∠2 are complementary	1. _____
2. _____	2. Def. of comp. ∠'s
3. _____	3. Given
4. ∠1 = ∠3	4. _____
5. _____	5. Given
6. ∠2 = ∠4	6. _____
7. ∠3 + ∠4 = 90°	7. Substitution law
8. _____	8. Def. of comp. ∠'s

16. Prove that the bisectors of adjacent supplementary angles are perpendicular.

Section 2.3

17. Draw a line segment $\overline{PQ}$ approximately 2 inches in length, and line m containing point C. Construct $\overline{CD}$ on m such that CD is half PQ.

18. Draw an acute angle and construct its bisector.

19. Draw a line segment $\overline{AB}$ and construct the perpendicular bisector of $\overline{AB}$.

20. How many bisectors does segment $\overline{CD}$ have?

21. How many perpendicular bisectors does segment $\overline{CD}$ have?

22. Draw an acute angle, ∠PQR and ray $\overrightarrow{AB}$. Construct an acute ∠CAB so that ∠CAB = ∠PQR.

PRACTICE TEST

1. Give a direct proof of the following "theorem."

Premise 1: If I go on trial, then I will be convicted.
Premise 2: If I am arrested, then I will go on trial.
Premise 3: If I am convicted, then I will go to jail.
Premise 4: If I rob a bank, then I will be arrested.
Theorem: If I rob a bank, then I will go to jail.

2. Give the converse of the statement, "If it is milk, then it is white."

3. Give the negation of the statement, "The road to success is difficult."

4. Give the inverse of the statement, "If it is a collie, then it is a dog."

5. Give the contrapositive of the statement, "If the fruit is picked, then it is ripe."

6. *Given:* ∠1 = ∠2
   Prove: ∠3 = ∠4

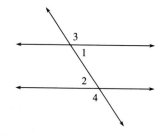

7. Draw a line segment $\overline{AB}$ and an acute angle ∠*PQR*. Construct an acute angle equal to ∠*PQR* with vertex at the midpoint *C* of $\overline{AB}$ and one side $\overrightarrow{CB}$.

8. Draw an obtuse angle and construct its bisector.

3

TRIANGLES

The triangle is the most important geometric figure we will study because of its applications in art and engineering as well as its influence in developing other areas of geometry. In this chapter, we classify triangles by their sides and angles and develop three ways to determine if two triangles have the same size and shape.

Properties of triangles give us the necessary tools for solving many applied problems. One of these is given below and solved in Example 2 of Section 3.3.

Mr. Wells owns a cabin on the south shore of Lake Pleasant. Directly north of the cabin on the north shore is a boat dock. To determine the distance from the cabin to the dock, Mr. Wells does the following: He first paces 40 yd due east from the cabin and drives a reference stake into the ground. He then paces another 40 yd due east, at which point he turns due south and walks until he meets the straight line formed by the dock and the stake. The total distance paced off in the southerly direction is 150 yd, and he concludes that the distance from the cabin to the dock is also 150 yd. Explain why this is correct.

3.1 CLASSIFYING TRIANGLES

Much of geometry involves studying figures such as those in Figure 3.1. The undefined terms *interior* and *exterior* used in conjunction with angles in Chapter 2 are also applicable to other geometric figures. Four additional undefined terms are *sides, closed, included,* and *opposite*. Figure 3.1(a) shows a closed three-sided figure; Figure 3.1(b) shows a closed six-sided figure with the point *G* in its interior and point *H* in its exterior; Figure 3.1(c) shows a four-sided figure that is not closed.

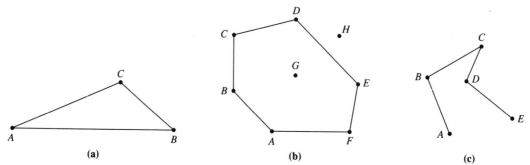

Figure 3.1   Geometric Figures

Perhaps the simplest but most important geometric figure is a *triangle*, a closed three-sided figure.

DEFINITION: TRIANGLE

Let *A*, *B*, and *C* be three points not on the same line. The figure formed by the three segments $\overline{AB}$, $\overline{BC}$, and $\overline{AC}$ is called a **triangle**, denoted by $\triangle ABC$. The three segments $\overline{AB}$, $\overline{BC}$, and $\overline{AC}$ form the **sides** of $\triangle ABC$, and the three points *A*, *B*, and *C* are the **vertices** (singular—**vertex**) of $\triangle ABC$.

How many different triangles can you find in the figure below?

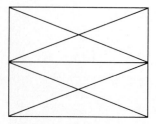

You might be amazed to discover that there are actually eighteen.

Figure 3.1(a) shows $\triangle ABC$. We say, for example, that side $\overline{BC}$ is opposite $\angle A$, and that $\angle A$ is included by the sides $\overline{AC}$ and $\overline{AB}$. Also, side $\overline{BC}$ is included by $\angle B$ and $\angle C$, and $\angle A$ is opposite side $\overline{BC}$. Notice that every triangle has six basic parts: three sides and three angles.

Triangles are often classified by their angles.

USING ANGLES TO CLASSIFY TRIANGLES

1. An **acute triangle** is a triangle in which all angles are acute.
2. A **right triangle** is a triangle in which one angle is a right angle. The side opposite the right angle is the **hypotenuse** of the triangle, and the other two sides are **legs**.

3. An **obtuse triangle** is a triangle in which one angle is obtuse.
4. An **equiangular triangle** is a triangle in which all three angles are equal.

Figure 3.2 shows four triangles classified by their angles.

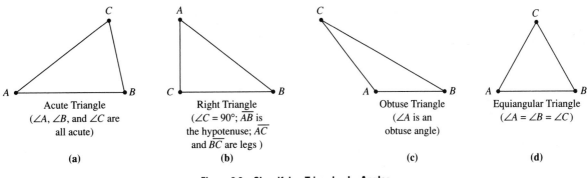

Acute Triangle
($\angle A$, $\angle B$, and $\angle C$ are all acute)
(a)

Right Triangle
($\angle C = 90°$; $\overline{AB}$ is the hypotenuse; $\overline{AC}$ and $\overline{BC}$ are legs)
(b)

Obtuse Triangle
($\angle A$ is an obtuse angle)
(c)

Equiangular Triangle
($\angle A = \angle B = \angle C$)
(d)

Figure 3.2   Classifying Triangles by Angles

Triangles can also be classified by their sides.

> USING SIDES TO CLASSIFY TRIANGLES
> 1. A **scalene triangle** is a triangle in which no two sides are equal.
> 2. An **isosceles triangle** is a triangle in which two sides are equal. The third side is its **base.**
> 3. An **equilateral triangle** is a triangle in which all three sides are equal.

Figure 3.3 shows three triangles classified by their sides. Notice that if a triangle is equilateral, then it is also isosceles. However, the converse is not true because an isosceles triangle need not be equilateral.

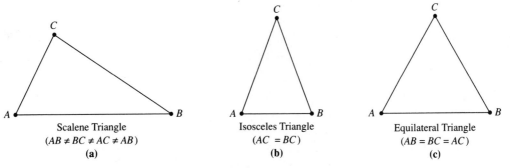

Scalene Triangle
($AB \ne BC \ne AC \ne AB$)
(a)

Isosceles Triangle
($AC = BC$)
(b)

Equilateral Triangle
($AB = BC = AC$)
(c)

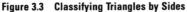

Figure 3.3   Classifying Triangles by Sides

We can also describe triangles with a combination of terms involving both angles and sides as shown in Figure 3.4.

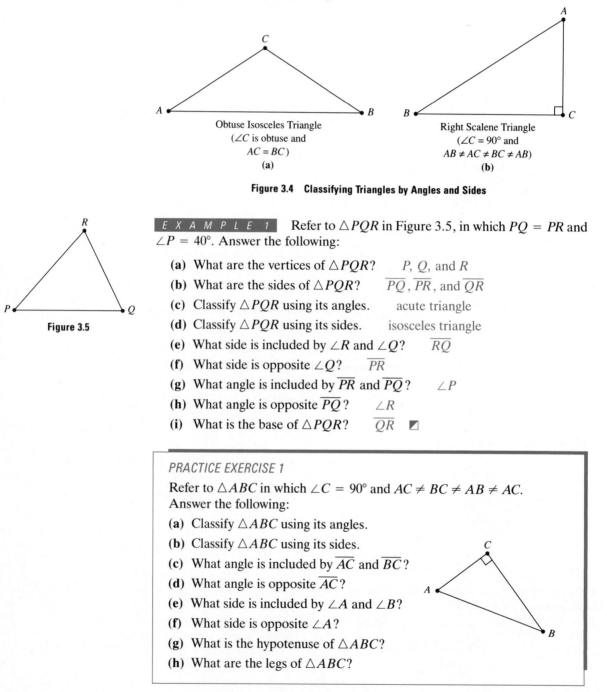

Obtuse Isosceles Triangle
($\angle C$ is obtuse and
$AC = BC$)
(a)

Right Scalene Triangle
($\angle C = 90°$ and
$AB \neq AC \neq BC \neq AB$)
(b)

Figure 3.4   Classifying Triangles by Angles and Sides

E X A M P L E 1   Refer to $\triangle PQR$ in Figure 3.5, in which $PQ = PR$ and $\angle P = 40°$. Answer the following:

Figure 3.5

(a) What are the vertices of $\triangle PQR$?   *P, Q,* and *R*

(b) What are the sides of $\triangle PQR$? $\overline{PQ}, \overline{PR},$ and $\overline{QR}$

(c) Classify $\triangle PQR$ using its angles.   acute triangle

(d) Classify $\triangle PQR$ using its sides.   isosceles triangle

(e) What side is included by $\angle R$ and $\angle Q$? $\overline{RQ}$

(f) What side is opposite $\angle Q$? $\overline{PR}$

(g) What angle is included by $\overline{PR}$ and $\overline{PQ}$? $\angle P$

(h) What angle is opposite $\overline{PQ}$? $\angle R$

(i) What is the base of $\triangle PQR$? $\overline{QR}$ ◪

PRACTICE EXERCISE 1

Refer to $\triangle ABC$ in which $\angle C = 90°$ and $AC \neq BC \neq AB \neq AC$. Answer the following:

(a) Classify $\triangle ABC$ using its angles.

(b) Classify $\triangle ABC$ using its sides.

(c) What angle is included by $\overline{AC}$ and $\overline{BC}$?

(d) What angle is opposite $\overline{AC}$?

(e) What side is included by $\angle A$ and $\angle B$?

(f) What side is opposite $\angle A$?

(g) What is the hypotenuse of $\triangle ABC$?

(h) What are the legs of $\triangle ABC$?

A useful number associated with every triangle is its *perimeter.*

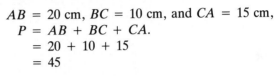

DEFINITION: PERIMETER OF A TRIANGLE

The **perimeter** P of $\triangle ABC$ is the sum of the lengths of its sides. That is,

$$P = AB + BC + CA.$$

E X A M P L E   2   Find the perimeter of $\triangle ABC$ shown in Figure 3.6. Because the lengths of the sides are

$$AB = 20 \text{ cm}, BC = 10 \text{ cm}, \text{ and } CA = 15 \text{ cm},$$
$$P = AB + BC + CA.$$
$$= 20 + 10 + 15$$
$$= 45$$

Figure 3.6

Thus, the perimeter of $\triangle ABC$ is 45 cm.   ◪

The next example requires the ability to solve a simple equation using techniques learned in beginning algebra.

E X A M P L E   3   The base of an isosceles triangle is 12 inches in length and its perimeter is 40 inches. What is the length of each of its equal sides?

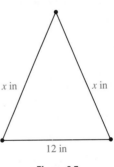

Figure 3.7

First make a sketch of the triangle as shown in Figure 3.7 with the unknown sides labeled x inches. Because the perimeter is 40 inches, we have

$$40 = x + x + 12.$$
$$40 = 2x + 12 \qquad \text{Combine terms}$$
$$28 = 2x \qquad \text{Subtract 12 from both sides}$$
$$14 = x \qquad \text{Divide both sides by 2}$$

Thus, each equal side is 14 inches long.   ◪

Because of their properties of strength as well as aesthetic appeal, triangles are often used in the construction of buildings such as the Cadet Chapel at the Air Force Academy outside Colorado Springs, Colorado.

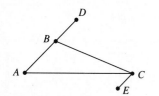

Figure 3.8   Exterior and Interior Angles

PRACTICE EXERCISE 2

The perimeter of an equilateral triangle is 75 ft. What is the length of each side?

Several other types of angles can be associated with a given triangle. Refer to Figure 3.8 for the following definition.

> **DEFINITION: EXTERIOR AND INTERIOR ANGLES**
>
> Given $\triangle ABC$ and point D on $\overleftrightarrow{AB}$ with B between A and D. Then $\angle DBC$ is an **exterior angle** of $\triangle ABC$, and $\angle A$, $\angle ABC$, and $\angle BCA$ are **interior angles** of $\triangle ABC$. Also, $\angle A$ and $\angle BCA$ are **remote interior angles** relative to $\angle DBC$, and $\angle ABC$ is an **adjacent interior angle** relative to $\angle DBC$.

NOTE: In Figure 3.8, $\angle ECA$ is *not* an exterior angle of $\triangle ABC$ because E is not on $\overleftrightarrow{AC}$ or $\overleftrightarrow{BC}$.   □

ANSWERS TO PRACTICE EXERCISES: 1. (a) right triangle **(b)** scalene **(c)** $\angle C$ **(d)** $\angle B$ **(e)** $\overline{AB}$ **(f)** $\overline{BC}$ **(g)** $\overline{AB}$ **(h)** $\overline{AC}$ and $\overline{BC}$ **2.** 25 ft

3.1 EXERCISES

Exercises 1–10 refer to $\triangle DEF$ shown below, in which $DF = EF$ and $\angle DFE = 120°$.

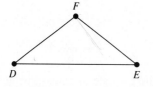

1. What are the vertices of $\triangle DEF$?

2. What are the sides of $\triangle DEF$?

3. Classify $\triangle DEF$ using its angles.

4. Classify $\triangle DEF$ using its sides.

5. What side is included by $\angle D$ and $\angle E$?

6. What side is opposite $\angle E$?

7. What angle is included by $\overline{DF}$ and $\overline{EF}$?

8. What angle is opposite $\overline{FE}$?

9. What is the base of $\triangle DEF$?

10. What is the hypotenuse of $\triangle DEF$?

Exercises 11–20 refer to $\triangle ABC$ shown below, in which $AB = BC = AC$ and $\angle A = \angle B = \angle C$.

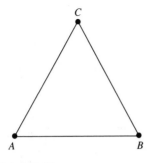

11. What are the sides of $\triangle ABC$?

12. What are the vertices of $\triangle ABC$?

13. Classify $\triangle ABC$ using its sides.

14. Classify $\triangle ABC$ using its angles.

15. What side is opposite $\angle A$?

16. What side is included by $\angle A$ and $\angle C$?

17. What angle is opposite $\overline{BC}$?

18. What angle is included by $\overline{CA}$ and $\overline{BC}$?

19. What is the hypotenuse of $\triangle ABC$?

20. Is $\triangle ABC$ is an isosceles triangle?

21. Find the perimeter of a triangle with sides 20 cm, 30 cm, and 40 cm.

22. Find the perimeter of an equilateral triangle with sides 18 ft.

23. Find the perimeter of an isosceles triangle with base 8 inches and equal sides 12.5 inches.

24. A triangle with sides 14 cm and 22 cm has a perimeter 66 cm. Find the length of the third side.

25. The perimeter of an equilateral triangle is 69 ft. Find the length of each side.

26. The base of an isosceles triangle is 13 inches and the perimeter is 47 inches. Find the length of its equal sides.

27. The base of an isosceles triangle is one-third the length of each equal side. If the perimeter is 105 cm, find the length of the base and each side.

28. The base of an isosceles triangle is half the length of each equal side. If the perimeter is 30 ft, find the length of the base and each side.

Exercises 29–32 refer to the figure below. Name all triangles that have the given part.

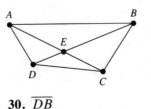

29. $\overline{AE}$ **30.** $\overline{DB}$ **31.** $\angle ECD$ **32.** $\angle ABD$

Exercises 33–40 refer to the figure below. Answer true or false in each.

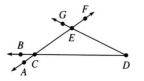

33. $\angle CED$ is an interior angle of $\triangle EDC$.

34. $\angle FED$ is an exterior angle of $\triangle EDC$.

35. $\angle ACB$ is an exterior angle of $\triangle EDC$.

36. $\angle D$ and $\angle CED$ are remote interior angles relative to $\angle ACD$.

37. $\angle CED$ is an interior adjacent angle to $\angle GEF$.

38. $\angle ACB = \angle ECD$.

39. Exterior angle $\angle BCE$ is supplementary to interior angle $\angle ECD$.

40. Exterior angle $\angle FED$ is complementary to interior angle $\angle CED$.

41. Can a scalene triangle be an obtuse triangle?

42. Can a scalene triangle be an isosceles triangle?

43. Can a right triangle be an isosceles triangle?

44. Can an isosceles triangle be an acute triangle?

45. Can an equilateral triangle be an obtuse triangle?

3.2 CONGRUENT TRIANGLES

In this section, we compare triangles that are the same size and shape. The term **congruent,** from the Latin words *con* (with) and *gruere* (to agree), is applied to such figures and literally means "in agreement with." Intuitively, congruent triangles can be made to coincide by placing one on top of the other either directly or by *flipping* over one of them.

Because every triangle has six **parts,** three angles and three sides, two congruent triangles have equal parts that can be made to coincide when one is placed on top of the other. In Figure 3.9, $\triangle ABC$ and $\triangle DEF$ are congruent because $\triangle ABC$ can be placed on top of $\triangle DEF$. Also, $\triangle ABC$ and $\triangle GHI$ are congruent because if $\triangle GHI$ were flipped over, it could be made to coincide with $\triangle ABC$. On the other hand, $\triangle ABC$ and $\triangle JKL$ are not congruent because they cannot be made to coincide either directly or by flipping over one of them.

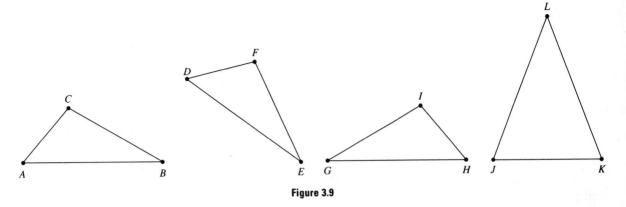

Figure 3.9

The parts of congruent triangles that coincide when one is placed on top of the other are called **corresponding parts.** For example, congruent triangles $\triangle ABC$ and $\triangle DEF$ in Figure 3.9 have six pairs of corresponding parts:

$\angle A$ corresponds to $\angle D$ $\overline{AB}$ corresponds to $\overline{DE}$
$\angle B$ corresponds to $\angle E$ $\overline{BC}$ corresponds to $\overline{EF}$
$\angle C$ corresponds to $\angle F$ $\overline{AC}$ corresponds to $\overline{DF}$

Notice that corresponding angles are opposite corresponding sides, and corresponding sides are opposite corresponding angles. We see that two triangles are congruent when their corresponding parts are equal.

DEFINITION: CONGRUENT TRIANGLES

Two triangles, $\triangle ABC$ and $\triangle DEF$ are **congruent,** written $\triangle ABC \cong \triangle DEF$, whenever $\angle A = \angle D$, $\angle B = \angle E$, $\angle C = \angle F$, $AB = DE$, $BC = EF$, and $AC = DF$.

To avoid giving specific lengths and angle measures, we often mark the parts in working with figures of congruent triangles to indicate those that are equal. Figure 3.10 illustrates this and shows that $\angle A = \angle D$, $\angle B = \angle E$, $\angle C = \angle F$, $AB = DE$, $BC = EF$, and $AC = DF$ in congruent triangles $\triangle ABC$ and $\triangle DEF$.

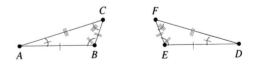

Figure 3.10 Labeling Congruent Triangles

According to our definition, in order to prove that two triangles are congruent, we would need to show that *all* six parts of one are equal to *all* six parts of the other. Fortunately, it is necessary to show congruence using only *three* pairs of corresponding parts as shown in the next three postulates.

> **POSTULATE 3.1   SAS = SAS**
>
> If two sides and the included angle of one triangle are equal, respectively, to two sides and the included angle of a second triangle, then the triangles are congruent.

EXAMPLE 1   Verify that $\triangle ABC \cong \triangle DEF$ in Figure 3.11.

Because AC and DE are both 12 cm, $\angle C$ and $\angle E$ both measure 24°, and BC and EF are both 13 cm, $\triangle ABC \cong \triangle DEF$ by Postulate 3.1, SAS = SAS.   ▨

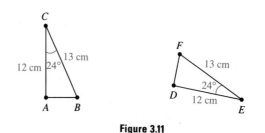

Figure 3.11

> **POSTULATE 3.2   ASA = ASA**
>
> If two angles and the included side of one triangle are equal, respectively, to two angles and the included side of a second triangle, then the triangles are congruent.

EXAMPLE 2   Use the information given in Figure 3.12 to verify that $\triangle PQR \cong \triangle MNO$.

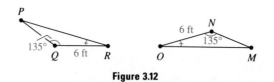

Figure 3.12

Because $QR = 6$ ft and $ON = 6$ ft, $QR = ON$. Also, $\angle Q = \angle N$ because both measure 135°, and we are given that $\angle R = \angle O$. Thus, $\triangle PQR \cong \triangle MNO$ by Postulate 3.2, ASA = ASA.   ▨

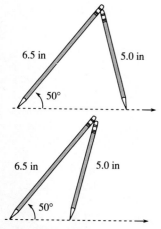

The figure above shows why we *do not* have a SSA postulate for proving triangles congruent. Two noncongruent triangles can be formed with sides measuring 6.5 inches, 5.0 inches, and a nonincluded angle of 50°.

POSTULATE 3.3   SSS = SSS

If three sides of one triangle are equal, respectively, to three sides of a second triangle, then the triangles are congruent.

PRACTICE EXERCISE 1

Use the information given to verify that $\triangle ABC \cong \triangle STU$.

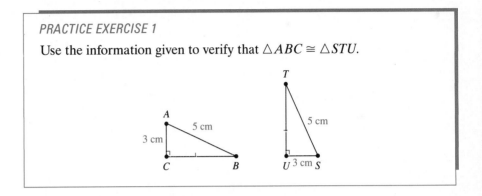

To prove two triangles congruent, we sometimes need to derive information about equal parts using our knowledge about vertical angles or the fact that two triangles share a common side. The following two examples illustrate.

E X A M P L E 3   Refer to Figure 3.13.

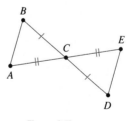

Figure 3.13

Given: BC = CD
     AC = CE
Prove: $\triangle ABC \cong \triangle CDE$
Proof:

STATEMENTS	REASONS
1. $BC = CD$	1. Given
2. $AC = CE$	2. Given
3. $\angle BCA = \angle ECD$	3. Vert. ∠'s are =
4. $\triangle ABC \cong \triangle CDE$	4. SAS = SAS

EXAMPLE 4   Refer to Figure 3.14.

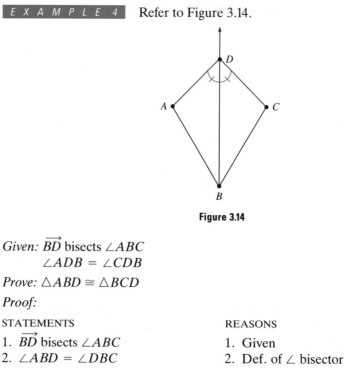

Figure 3.14

Given: $\overrightarrow{BD}$ bisects $\angle ABC$
        $\angle ADB = \angle CDB$

Prove: $\triangle ABD \cong \triangle BCD$

Proof:

STATEMENTS	REASONS
1. $\overrightarrow{BD}$ bisects $\angle ABC$	1. Given
2. $\angle ABD = \angle DBC$	2. Def. of $\angle$ bisector
3. $\angle ADB = \angle CDB$	3. Given
4. $BD = BD$	4. Reflexive law
5. $\triangle ABD \cong \triangle BCD$	5. ASA = ASA ◪

NOTE: Although two segments or two angles may appear to be equal, never make this assumption simply because they look the same. When writing proofs, use only given information together with known facts from previously proved theorems, postulates, or definitions.   ☐

The proof of the next theorem is requested in the exercises.

THEOREM 3.1   TRANSITIVE LAW FOR CONGRUENT TRIANGLES

If $\triangle ABC \cong \triangle DEF$ and $\triangle DEF \cong \triangle GHI$, then $\triangle ABC \cong \triangle GHI$.

Thales (640–546 B.C.)

Thales of Miletus was a wealthy merchant who became interested in the practical aspects of geometry. In Greece, he taught geometry to many of his friends. His most noteworthy pupil was Pythagoras. Thales has been called the "father of Greek mathematics."

Simply stated, Theorem 3.1 says that two triangles congruence to the same triangle are congruent to each other.

ANSWER TO PRACTICE EXERCISE: **1.** The desired congruence can be shown in two ways. Because $AC = 3$ cm and $US = 3$ cm, $AC = US$. Also, we are given that $CB = UT$ and $\angle C = \angle U$ (both are right angles). Thus, $\triangle ABC \cong \triangle STU$ by SAS = SAS. Alternatively, because $AC = US$, $AB = ST$, and $BC = UT$, $\triangle ABC \cong \triangle STU$ by SSS = SSS.

3.2 EXERCISES

Decide whether the triangles given in Exercises 1–6 are congruent. If so, state why.

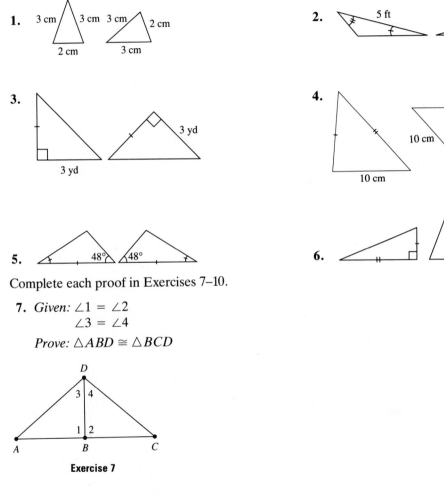

1. 3 cm 3 cm 3 cm 2 cm 2 cm 3 cm

2. 5 ft 5 ft

3. 3 yd 3 yd

4. 10 cm 10 cm

5. 48° 48°

6.

Complete each proof in Exercises 7–10.

7. *Given:* $\angle 1 = \angle 2$
    $\angle 3 = \angle 4$

 Prove: $\triangle ABD \cong \triangle BCD$

Exercise 7

Proof:

STATEMENTS	REASONS
1. $\angle 1 = \angle 2$	1. _____
2. _____	2. Given
3. _____	3. Reflexive law
4. $\triangle ABD \cong \triangle BCD$	4. _____

8. *Given:* C is the midpoint of $\overline{AE}$
$\qquad \angle E = \angle A$

Prove: $\triangle ABC \cong \triangle CED$

Proof:

STATEMENTS	REASONS
1. C is the midpoint of $\overline{AE}$	1. _____
2. $AC = CE$	2. _____
3. _____	3. Given
4. $\angle ACB = \angle DCE$	4. _____
5. $\triangle ABC \cong \triangle CED$	5. _____

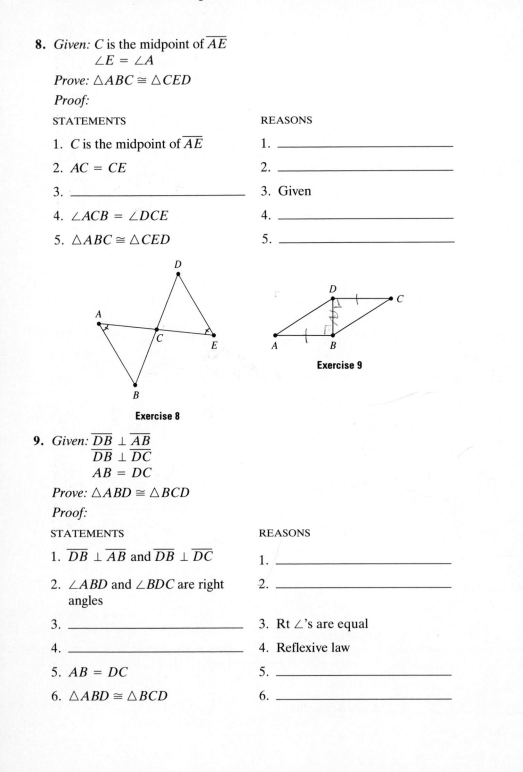

Exercise 8

Exercise 9

9. *Given:* $\overline{DB} \perp \overline{AB}$
$\qquad \overline{DB} \perp \overline{DC}$
$\qquad AB = DC$

Prove: $\triangle ABD \cong \triangle BCD$

Proof:

STATEMENTS	REASONS
1. $\overline{DB} \perp \overline{AB}$ and $\overline{DB} \perp \overline{DC}$	1. _____
2. $\angle ABD$ and $\angle BDC$ are right angles	2. _____
3. _____	3. Rt $\angle$'s are equal
4. _____	4. Reflexive law
5. $AB = DC$	5. _____
6. $\triangle ABD \cong \triangle BCD$	6. _____

10. *Given:* $AB = CD$
$AC = BD$

Prove: $\triangle ABC \cong \triangle BCD$

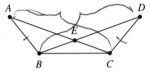

Exercise 10

Proof:

STATEMENTS	REASONS
1. $AB = CD$	1. _____
2. _____	2. Given
3. _____	3. Reflexive law
4. $\triangle ABC \cong \triangle BCD$	4. _____

Write a two-column proof in Exercises 11–18.

11. *Given:* $\overline{AD}$ bisects $\overline{BE}$
$\overline{BE}$ bisects $\overline{AD}$

Prove: $\triangle ABC \cong \triangle CDE$

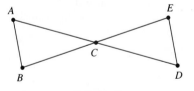

Exercise 11

12. *Given:* $\angle B$ and $\angle E$ are right angles
$\overline{AD}$ bisects $\overline{BE}$

Prove: $\triangle ABC \cong \triangle CDE$

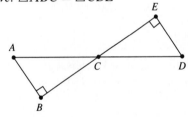

Exercise 12

13. *Given:* $AD = BD$
$AE = BC$

Prove: $\triangle ACD \cong \triangle BDE$

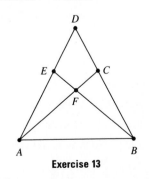

Exercise 13

14. *Given:* $\overline{DB} \perp \overline{AC}$
$\overline{DB}$ bisects $\overline{AC}$

Prove: $\triangle ABD \cong \triangle BCD$

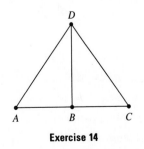

Exercise 14

15. *Given:* $AB = BC$
       $\angle 1 = \angle 2$
  Prove: $\triangle ABE \cong \triangle BCD$

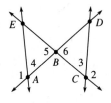

Exercise 15

16. *Given:* $\triangle ABC$ is equilateral
       $\triangle BDC$ is equilateral
  Prove: $\triangle ABC \cong \triangle BDC$

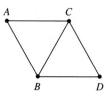

Exercise 16

17. *Given:* $\triangle ACD$ is equilateral
       $\angle 1 = \angle 2$
       $BC = DE$
  Prove: $\triangle ABC \cong \triangle ADE$

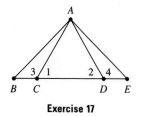

Exercise 17

18. *Given:* $\triangle ACD$ is isosceles with base $\overline{CD}$
       $\angle 1 = \angle 2$
       $BD = CE$
  Prove: $\triangle ABC \cong \triangle ADE$

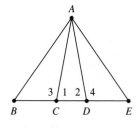

Exercise 18

Use the figure below in Exercises 19–30. You will need to recall what you learned in beginning algebra to solve several equations. Assume that $\triangle ABC \cong \triangle PQR$, $AC = x + 1$, $PR = 3x - 5$, $\angle B = (100 + y)°$, and $\angle Q = (5y + 20)°$.

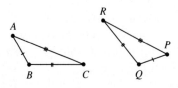

19. Find the value of x.

20. Find the value of y.

21. Find AC.

22. Find PR.

23. Find the measure of $\angle B$.

24. Find the measure of $\angle Q$.

25. $\overline{AC}$ corresponds to which side in $\triangle PQR$?

26. $\overline{QP}$ corresponds to which side in $\triangle ABC$?

27. $\overline{BC}$ corresponds to which side in $\triangle PQR$?

28. $\angle Q$ corresponds to which angle in $\triangle ABC$?

29. $\angle A$ corresponds to which angle in $\triangle PQR$?

30. $\angle R$ corresponds to which angle in $\triangle ABC$?

31. Prove Theorem 3.1.

3.3   PROOFS INVOLVING CONGRUENCE

Sometimes we must prove that two segments or two angles are equal. This is often done by showing that they are corresponding parts of congruent triangles. The reason given for this statement in a proof will be abbreviated **cpoctae** (corresponding parts of congruent triangles are equal).

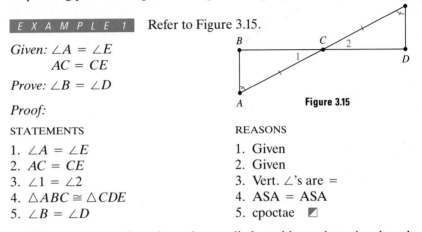

Figure 3.15

E X A M P L E   1   Refer to Figure 3.15.

Given: $\angle A = \angle E$
$AC = CE$

Prove: $\angle B = \angle D$

Proof:

STATEMENTS	REASONS
1. $\angle A = \angle E$	1. Given
2. $AC = CE$	2. Given
3. $\angle 1 = \angle 2$	3. Vert. $\angle$'s are $=$
4. $\triangle ABC \cong \triangle CDE$	4. ASA = ASA
5. $\angle B = \angle D$	5. cpoctae ☑

The next example solves the applied problem given in the chapter introduction.

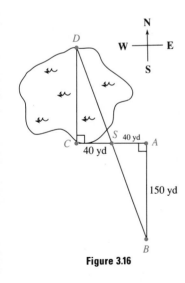

Figure 3.16

E X A M P L E   2   Mr. Wells owns a cabin on the south shore of Lake Pleasant. Directly north of the cabin on the north shore is a boat dock. To determine the distance from the cabin to the dock, Mr. Wells does the following: He first paces 40 yd due east from the cabin and drives a reference stake into the ground. He then paces another 40 yd due east, at which point he turns due south and walks until he meets the straight line formed by the dock and the stake. The total distance paced off in the southerly direction is 150 yd, and he concludes that the distance from the cabin to the dock is also 150 yd. Explain why this is correct.

We'll make a sketch with point D corresponding to the boat dock, point C corresponding to the cabin, and point S corresponding to the stake as in Figure 3.16. Because $\angle DCS$ and $\angle SAB$ are right angles, $\angle DCS = \angle SAB$. Because $\angle DSC$ and $\angle ASB$ are vertical angles, $\angle DSC = \angle ASB$. And because $CS = 40$ yd and $SA = 40$ yd, $CS = SA$. Thus, $\triangle CSD \cong \triangle SAB$ by ASA = ASA. Because $\overline{DC}$ and $\overline{AB}$ are corresponding parts of congruent triangles, $DC = AB$. Thus, CD, the distance from the cabin to the boat dock is the same as AB which is 150 yd. ☑

In Chapter 2 we showed how to construct the (perpendicular) bisector of a line segment. Congruent triangles can illustrate that our construction gives the desired result.

THEOREM 3.2   SEGMENT BISECTOR THEOREM

Construction 2.3 gives the perpendicular bisector of a given line segment.

When constructing a roof, bridge, or other type of structure that must support heavy weight, a triangular-shaped frame called a truss is used. The SSS postulate guarantees that a triangle with given sides cannot be distorted in shape. It is this fact that gives a truss its strength.

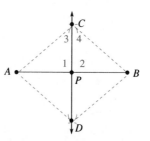

Figure 3.17   Perpendicular Bisector

Given: $AC = BC$ (by construction)
$\quad\quad\quad AD = BD$ (by construction)
$\quad\quad\quad$ (See Figure 3.17.)
Prove: $\overleftrightarrow{CD}$ is the perpendicular bisector of $\overline{AB}$

Proof:

STATEMENTS	REASONS
1. $AC = BC$ and $AD = BD$	1. Given
2. $CD = CD$	2. Reflexive law
3. $\triangle ACD \cong \triangle BCD$	3. SSS = SSS
4. $\angle 3 = \angle 4$	4. cpoctae
5. $CP = CP$	5. Reflexive law
6. $\triangle ACP \cong \triangle BPC$	6. SAS = SAS
7. $AP = PB$	7. cpoctae
8. P is the midpoint of $\overline{AB}$	8. Def. of midpoint
9. $\overleftrightarrow{CD}$ bisects $\overline{AB}$	9. Def. of bisector
10. $\angle 1$ and $\angle 2$ are supplementary	10. Adj. $\angle$'s whose noncommon sides are in line are supp.
11. $\angle 1 = \angle 2$	11. cpoctae
12. $\overleftrightarrow{CD} \perp \overline{AB}$	12. Def. of $\perp$
13. $\overleftrightarrow{CD}$ is the perpendicular bisector of $\overline{AB}$	13. Combine 9 and 12

The next theorem is closely related to Theorem 3.2, and its proof is requested in the exercises.

> **THEOREM 3.3**
> Every point on the perpendicular bisector of a segment $\overline{AB}$ is equidistant from A and B.

We can also show that the method used to construct the bisector of an angle gives the desired result.

> **THEOREM 3.4   ANGLE BISECTOR THEOREM**
> Construction 2.6 gives the bisector of a given angle.

Refer to Figure 3.18 for the proof of Theorem 3.4.

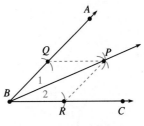

Figure 3.18   Angle Bisector

Given: $BQ = BR$ (by construction)
$\qquad QP = RP$ (by construction)
Prove: $\overrightarrow{BP}$ bisects $\angle ABC$
Proof:

Why does the frame of a bicycle maintain its shape?

STATEMENTS	REASONS
1. $BQ = BR$	1. Given
2. $QP = RP$	2. Given
3. $BP = BP$	3. Reflexive law
4. $\triangle BQP \cong \triangle BPR$	4. SSS = SSS
5. $\angle 1 = \angle 2$	5. cpoctae
6. $\overrightarrow{BP}$ bisects $\angle ABC$	6. Def. of $\angle$ bisector

Sometimes it is necessary to prove that several pairs of triangles are congruent to show that two segments or angles are equal. The following example illustrates.

E X A M P L E 3 Refer to Figure 3.19.

Given: $\angle 1 = \angle 2$
$\angle 3 = \angle 4$

Prove: $CB = CD$

Figure 3.19

Proof:

STATEMENTS	REASONS
1. $\angle 1 = \angle 2$	1. Given
2. $\angle 3 = \angle 4$	2. Given
3. $\angle BEA$ is supplementary to $\angle 3$ and $\angle DEA$ is supplementary to $\angle 4$	3. Adj. $\angle$'s whose noncommon sides are in line are supp.
4. $\angle BEA = \angle DEA$	4. Supp. of $=$ $\angle$'s are $=$
5. $AE = AE$	5. Reflexive law
6. $\triangle BEA \cong \triangle DEA$	6. ASA $=$ ASA
7. $BE = DE$	7. cpoctae
8. $CE = CE$	8. Reflexive law
9. $\triangle CEB \cong \triangle CED$	9. SAS $=$ SAS
10. $CB = CD$	10. cpoctae ◼

NOTE: As we have seen, many congruency proofs involve triangles that share a common angle or a common side. For example, in Figure 3.19 $\triangle ABE$ and $\triangle ADE$ share a common side $\overline{AE}$. In such cases, the triangles are sometimes called **overlapping.** ☐

3.3 EXERCISES

Exercises 1–4 refer to the congruent triangles given below in which $\angle A$ and $\angle F$ correspond as do $\angle C$ and $\angle E$.

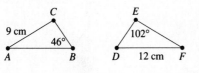

1. Find EF.

2. Find AB.

3. Find the measure of $\angle D$.

4. Find the measure of $\angle C$.

Complete each proof in Exercises 5–8.

5. *Given:* $BG = CE$
$\quad\quad\quad \angle 1 = \angle 2$

Prove: $\angle G = \angle E$

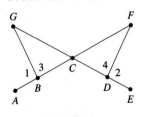

Exercise 5

Proof:

STATEMENTS	REASONS
1. $BG = CE$	1. _____
2. _____	2. Given
3. $\angle 1$ and $\angle GBC$ are supplementary	3. _____
4. _____	4. Adj. $\angle$'s whose noncommon sides are in a line are supp.
5. $\angle GBC = \angle ECB$	5. _____
6. $BC = BC$	6. _____
7. $\triangle GBC \cong \triangle ECB$	7. _____
8. _____	8. cpoctae

6. *Given:* $BC = CD$
$\quad\quad\quad \angle 1 = \angle 2$

Prove: $GC = CF$

G   F
1 3 C 4 2
  B   D
A   E

Exercise 6

Proof:

STATEMENTS	REASONS
1. $BC = CD$	1. _____
2. _____	2. Given

3. ∠1 and ∠3 are supplementary

3. _____

4. _____

4. Adj. ∠'s whose noncommon sides are in a line are supp.

5. _____

5. Supp. of = ∠'s are =

6. ∠*GCB* and ∠*FCD* are vertical angles

6. _____

7. _____

7. Vert. ∠'s are =

8. _____

8. ASA = ASA

9. *GC = CF*

9. _____

7. *Given: AB = CD*
    ∠A = ∠D
    ∠1 = ∠2
   Prove: AF = ED

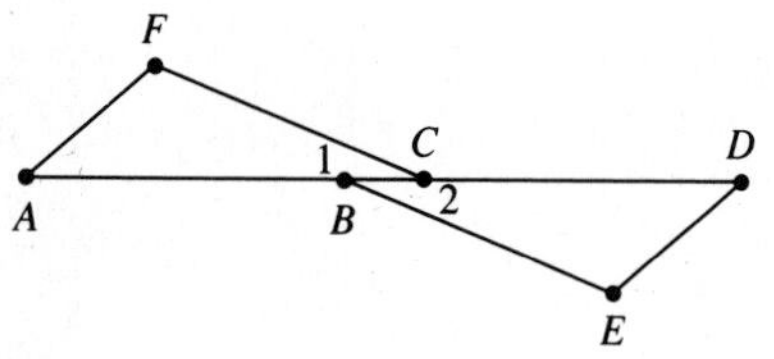

Exercise 7

Proof:

STATEMENTS	REASONS
1. ∠*A* = ∠*D*	1. _____
2. ∠1 = ∠2	2. _____
3. _____	3. Given
4. *AC = AB + BC*	4. _____
5. _____	5. Seg. add. post.
6. *AB + BC = CD + BC*	6. _____
7. *AC = BD*	7. _____
8. _____	8. ASA = ASA
9. *AF = ED*	9. _____

8. *Given:* $\overline{BD} \perp \overline{AC}$
$BC = BE$
$\angle 1 = \angle C$

Prove: $\angle A = \angle D$

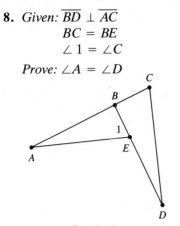

Exercise 8

Proof:

STATEMENTS	REASONS
1. $\angle 1 = \angle C$	1. _____
2. $\overline{BD} \perp \overline{AC}$	2. _____
3. _____	3. $\perp$ lines form $=$ rt $\angle$'s
4. _____	4. Given
5. $\triangle ABE \cong \triangle BCD$	5. _____
6. _____	6. cpoctae

Write a two-column proof in Exercises 9–16.

9. *Given:* $AC = CE$
$DC = CB$

Prove: $\angle A = \angle E$

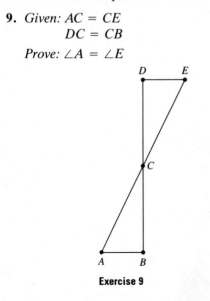

Exercise 9

10. *Given:* $\overrightarrow{AC}$ bisects $\angle BAD$
$\overrightarrow{CA}$ bisects $\angle BCD$

Prove: $\angle B = \angle D$

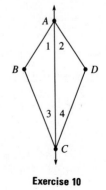

Exercise 10

11. *Given:* $BC = CD$
   $\angle 1 = \angle 2$
   Prove: $AB = AD$

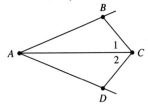

Exercise 11

12. *Given:* $\overrightarrow{DB}$ bisects $\angle ADC$
   $AD = CD$
   Prove: $\overline{DB} \perp \overline{AC}$

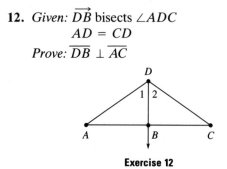

Exercise 12

13. *Given:* $\angle 1 = \angle 2$
   $\angle 3 = \angle 4$
   Prove: $\angle A = \angle C$

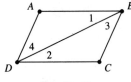

Exercise 13

14. *Given:* $\overline{BC} \perp \overline{AB}$
   $\overline{AD} \perp \overline{DC}$
   $\angle 1 = \angle 2$
   Prove: $AB = DC$

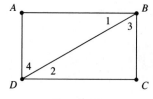

Exercise 14

15. *Given:* $\overline{GB} \perp \overline{AF}$
   $\overline{FD} \perp \overline{GE}$
   $GD = BF$
   $GB = FD$
   Prove: $BC = DC$
   [Hint: Several pairs of triangles must be shown congruent.]

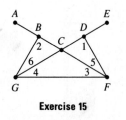

Exercise 15

16. *Given:* $\angle A = \angle C$
   $\angle 1 = \angle 2$
   B is the midpoint of $\overline{AC}$
   Prove: $GF = DE$
   [Hint: Several pairs of triangles must be shown congruent.]

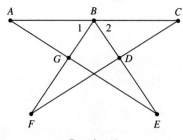

Exercise 16

17. To measure the distance between two points A and B on opposite sides of a lake, a ranger first sets a stake at point C as shown in the Exercise 17 figure. He then sets a stake at point D in line with B and C so that $BC = CD$. Next he sets a stake at point E in line with A and C so that $AC = CE$. He measures the distance between D and E, 105 yd, and concludes that the distance between A and B is also 105 yd. Why is this true?

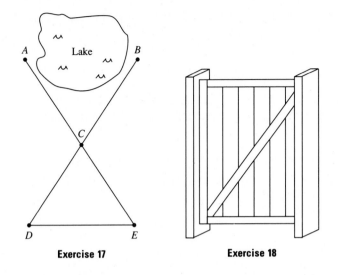

Exercise 17   Exercise 18

18. Why do we put a diagonal brace on a gate as shown in the figure above? [Hint: Your answer will involve one of the postulates for congruent triangles.]

19. The figure below shows a bridge. Why is it composed of triangles? Can the bridge change in shape without breaking?

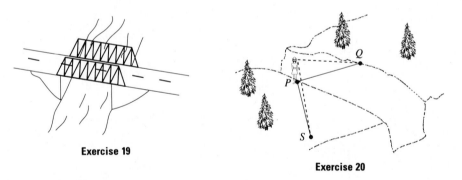

Exercise 19

Exercise 20

20. To measure the distance across a canyon, Diana stood at the edge of the canyon at point P and sighted a point Q on the far rim. Without raising or lowering her eyes, she turned around and sighted a point S on the ground. She then measured the distance between P and S and found it to be 45 ft. She concluded that the width of the canyon is also 45 ft. Why is this true?

21. Prove that Construction 2.4 gives the desired result.

22. Prove that Construction 2.5 gives the desired result.

23. Prove Theorem 3.3.

3.4 ISOSCELES TRIANGLES, MEDIANS, AND ALTITUDES

Recall that an isosceles triangle is a triangle having two equal sides with the third side called the base of the triangle. The angle included between the equal sides of an isosceles triangle is the **vertex angle,** and the remaining angles are the **base angles.** The next theorem shows that the base angles of an isosceles triangle are equal.

THEOREM 3.5

If two sides of a triangle are equal, then the angles opposite these sides are also equal.

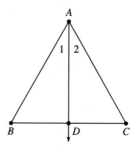

Figure 3.20

Given: $\triangle ABC$

     $AB = AC$ (See Figure 3.20.)

Prove: $\angle C = \angle B$

Construction: Construct the bisector $\overrightarrow{AD}$ of $\angle BAC$

Proof:

STATEMENTS	REASONS
1. $\triangle ABC$ with $AB = AC$	1. Given
2. $\overrightarrow{AD}$ bisects $\angle BAC$	2. An $\angle$ can be bisected
3. $\angle 1 = \angle 2$	3. Def. of $\angle$ bisector
4. $AD = AD$	4. Reflexive law
5. $\triangle ABD \cong \triangle ADC$	5. SAS = SAS
6. $\angle C = \angle B$	6. cpoctae

The next corollary follows directly from Theorem 3.5, and its proof is requested in the exercises.

COROLLARY 3.6

If a triangle is equilateral, then it is equiangular.

A theorem of the form $P \longrightarrow Q$, that is, *if P then Q*, has a converse $Q \longrightarrow P$ that may or may not be a theorem. In the case of Theorem 3.5, its converse is also true.

THEOREM 3.7

If two angles of a triangle are equal, then the sides opposite these angles are also equal.

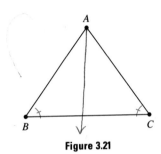

Figure 3.21

Given: $\triangle ABC$

$\angle B = \angle C$ (See Figure 3.21.)

Prove: $AB = AC$

Proof:

STATEMENTS	REASONS
1. $\angle B = \angle C$	1. Given
2. $\angle C = \angle B$	2. Symmetric law
3. $BC = CB$	3. Reflexive law
4. $\triangle ABC \cong \triangle ACB$	4. ASA = ASA
5. $AB = AC$	5. cpoctae

In the proof of Theorem 3.7 we showed that a triangle is congruent to itself, and as a result, we found equal sides. This may seem strange at first because in all previous cases we had two distinct triangles in a congruence proof. Actually, we are thinking of $\triangle ABC$ and $\triangle ACB$ as overlapping triangles much like the example in which two triangles shared a common side or common angle.

The next corollary follows directly from Theorem 3.7.

COROLLARY 3.8

If a triangle is equiangular, then it is equilateral.

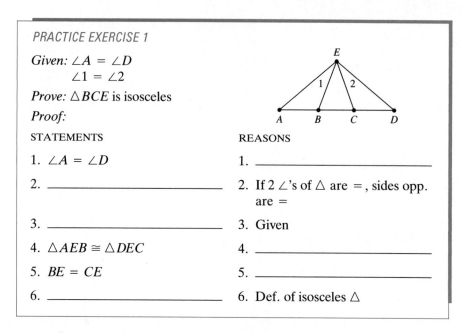

PRACTICE EXERCISE 1

Given: $\angle A = \angle D$
        $\angle 1 = \angle 2$
Prove: $\triangle BCE$ is isosceles
Proof:

STATEMENTS	REASONS
1. $\angle A = \angle D$	1. _____
2. _____	2. If 2 $\angle$'s of $\triangle$ are =, sides opp. are =
3. _____	3. Given
4. $\triangle AEB \cong \triangle DEC$	4. _____
5. $BE = CE$	5. _____
6. _____	6. Def. of isosceles $\triangle$

Some proofs require the use of **auxiliary lines** or **segments,** which we illustrated in the proof of Theorem 3.5 when we constructed the bisector of an angle in a triangle. Often these lines are drawn in a figure using a dashed line because they are not part of the original problem. The following example illustrates.

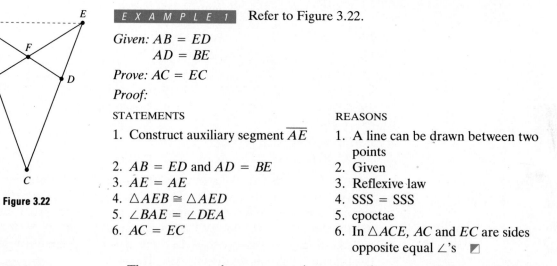

Figure 3.22

E X A M P L E   1   Refer to Figure 3.22.

Given: $AB = ED$
        $AD = BE$
Prove: $AC = EC$
Proof:

STATEMENTS	REASONS
1. Construct auxiliary segment $\overline{AE}$	1. A line can be drawn between two points
2. $AB = ED$ and $AD = BE$	2. Given
3. $AE = AE$	3. Reflexive law
4. $\triangle AEB \cong \triangle AED$	4. SSS = SSS
5. $\angle BAE = \angle DEA$	5. cpoctae
6. $AC = EC$	6. In $\triangle ACE$, AC and EC are sides opposite equal $\angle$'s ◪

The next two theorems are important theorems relating to isosceles triangles.

THEOREM 3.9

The bisector of the vertex angle of an isosceles triangle is the perpendicular bisector of the base of the triangle.

Given: Isosceles $\triangle ABC$ with $AB = AC$
$\overrightarrow{AD}$ bisects $\angle BAC$
(See Figure 3.23.)

Prove: $\overline{AD}$ is the $\perp$ bisector of $\overline{BC}$

Proof:

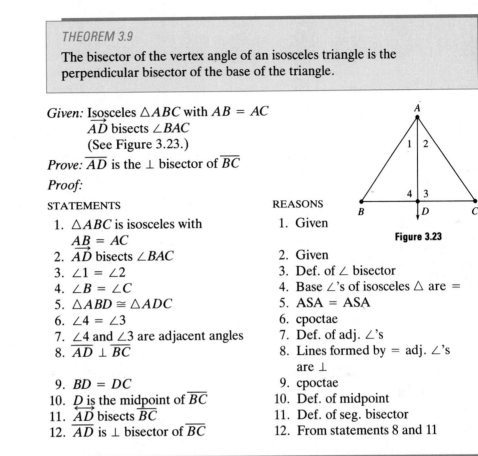

Figure 3.23

STATEMENTS	REASONS
1. $\triangle ABC$ is isosceles with $AB = AC$	1. Given
2. $\overrightarrow{AD}$ bisects $\angle BAC$	2. Given
3. $\angle 1 = \angle 2$	3. Def. of $\angle$ bisector
4. $\angle B = \angle C$	4. Base $\angle$'s of isosceles $\triangle$ are $=$
5. $\triangle ABD \cong \triangle ADC$	5. ASA = ASA
6. $\angle 4 = \angle 3$	6. cpoctae
7. $\angle 4$ and $\angle 3$ are adjacent angles	7. Def. of adj. $\angle$'s
8. $\overline{AD} \perp \overline{BC}$	8. Lines formed by $=$ adj. $\angle$'s are $\perp$
9. $BD = DC$	9. cpoctae
10. D is the midpoint of $\overline{BC}$	10. Def. of midpoint
11. $\overleftrightarrow{AD}$ bisects $\overline{BC}$	11. Def. of seg. bisector
12. $\overline{AD}$ is $\perp$ bisector of $\overline{BC}$	12. From statements 8 and 11

THEOREM 3.10

The perpendicular bisector of the base of an isosceles triangle passes through the vertex of its vertex angle and bisects the vertex angle.

Given: Isosceles $\triangle ABC$ with
$AB = AC$ and base BC
$\overline{DF}$ is $\perp$ bisector of $\overline{BC}$
(See Figure 3.24.)

Prove: $\overline{DF}$ passes through A and $\overline{DF}$ bisects $\angle A$

Construction: Construct the bisector $\overrightarrow{AE}$ of vertex $\angle BAC$.

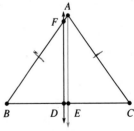

Figure 3.24

Proof:

STATEMENTS	REASONS
1. $\triangle ABC$ is isosceles with $AB = AC$	1. Given

Cut a triangle from a piece of cardboard, draw the medians of the triangle. If you are careful, the medians will intersect at a point. Insert the point of a sharp pencil through this point. Does the triangle balance without tilting? Why do you suppose the intersection of the medians is called the *centroid* or *balance point* of the triangle?

2. $\overrightarrow{AE}$ is the bisector of $\angle BAC$
3. $\overline{AE}$ is the $\perp$ bisector of $\overline{BC}$

4. $\overline{DF}$ is the $\perp$ bisector of $\overline{BC}$
5. $\overline{DF} = \overline{AE}$
6. $\overline{DF}$ passes through A

7. $\overline{DF}$ bisects $\angle A$

2. We can construct an $\angle$ bisector
3. Bisector of vertex $\angle$ in isos. $\triangle$ is $\perp$ bisector of base
4. Given
5. The $\perp$ bisector of a line is unique
6. $\overline{DF}$ and $\overline{AE}$ are same and A is on $\overrightarrow{AE}$
7. $\overrightarrow{AE}$ bisects $\angle A$ and $\overline{DF} = \overline{AE}$

The preceding proof shows how to prove that two lines are the same, or **coincide.** Proofs such as this are often called **coincidence proofs.**

In addition to its three sides, every triangle has nine segments associated with it: three *medians*, three *altitudes,* and three *angle bisectors.*

DEFINITION: MEDIAN OF A TRIANGLE

A **median** of a triangle is the segment joining a vertex to the midpoint of the side opposite that vertex.

In $\triangle ABC$ shown in Figure 3.25, D is the midpoint of $\overline{BC}$, and $\overline{AD}$ is a median of $\triangle ABC$. Notice that $\triangle ABC$ has three medians, one from each of the three vertices.

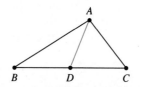

Figure 3.25   Median of a Triangle

DEFINITION: ALTITUDE OF A TRIANGLE

An **altitude** of a triangle is a line segment from a vertex perpendicular to the side opposite that vertex (possibly extended).

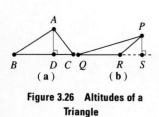

Figure 3.26   Altitudes of a Triangle

In $\triangle ABC$ shown in Figure 3.26(a), $\overline{AD}$ is perpendicular to $\overline{BC}$, and $\overline{AD}$ is the altitude of $\triangle ABC$ from vertex A. In $\triangle PQR$ shown in Figure 3.26(b), $\overline{PS}$ is perpendicular to the extension of $\overline{QR}$, that is $\overline{QS}$, and $\overline{PS}$ is the altitude of $\triangle PQR$ from vertex P. Notice that any triangle has three altitudes, one from each of the three vertices.

DEFINITION: ANGLE BISECTOR OF A TRIANGLE

An **angle bisector** of a triangle is a line segment from a vertex that bisects the angle determined by that vertex.

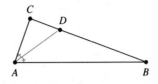

Figure 3.27   Angle Bisector of a Triangle

In $\triangle ABC$ shown in Figure 3.27, $\overline{AD}$ is the bisector of $\angle CAB$ because $\angle CAD = \angle DAB$. Notice that $\triangle ABC$ has three angle bisectors, one for each angle in the triangle.

In view of Theorems 3.9 and 3.10, we have the following corollary for isosceles triangles.

COROLLARY 3.11

The bisector of the vertex angle of an isosceles triangle coincides with the altitude and median drawn from that vertex.

ANSWER TO PRACTICE EXERCISE: **1.** 1. Given 2. $EA = ED$ 3. $\angle 1 = \angle 2$ 4. ASA = ASA 5. cpoctae 6. $\triangle BCE$ is isosceles

3.4 EXERCISES

Exercises 1–6 refer to the figure below in which $\angle 1 = \angle 2$, $BD = DC$, and $\overline{CF} \perp \overline{BF}$. Answer true or false.

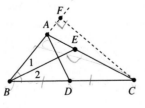

1. $\overline{AD}$ is a median of $\triangle ABC$.

2. $\overline{CF}$ is an altitude of $\triangle ABC$.

3. $\overline{BE}$ is an altitude of $\triangle ABC$.

4. $\overline{BE}$ is an angle bisector of $\triangle ABC$.

5. $\overline{AD}$ is an angle bisector of $\triangle ABC$.

6. $\overline{CF}$ is a median of $\triangle ABC$.

Complete each proof in Exercises 7–10.

7. *Given:* $AB = AD$
$\overline{AC}$ bisects $\angle BAD$

Prove: $\triangle BDE$ is isosceles

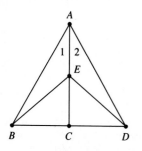

Exercise 7

Proof:

STATEMENTS	REASONS
1. $AB = AD$	1. _____
2. _____	2. Given
3. $\angle 1 = \angle 2$	3. _____
4. $AE = AE$	4. _____
5. _____	5. SAS = SAS
6. $BE = ED$	6. _____
7. _____	7. Def. of isos. $\triangle$

8. *Given:* $\angle 1 = \angle 2$
$AB = AE$

Prove: $\angle 3 = \angle 4$

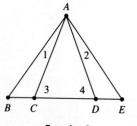

Exercise 8

Proof:

STATEMENTS	REASONS
1. $\angle 1 = \angle 2$	1. _____

2. _____ 2. Given

3. $\angle B = \angle E$ 3. _____

4. $\triangle ABC \cong \triangle AED$ 4. _____

5. $AC = AD$ 5. _____

6. _____ 6. $\angle$'s opp. = sides are =

9. *Given:* Isosceles $\triangle ACD$ with base $\overline{CD}$
B is the midpoint of $\overline{AC}$
E is the midpoint of $\overline{AD}$

Prove: $\angle 1 = \angle 2$

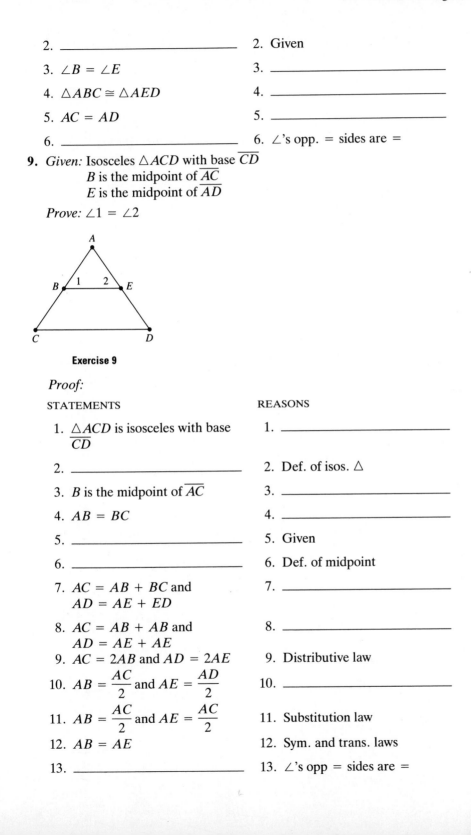

Exercise 9

Proof:

STATEMENTS	REASONS
1. $\triangle ACD$ is isosceles with base $\overline{CD}$	1. _____
2. _____	2. Def. of isos. $\triangle$
3. B is the midpoint of $\overline{AC}$	3. _____
4. $AB = BC$	4. _____
5. _____	5. Given
6. _____	6. Def. of midpoint
7. $AC = AB + BC$ and $AD = AE + ED$	7. _____
8. $AC = AB + AB$ and $AD = AE + AE$	8. _____
9. $AC = 2AB$ and $AD = 2AE$	9. Distributive law
10. $AB = \dfrac{AC}{2}$ and $AE = \dfrac{AD}{2}$	10. _____
11. $AB = \dfrac{AC}{2}$ and $AE = \dfrac{AC}{2}$	11. Substitution law
12. $AB = AE$	12. Sym. and trans. laws
13. _____	13. $\angle$'s opp = sides are =

10. *Given: AC = AD*
　　　　BD = CE
　　Prove: AB = AE

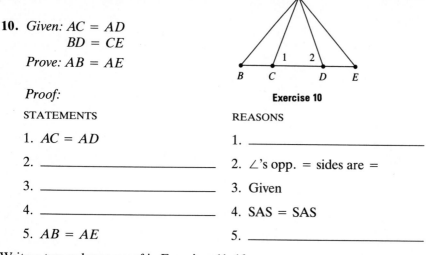

Exercise 10

Proof:

STATEMENTS	REASONS
1. *AC = AD*	1. _____
2. _____	2. ∠'s opp. = sides are =
3. _____	3. Given
4. _____	4. SAS = SAS
5. *AB = AE*	5. _____

Write a two-column proof in Exercises 11–16.

11. *Given: ∠1 = ∠2*
　　　　∠3 = ∠4
　　Prove: ∠A = ∠D

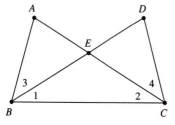

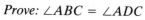

Exercise 11

12. *Given: AB = AE*
　　　　BC = DE
　　Prove: ∠1 = ∠2

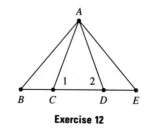

Exercise 12

13. *Given: AB = AD*
　　　　BC = DC
　　Prove: ∠ABC = ∠ADC

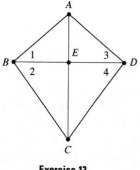

Exercise 13

14. *Given: AD = BC*
　　　　AC = BD
　　Prove: ∠A = ∠B
　　　[Hint: Draw an auxiliary line
　　　joining *D* and *C*.]

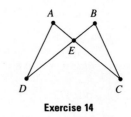

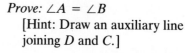

Exercise 14

15. *Given:* △*ABC* is isosceles with base $\overline{BC}$
           △*DBC* is isosceles with base $\overline{BC}$

   Prove: ∠1 = ∠2

16. *Given:* ∠1 = ∠2
           ∠3 = ∠4

   Prove: *AD* = *BC*

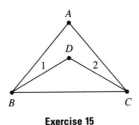

Exercise 15

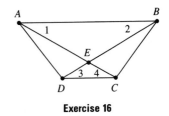

Exercise 16

17. Prove that the bisectors (terminated by the opposite side) of two corresponding angles of two congruent triangles are equal.

18. Prove that the corresponding medians of two congruent triangles are equal.

19. Prove that the line segments joining the midpoints of the equal sides of an isosceles triangle with the midpoint of the base are equal.

20. Prove that the medians to the equal sides of an isosceles triangle are equal.

21. Prove that the perimeters of congruent triangles are equal.

3.5 CONSTRUCTIONS INVOLVING TRIANGLES

The first construction we'll consider involves duplicating a given triangle.

CONSTRUCTION 3.1

Construct a triangle that is congruent to a given triangle.

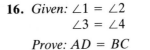

Given: △*ABC* (See Figure 3.28.)

To Construct: △*DEF* such that △*DEF* ≅ △*ABC*

Construction:

1. Draw line ℓ, choose point *D* on ℓ, and use Construction 2.1 to duplicate $\overline{AB}$ ($\overline{DE}$ in Figure 3.28).

2. Set the compass at the length *AC*, place the point at *D* and draw an arc. Then set the compass at the length *BC*, place the point at *E*, and draw an arc that intersects the first arc in point *F*.

3. Use a straightedge to draw $\overline{DF}$ and $\overline{EF}$. The resulting △*DEF* is congruent to △*ABC* by SSS = SSS because *AB* = *DE*, *AC* = *DF*, and *BC* = *EF*.

Figure 3.28

One way to view Construction 3.1 is to construct a triangle given the three sides. Next we construct a triangle when two sides and the included angle are given.

CONSTRUCTION 3.2

Construct a triangle with two sides and the included angle given.

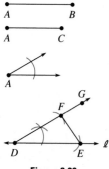

Figure 3.29

Given: Sides $\overline{AB}$ and $\overline{AC}$ and $\angle A$ (See Figure 3.29.)

To Construct: The triangle with these parts

Construction:

1. Draw line ℓ, choose point D on ℓ, and use Construction 2.1 to duplicate $\overline{AB}$ ($\overline{DE}$ in Figure 3.29).
2. Use Construction 2.2 to duplicate $\angle A$ with vertex at D ($\angle GDE$ in Figure 3.29).
3. On $\overrightarrow{DG}$ use Construction 2.1 to duplicate $\overline{AC}$ ($\overline{DF}$ in Figure 3.29).
4. Use a straightedge to draw $\overline{EF}$ forming $\triangle DEF$ with the required properties. By SAS = SAS, $\triangle DEF$ is the triangle that satisfies the given conditions.

Next we show how to construct a triangle when two angles and the included side are given.

CONSTRUCTION 3.3

Construct a triangle with two angles and the included side given.

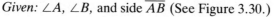

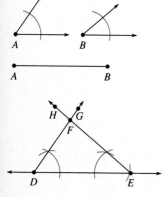

Figure 3.30

Given: $\angle A$, $\angle B$, and side $\overline{AB}$ (See Figure 3.30.)

To Construct: The triangle with these parts

Construction:

1. Draw line ℓ, choose point D on ℓ, and use Construction 2.1 to duplicate $\overline{AB}$ ($\overline{DE}$ in Figure 3.30).
2. Use Construction 2.2 to duplicate $\angle A$ with vertex at D ($\angle GDE$) and $\angle B$ with vertex at E ($\angle HED$).
3. Locate point F at the intersection of rays $\overrightarrow{DG}$ and $\overrightarrow{EH}$ forming sides $\overline{DF}$ and $\overline{EF}$ of the desired triangle $\triangle DEF$. By ASA = ASA, $\triangle DEF$ is the triangle that satisfies the given conditions.

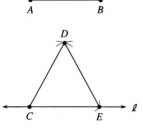

Plato (430–349 B.C.)

The Greek philosopher Plato, perhaps best known as the author of *The Republic,* also studied geometry. Much of the mathematical work done in the fourth century (*B.C.*) was done by his students and friends. Plato is credited with the concept that geometric constructions should be formed only with a compass and straightedge. Over the door to his school, The Academy, was found the statement "Let no one ignorant of geometry enter here."

E X A M P L E   1   Construct an isosceles triangle with base $\overline{AB}$ and sides $\overline{CD}$, given in Figure 3.31.

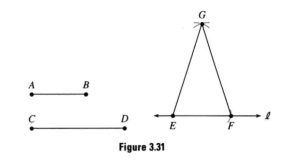

Figure 3.31

The construction is similar to Construction 3.1. Draw line ℓ and duplicate $\overline{AB}$ on ℓ, call it $\overline{EF}$. Set the compass at length CD, place the point at E and draw an arc, then at F and draw another arc intersecting the first one. This determines point G. Use a straightedge and draw $\overline{EG}$ and $\overline{FG}$ to form the desired isosceles triangle, $\triangle EFG$.   ◾

PRACTICE EXERCISE 1

Construct an isosceles triangle with vertex angle $\angle A$ and sides $\overline{BC}$, given in the figure below.

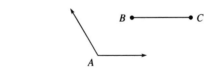

Equilateral triangles are easy to construct.

CONSTRUCTION 3.4

Construct an equilateral triangle when given a single side.

Given: Side $\overline{AB}$ (See Figure 3.32.)

To Construct: An equilateral triangle with sides of length AB

Construction:

1. Draw line ℓ and duplicate $\overline{AB}$ on ℓ $(\overline{CE})$.
2. Using the compass set at length AB, make two arcs by placing the point at C and at E. The point of intersection of the arcs is D.

Figure 3.32

3. Use a straightedge and draw $\overline{CD}$ and $\overline{ED}$. Then $\triangle CDE$ is the desired equilateral triangle.

Constructing an altitude of a triangle is a direct application of Construction 2.5, constructing a line perpendicular to a given line from a point not on that line.

CONSTRUCTION 3.5

Construct an altitude of a given triangle.

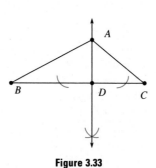

Figure 3.33

Given: $\triangle ABC$ (See Figure 3.33.)

To Construct: The altitude of $\triangle ABC$ from vertex A

Construction:

1. Use Construction 2.5 to construct the line $\overleftrightarrow{AD}$ such that $\overleftrightarrow{AD} \perp \overline{BC}$. To do this, we sometimes need to extend $\overline{BC}$ and consider $\overrightarrow{BC}$.

2. Then $\overline{AD}$ is the desired altitude from vertex A.

NOTE: In Construction 3.5 we said that $\overline{AD}$ is *the* desired altitude from vertex A implying that it is also the *only* such altitude from A. This is true because there is exactly one line perpendicular to a given line passing through a given point not on the line by Postulate 2.4.   □

Constructing a median of a triangle uses Construction 2.3 to find the midpoint of a side.

CONSTRUCTION 3.6

Construct a median of a given triangle.

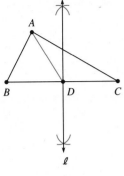

Figure 3.34

Given: △*ABC* (See Figure 3.34.)

To Construct: The median of △*ABC* from vertex *A*

Construction:

1. Use Construction 2.3 to construct line ℓ, the (perpendicular) bisector of $\overline{BC}$, to determine the midpoint *D* of $\overline{BC}$.

2. Use a straightedge to draw $\overline{AD}$, the desired median from vertex *A*. We say *the* median because by the midpoint postulate, $\overline{BC}$ has exactly one midpoint *D,* and by Postulate 1.1, the two points *A* and *D* determine exactly one line.

NOTE: Some students might wonder why we do not specifically give the construction of an angle bisector of a triangle similar to that for medians and altitudes. This is because we already know how to bisect an angle by Construction 2.6, and this procedure works even when the given angle is part of a triangle. □

ANSWER TO PRACTICE EXERCISE: **1.** Construct the desired triangle using Construction 3.2 with sides $\overline{BC}$ and included angle ∠*A*.

3.5 *EXERCISES*

1. Draw an acute scalene triangle and construct a triangle that is congruent to it.

2. Draw an obtuse scalene triangle and construct a triangle that is congruent to it.

3. Draw two segments, one about 2 inches long and the other about 3 inches long. Draw an obtuse angle, and construct the triangle with these sides and included angle.

4. Draw two segments, one about 2 inches long and the other about 3 inches long. Draw an acute angle, and construct the triangle with these sides and included angle.

5. Draw a segment about 2 inches long. Draw two angles, both acute, and construct the triangle with these angles and included side.

6. Draw a segment about 3 inches long. Draw an angle that is about 90° and another acute angle. Construct the triangle with these angles and included side.

7. Give an example to show that two obtuse angles and a segment cannot be used in Construction 3.3.

8. Give examples to show that an obtuse angle, an acute angle, and a segment may or may not yield a triangle using Construction 3.3.

9. Draw a segment about 2 inches long and an obtuse angle. Construct the isosceles triangle with these as sides and the vertex angle.

10. Draw a segment about 2 inches long and a second segment 5 inches long. Construct an isosceles triangle with the base as the 2-inch segment and each side as the 5-inch segment.

11. What happens in Exercise 10 if the base is a 5-inch segment and each side is a 2-inch segment?

12. Construct a right triangle with legs about 2 inches and 3 inches in length, respectively.

13. Construct a right isosceles triangle with legs that are about 2 inches in length.

14. Construct a right angle and bisect it to obtain an acute angle, $\angle A$.

15. Use $\angle A$ in Exercise 14, a segment about 3 inches in length, and a segment about 2.5 inches in length. Try to construct a triangle using $\angle A$, the 3-inch segment as a side of $\angle A$, and the 2.5-inch segment as the side opposite $\angle A$. What happens?

16. Repeat Exercise 15 using a 1-inch segment instead of the 2.5-inch segment. What happens?

17. Draw a scalene acute triangle and construct its three altitudes.

18. Draw a scalene obtuse triangle and construct its three altitudes.

19. Draw a scalene obtuse triangle and construct its three medians.

20. Draw a scalene acute triangle and construct its three medians.

21. Draw a scalene acute triangle and construct its three angle bisectors.

22. Draw a scalene obtuse triangle and construct its three angle bisectors.

23. What common relationship do the altitudes of a triangle seem to have in view of Exercises 17 and 18?

24. What common relationship do the medians of a triangle seem to have in view of Exercises 19 and 20?

25. What common relationship do the angle bisectors of a triangle seem to have in view of Exercises 21 and 22?

26. Construct the perpendicular bisectors of the sides of an acute triangle. What common relationship do these lines seem to have?

27. Construct the perpendicular bisectors of the sides of an obtuse triangle. What common relationships do these lines seem to have?

28. Starting with three given line segments, can you always construct a triangle (using the method in Construction 3.1) with these segments as its sides?

29. Construct an equilateral triangle with sides about 3 inches in length.

30. Use the triangle constructed in Exercise 29 and construct its three medians, three altitudes, and three angle bisectors. What happens?

KEY TERMS AND SYMBOLS

3.1 sides, p. 62
  closed, p. 62
  included, p. 62
  opposite, p. 62
  triangle, p. 62
  vertex, p. 62
  acute triangle, p. 62
  right triangle, p. 62
  hypotenuse, p. 62
  legs, p. 62
  obtuse triangle, p. 63
  equiangular triangle, p. 63
  scalene triangle, p. 63
  isosceles triangle, p. 63

base, p. 63
equilateral triangle, p. 63
perimeter, p. 65
exterior angle, p. 66
interior angle, p. 66
remote interior angles,
  p. 66
adjacent interior angles,
  p. 66
3.2 congruent ($\cong$) parts (of a
  triangle), p. 68
corresponding parts, p. 69
3.3 cpoctae, p. 77
overlapping triangles, p. 80

3.4 vertex angles (of an
  isosceles triangle), p. 86
base angles (of an isosceles
  triangle), p. 86
auxiliary lines or segments,
  p. 88
coincide, p. 90
coincidence proof, p. 90
median, p. 90
altitude, p. 90
angle bisector (of a
  triangle), p. 91

PROOF TECHNIQUES

To Prove:

Two Triangles Congruent

1. Show each is congruent to a third triangle. (Theorem 3.1, p. 72)
2. Show two sides and the included angle of one are equal, respectively, to two sides and the included angle of the other. (Postulate 3.1, p. 70)
3. Show two angles and the included side of one are equal, respectively, to two angles and the included side of the other. (Postulate 3.2, p. 70)
4. Show three sides of one are equal, respectively, to three sides of the other. (Postulate 3.3, p. 71)

Two Segments Equal

1. Show they are corresponding parts of congruent triangles.
2. Show they are drawn from any point on the perpendicular bisector of a segment to the endpoints of that segment. (Theorem 3.3, p. 79)
3. Show they are sides opposite equal angles in a triangle. (Theorem 3.7, p. 87)
4. Show they are segments on the base of an isosceles triangle formed by the bisector of the vertex angle. (Theorem 3.9, p. 89)

Two Angles Equal

1. Show they are corresponding parts of congruent triangles.

2. Show they are angles formed by the bisector of an angle. (Theorem 3.4, p. 79)

3. Show they are angles opposite the equal sides of a triangle. (Theorem 3.5, p. 86)

4. Show they are formed by the perpendicular bisector of the base of an isosceles triangle relative to the vertex angle. (Theorem 3.10, p. 89)

REVIEW EXERCISES

Section 3.1

Exercises 1–8 refer to $\triangle ABC$ given below in which all angles are unequal and all sides are unequal.

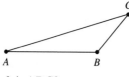

1. What are the vertices of $\triangle ABC$?   2. What are the sides of $\triangle ABC$?

3. Classify $\triangle ABC$ using its sides.   4. Classify $\triangle ABC$ using its angles.

5. What side is included by $\angle A$ and $\angle C$?   6. What angle is included by $\overline{AB}$ and $\overline{BC}$?

7. What side is opposite $\angle A$?   8. What angle is opposite $\overline{AB}$?

9. Find the perimeter of an isosceles triangle with base 12 cm and equal sides 16 cm.

10. The base of an isosceles triangle is half the length of each equal side. If the perimeter is 65 ft, find the length of the base and each side.

Exercises 11–14 refer to the figure below. Answer true or false in each.

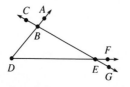

11. $\angle D$ is an interior angle of $\triangle BDE$.   12. $\angle BEF$ is an exterior angle of $\triangle BDE$.

13. $\angle FEG$ is an exterior angle of $\triangle BDE$.

14. $\angle D$ and $\angle BED$ are remote interior angles relative to $\angle ABE$.

15. Can an isosceles triangle be an obtuse triangle?

Section 3.2

In Exercises 16–17, state why the given pair of triangles are congruent.

16. 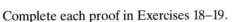 2 cm   2 cm

3 cm   3 cm

17. 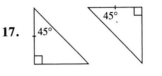 45°   45°

Complete each proof in Exercises 18–19.

18. *Given:* $\angle C = \angle E$

$AC = AE$

Prove: $\triangle ACF \cong \triangle AEB$

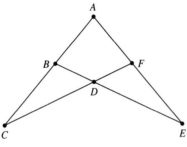

Exercise 18

Proof:

STATEMENTS	REASONS
1. $\angle C = \angle E$	1. _____
2. _____	2. Given
3. _____	3. Reflexive law
4. _____	4. ASA = ASA

19. *Given:* $AB = AE$

$\overline{AC}$ bisects $\angle BAD$

Prove: $\triangle ABC \cong \triangle AEC$

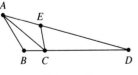

Exercise 19

Proof:

STATEMENTS

1. $AB = AE$

2. _____

3. _____

4. $AC = AC$

5. _____

REASONS

1. _____

2. Given

3. Def. of ∠ bisector

4. _____

5. SAS = SAS

Write a two-column proof in Exercises 20–21.

20. *Given:* $AD = BC$
  $AB = DC$
  Prove: $\triangle ADB \cong \triangle BCD$

21. *Given:* $\overline{AC} \perp \overline{BD}$
  $\angle 1 = \angle 2$
  Prove: $AB = AD$

Exercise 20

Exercise 21

Use the figure below in Exercises 22–27. Assume that $\triangle ABC \cong \triangle DEF$, $AC = x + 2$, $EF = 4x - 4$, $\angle E = (y + 10)°$, and $\angle A = (2y - 15)°$.

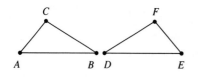

22. Find the value of x.

23. Find the value of y.

24. Find AC.

25. Find the measure of $\angle A$.

26. $\overline{DF}$ corresponds to which side in $\triangle ABC$?

27. $\angle B$ corresponds to which angle in $\triangle DEF$?

Section 3.3

28. Complete the following proof.

  Given: $\angle DAB = \angle DBA$
    $\overline{AC}$ bisects $\angle DAB$
    $\overline{BE}$ bisects $\angle DBA$
  Prove: $AC = BE$

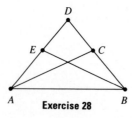

Exercise 28

Proof:

STATEMENTS	REASONS
1. $\angle DAB = \angle DBA$	1. _____
2. _____	2. Given
3. $\angle CAB = \angle DAC$	3. _____
4. $\overline{BE}$ bisects $\angle DBA$	4. _____
5. _____	5. Def. of $\angle$ bisector
6. $\angle CAB + \angle DAC = \angle DAB$ and $\angle EBA + \angle DBE = \angle DBA$	6. _____
7. $\angle CAB + \angle CAB = \angle DAB$ and $\angle EBA + \angle EBA = \angle DBA$	7. _____
8. $2\angle CAB = \angle DAB$ and $2\angle EBA = \angle DBA$	8. Distributive law
9. $2\angle CAB = 2\angle EBA$	9. _____
10. $\angle CAB = \angle EBA$	10. _____
11. $AB = AB$	11. _____
12. $\triangle ACB \cong \triangle BEA$	12. _____
13. _____	13. cpoctae

Write a two-column proof in Exercises 29–31.

29. *Given:* E is the midpoint of $\overline{AC}$
   E is the midpoint of $\overline{BD}$

Prove: $AB = CD$

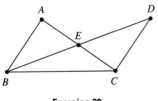

Exercise 29

Section 3.4

30. *Given:* $AB = CD$
$\qquad\quad BC = DE$
$\qquad\quad \angle CAE = \angle CEA$
$\quad$ *Prove:* $\angle B = \angle D$

31. *Given:* $\triangle ABD$ is isosceles with base $\overline{BD}$
$\qquad\quad \triangle BDE$ is isosceles with base $\overline{BD}$
$\quad$ *Prove:* $\angle 1 = \angle 2$

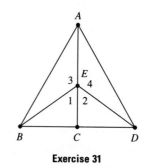

Exercise 30

Exercise 31

Section 3.5

32. Draw an obtuse triangle and construct a triangle that is congruent to it.

33. Draw a segment 3 inches long and an acute angle. Construct a right triangle with the 3-inch segment as the leg included between the right angle and the acute angle.

34. Construct the median from the vertex at the right angle in the triangle you constructed in Exercise 33.

35. Draw a segment about 2 inches long and an obtuse angle. Construct an isosceles triangle with the two sides as the 2-inch segment and vertex angle as the obtuse angle.

PRACTICE TEST

Refer to the figure below in problems 1–10.

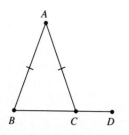

1. Classify $\triangle ABC$ using its angles.

2. Classify $\triangle ABC$ using its sides.

3. What is the vertex angle?

4. What can be said about $\angle B$ and $\angle C$?

5. What angle is opposite $\overline{AC}$?

6. What side is opposite $\angle A$?

7. Is $\angle B$ an interior angle of $\triangle ABC$?

8. Is $\angle ACD$ an exterior angle of $\triangle ABC$?

9. $\angle A$ and $\angle B$ are remote interior angles to what angle?

10. If $AB = 5$ cm and $BC = 3$ cm, what is the perimeter of $\triangle ABC$?

11. What is the perimeter of an equilateral triangle with one side measuring 11 inches?

12. Suppose $\overline{AB}$ and $\overline{EF}$ are corresponding sides of congruent triangles with $\angle C$ opposite $\overline{AB}$ and $\angle D$ opposite $\overline{EF}$. What is the value of x if $\angle C = 48°$ and $\angle D = (x + 20)°$?

Write a two-column proof in problems 13–14.

13. *Given:* $AC = AE$
    D is the midpoint of $\overline{CE}$
    $\angle 1 = \angle 2$
    Prove: $BD = FD$

14. *Given:* $\angle 3 = \angle 4$
    $BC = DE$
    Prove: $\angle 5 = \angle 6$

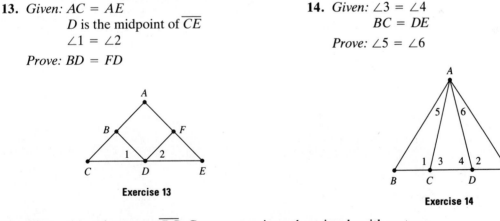

Exercise 13

Exercise 14

15. Given $\angle A$ and segment $\overline{BC}$. Construct an isosceles triangle with vertex angle equal to $\angle A$ and sides of length BC. Then construct the altitude from one of the base angles and the median from the vertex angle.

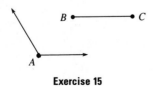

Exercise 15

4

PARALLEL LINES AND POLYGONS

This chapter begins with a discussion of another type of proof, an indirect proof. Such proofs are especially important when we want to show that two lines in a plane never intersect, that is, when the two lines are parallel. The study of parallel lines played a major role in the development of geometry and centers on one of the more controversial postulates, the parallel postulate. We will discover that geometric figures involving parallel line segments have numerous properties and applications in such diverse areas as navigation, physics, geology, construction, and architecture. One of these applications, presented below, is solved later in the chapter in Example 1 of Section 4.4.

The wind is blowing due north at 50 mph. The pilot of an airplane wants to fly due west. Because the wind will take the airplane off course toward the north, the pilot must set a course to the south of due west to maintain a true westerly resultant direction. If the velocity of the airplane is 500 mph, how can a parallelogram be used to assist the pilot in determining the correct course?

4.1 INDIRECT PROOF AND THE PARALLEL POSTULATE

In a direct proof of a geometry theorem, the reason given for each statement is either a definition, a postulate, or a theorem that has already been proven. Although direct proofs are the most common type of proof, some theorems are more easily proved using the format of an *indirect proof.* If $P \longrightarrow Q$ is a theorem we are to prove and we start by assuming P is true, then we must show that Q is true. Now we know that there are two possibilities for Q: either Q is true or Q is false. If we can show that the assumption that Q is false leads to a contradiction of a known fact, such as P, then we are forced to conclude that Q *cannot* be false, or that Q is indeed true (the only remaining possibility). This is the basis of an indirect proof. The following example illustrates this type of reasoning by solving a classic puzzle involving truth-telling and lying.

EXAMPLE 1 An explorer is in a primitive region inhabited by two tribes of natives. The members of one tribe always lie, and the members of the other always tell the truth. The explorer meets two natives on a trail, He asks the taller one "Are you a truth teller?" The native replies "Wagoo." "Him say 'yes' " explains the shorter native who fortunately speaks English, "but him a liar." Give an indirect proof to show that the shorter native is a member of the tribe of truth-tellers.

We know that the shorter native is either a truth-teller or a liar. Suppose we assume he is a liar. If we can show this assumption is wrong, then we are forced to conclude that he is a truth-teller, the desired result.

Note that the taller native must answer "yes" to the question "Are you a truth teller?" regardless of the tribe to which he belongs. That is, if he is a liar he must say "yes" and lie, and if he is a truth-teller he must say "yes" and tell the truth. Thus, assuming the shorter native is a liar, he is forced to tell the explorer that the taller native said "no." Because this is a contradiction to what actually happened, we must discard the assumption that the smaller native is a liar and conclude that the smaller native is a truth-teller. ◪

Indirect proofs require the ability to form the negation of a statement. Recall that if Q is the statement

"Line m is perpendicular to line n,"

then the negation of Q, $\sim Q$, is

"Line m is *not* perpendicular to line n."

In an indirect proof of the theorem $P \longrightarrow Q$, if we were to assume that the theorem is not true, we would be assuming that P is true and that Q is false. This is the same as assuming that P is true and that $\sim Q$ is also true. With this notation we can outline the format of an indirect proof.

INDIRECT PROOF OF P $\longrightarrow$ *Q*

Suppose we already know that $\sim Q \longrightarrow Q_1$, $Q_1 \longrightarrow Q_2$, $Q_2 \longrightarrow Q_3$, and $Q_3 \longrightarrow \sim P$ are accepted or previously proved statements, then the format used to write an **indirect proof** of $P \longrightarrow Q$ is:

Given: P

Prove: Q

Proof:

STATEMENTS	REASONS
1. P	1. Given
2. Assume $\sim Q$ is true	2. Assumption
3. Q_1	3. $\sim Q \longrightarrow Q_1$
4. Q_2	4. $Q_1 \longrightarrow Q_2$
5. Q_3	5. $Q_2 \longrightarrow Q_3$
6. $\sim P$	6. $Q_3 \longrightarrow \sim P$

But this is a contradiction because we were given that P is true and now we have $\sim P$ also true. Thus, our assumption that $\sim Q$ was true is incorrect; so Q must be true.

$\therefore P \longrightarrow Q$.

NOTE: The format for an indirect proof can consist of any number of steps. Notice that the progression of statements uses the same deductive reasoning (for example, $\sim Q$ and $\sim Q \longrightarrow Q_1$, together give Q_1) as was used in direct proofs. □

A student was asked to prove a particular theorem in geometry class. Her "proof" took the following form: "The theorem is either true or false. If it is true, we are finished. If it is false, why bother to try and prove it? Thus, the theorem is true." Did this student have a good understanding of what constitutes a mathematical proof?

EXAMPLE 2   Consider the following information taken from a newspaper article. "If the governor has committed a crime, the House of Representatives will impeach him. If impeached by the House, he will stand trial in the Senate. If tried in the Senate, the governor will be found guilty. If guilty, he will be permanently removed from office by May."

Give an indirect proof of the statement "If the governor is still in office during the summer, then he did not commit a crime."

Given: The governor is still in office during the summer.

Prove: The governor did not commit a crime.

Proof:

STATEMENTS	REASONS
1. The governor is still in office during the summer.	1. Given

2. Assume that the governor committed a crime.
    2. Assumption

3. The House of Representatives will impeach the governor.
    3. First sentence in the article

4. The governor will be tried in the Senate.
    4. Second sentence in the article

5. The governor will be found guilty.
    5. Third sentence in the article

6. The governor will be permanently removed from office by May.
    6. Fourth sentence in the article

But Statement 6 contradicts the given Statement 1 "The governor is still in office during the summer." Thus, our assumption that "The governor committed a crime" was incorrect; so we must conclude that he did not commit a crime.
∴ If the governor is still in office during the summer, then he did not commit a crime. ◪

Both the direct and indirect methods of proof use a chain of conditional statements. In a direct proof of $P \longrightarrow Q$, we start with the hypothesis P and end with the conclusion Q. In an indirect proof, we start with P and $\sim Q$ and end when we reach a contradiction (often $\sim P$). The contradiction tells us that if P is true, then Q cannot be false which means that $P \longrightarrow Q$.

Up to now, we've worked mostly with lines that intersect and the angles formed by these lines. Now we'll work with lines that do not intersect.

DEFINITION: PARALLEL LINES

Two lines in a plane that do not intersect are called **parallel lines.**

If two lines ℓ and m are parallel, we write $\ell \parallel m$. We also say that two segments, two rays, or a segment and a ray are parallel when the lines containing them are parallel. For example in Figure 4.1, if $\ell \parallel m$, we also have $\overline{AB} \parallel \overline{CD}$, $\overrightarrow{AB} \parallel \overrightarrow{CD}$, and $\overrightarrow{AB} \parallel \overrightarrow{DC}$. Also, ℓ is not parallel to n, written $\ell \nparallel n$, because ℓ and n intersect at point B. Similarly, $\overline{CB} \nparallel \overline{CD}$ and $\overrightarrow{AB} \nparallel \overrightarrow{BC}$.

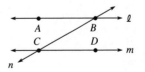

Figure 4.1   Parallel and Nonparallel Lines

The next postulate is perhaps the most famous and controversial of all postulates in geometry. Euclid tried to prove it, as did many mathematicians who followed him. All were unsuccessful. One of the major discoveries in mathematics history involved showing that this postulate *was* independent from the other postulates and *could not* be proved using them.

In the nineteenth century, the Russian mathematician Nicholas Lobachevsky (1793–1856) and a Hungarian colleague, John Bolyai (1802–1860), working independently, replaced Euclid's parallel postulate with the assumption that there are *infinitely many* lines through a point, not on a given line, parallel to the line. Somewhat later, Bernard Riemann (1826–1866), a German mathematician, assumed that there are *no* lines through a point *P*, not on a given line, parallel to the line. These assumptions led to new geometries, called non-Euclidean geometries, which have played a major role in the evolution of mathematics. Riemann's work was of extreme importance in Einstein's theory of relativity. In this text we follow the approach taken by Euclid.

POSTULATE 4.1 PARALLEL POSTULATE

For a given line ℓ and a point P not on ℓ, one and only one line through P is parallel to ℓ.

EXAMPLE 3 In Figure 4.2, $\ell \parallel m$ and P is a point on ℓ. How many lines through P are parallel to m? By the parallel postulate, there is one and

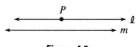

Figure 4.2

only one line through P parallel to m; that line must be ℓ. ∎

We need to be able to determine when two lines are parallel. The definition of parallel lines might be called a *negative definition;* it tells us that lines are parallel when they *do not* intersect. As a result, many proofs involving parallels are indirect proofs. To give an indirect proof of a theorem $P \longrightarrow Q$ we assume that P is true and that Q is false. Thus, when Q is the statement "two lines are parallel," we often assume that the lines are not parallel, that is, the lines intersect, and then see where the assumption leads. If we arrive at a contradictory fact, we can conclude that the lines do not intersect and therefore must be parallel. The proof of the next theorem illustrates and provides us with a more positive method for proving that two lines are parallel.

THEOREM 4.1

If two lines in a plane are both perpendicular to a third line, then they are parallel.

Given: Lines ℓ, m, and n with $m \perp \ell$ and $n \perp \ell$. (See Figure 4.3.)

Prove: $m \parallel n$

Proof:

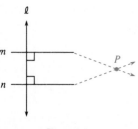

Figure 4.3

STATEMENTS	REASONS
1. Assume $m \nparallel n$	1. Assumption we wish to show is incorrect
2. m and n intersect at some point, say P	2. Def. of $\parallel$ lines
3. $m \perp \ell$ and $n \perp \ell$	3. Given

But this is a contradiction of Postulate 2.4 which says that there is only one line perpendicular to a given line passing through a point not on that line. Thus, the assumption $m \nparallel n$ is incorrect; so we must conclude that $m \parallel n$.

4.1 EXERCISES

Complete the indirect proof of each "theorem" in Exercises 1–2.

1. *Premise 1:* If Bob arrives on time for work, then he woke up on schedule.
   Premise 2: If he wakes up on time, then his alarm rang.
   Premise 3: If his alarm rings, then the power did not fail.
   Theorem: If the power fails, then Bob will be late for work.

   Given: The power fails.

   Prove: Bob will be late for work.

   Proof:

STATEMENTS	REASONS
1. The power fails.	1. Given
2. Assume Bob arrives on time for work.	2. _____
3. He woke up on schedule.	3. _____
4. His alarm rang.	4. _____
5. The power did not fail.	5. _____

   But not having a power failure contradicts the given Statement 1 "The power fails." Thus, our assumption in Statement 2 was incorrect; so we must conclude that Bob will be late for work. ∴ If the power fails, then Bob will be late for work.

2. *Premise 1:* If I gamble, then I will lose.
   Premise 2: If I am unhappy, then I will get a divorce.
   Premise 3: If I go to Las Vegas, then I will gamble.
   Premise 4: If I lose at gambling, then I will be unhappy.
   Theorem: If I stay married, then I did not go to Las Vegas.

   Given: I stay married.

   Prove: I did not go to Las Vegas.

   Proof:

STATEMENTS	REASONS
1. I stay married.	1. Given
2. _____	2. Assumption
3. I will gamble.	3. _____
4. I will lose.	4. _____
5. _____	5. Premise 4
6. I will get a divorce.	6. _____

   But this is a contradiction of Statement 1. Thus, we must conclude that our assumption in Statement 2 was incorrect. ∴ _____

Give an indirect proof of each "theorem" in Exercises 3–4.

3. *Premise 1:* If I don't exercise, then I'm not playing tennis regularly.
   Premise 2: If I'm unhealthy, then I don't exercise.
   Premise 3: If I don't play tennis regularly, then the weather is bad.
   Theorem: If the weather is nice, then I am healthy.

4. *Premise 1:* If it's not a parallelogram, then it's not a rectangle.
   Premise 2: If it's not a four-sided figure, then it's not a quadrilateral.
   Premise 3: If it's not a rectangle, then it's not a square.
   Premise 4: If it's not a quadrilateral, then it's not a parallelogram.
   Theorem: If it's a square, then it's a four-sided figure.

Use the figure below to answer each question in Exercises 5–14.

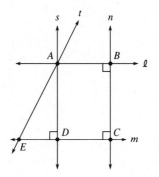

5. Is $\ell \parallel m$?

6. Is $s \parallel n$?

7. Is $\overline{AB} \parallel \overline{ED}$?

8. Is $\overline{AD} \parallel \overline{EC}$?

9. Is $\overrightarrow{EC} \parallel \overrightarrow{DA}$?

10. Is $\overrightarrow{AB} \parallel \overrightarrow{CE}$?

11. Can t be parallel to n?

12. Does there exist a line through E parallel to n?

13. How many lines through B are parallel to m?

14. Do $\overleftrightarrow{AE}$ and $\overleftrightarrow{BC}$ intersect?

Give an indirect proof of each Theorem in Exercises 15–16.

15. If two lines are parallel to a third line then they are parallel to each other.
    [Hint: Use Postulate 4.1.]

16. If a line intersects one of two parallel lines, then it intersects the other.
    [Hint: Use Postulate 4.1.]

In Exercises 17–22, assume that ℓ, m, and n are three distinct lines in a plane and P is a point in the plane.

17. If $\ell \parallel m$ and $\ell \parallel n$, is $m \parallel n$?

18. If $\ell \parallel m$ and P is a point on both m and n, is $n \parallel \ell$?

19. If $\ell \perp m$ and $\ell \perp n$, is $m \perp n$?

20. If $\ell \perp m$ and $\ell \perp n$, is $m \parallel n$?

21. If $\ell \parallel m$, $\ell \parallel n$, and P is on m, is P on n?

22. If $\ell \perp m$, $m \perp n$, and P is on ℓ, is P on n?

23. Give examples of parallel lines found in a classroom.

24. Give examples of perpendicular lines found in a classroom.

25. Give examples of perpendicular lines found on a football field.

26. Give examples of parallel lines found on a football field.

27. Can a triangle have two right angles?

28. Can a triangle have two sides that are parallel?

4.2 TRANSVERSALS AND ANGLES

One of the best ways to study two parallel lines is to consider the various angles that are formed when a third line intersects them.

> ### DEFINITION: TRANSVERSAL
>
> A **transversal** is a line that intersects two distinct lines in two distinct points.

Transversal
(a)

In Figure 4.4(a) ℓ is a transversal that **cuts** (intersects) lines m and n in the two points A and B, respectively. However, in Figure 4.4(b), s is not a transversal because it intersects t and u in only one point P.

When a transversal cuts two lines, several pairs of angles are formed.

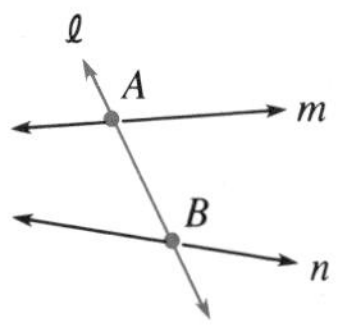

> ### DEFINITION: ANGLES FORMED BY A TRANSVERSAL
>
> Suppose two lines are cut by a transversal.
>
> 1. The nonadjacent angles on opposite sides of the transversal but on the interior of the two lines are called **alternate interior angles.**
> 2. The nonadjacent angles on the same side of the transversal and in the same corresponding positions with respect to the two lines are called **corresponding angles.**
> 3. The nonadjacent angles on opposite sides of the transversal and on the exterior of the two lines are called **alternate exterior angles.**

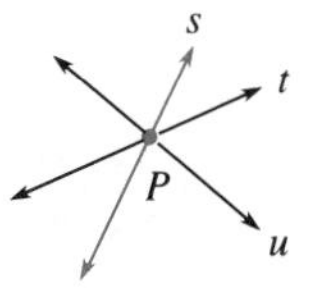

Not a Transversal
(b)

Figure 4.4

In Figure 4.5, ℓ is a transversal that cuts m and n. There are two pairs of alternate interior angles: $\angle 4$ and $\angle 6$; and $\angle 3$ and $\angle 5$. There are four pairs of cor-

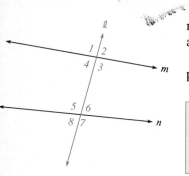

Figure 4.5

responding angles: $\angle 1$ and $\angle 5$; $\angle 2$ and $\angle 6$; $\angle 3$ and $\angle 7$; and $\angle 4$ and $\angle 8$. There are two pairs of alternate exterior angles: $\angle 1$ and $\angle 7$; and $\angle 2$ and $\angle 8$.

A pair of alternate interior angles can be used to show that two lines are parallel.

THEOREM 4.2

If two lines are cut by a transversal and a pair of alternate interior angles are equal, then the lines are parallel.

Given: Lines m and n cut by
transversal ℓ
$\angle 1 = \angle 2$ (See Figure 4.6.)

Prove: $m \parallel n$

Construction: Construct the midpoint
of $\overline{AB}$, C.
Construct $\overleftrightarrow{CD} \perp m$ through C.

Figure 4.6

Proof:

STATEMENTS	REASONS
1. m and n are lines with transversal ℓ	1. Given
2. C is the midpoint of $\overline{AB}$	2. Construction 2.3
3. $AC = CB$	3. Def. of midpoint
4. $\overleftrightarrow{CD} \perp m$	4. Construction 2.5
5. $\angle ADC$ is a right angle	5. $\perp$ lines form rt. $\angle$'s
6. $\angle DCA$ and $\angle BCE$ are vertical angles	6. Def. vert. $\angle$'s
7. $\angle DCA = \angle BCE$	7. Vert. $\angle$'s are $=$
8. $\angle 1 = \angle 2$	8. Given
9. $\triangle ACD \cong \triangle BEC$	9. ASA = ASA
10. $\angle ADC = \angle CEB$	10. cpoctae
11. $\angle CEB$ is a right angle	11. Substitution law
12. $\overleftrightarrow{CD} \perp n$	12. Lines forming rt $\angle$'s are $\perp$
13. $m \parallel n$	13. Two lines $\perp$ to third line are $\parallel$

Corresponding angles can also be used to show that lines are parallel.

THEOREM 4.3

If two lines are cut by a transversal and a pair of corresponding angles are equal, then the lines are parallel.

Given: Lines *m* and *n* cut by
   transversal ℓ
   ∠1 = ∠2 (See Figure 4.7.)

Prove: m ∥ n

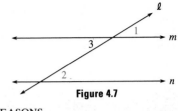

Figure 4.7

Proof:

STATEMENTS	REASONS
1. ∠1 = ∠2	1. Given
2. ∠1 and ∠3 are vertical angles	2. Def. vert. ∠'s
3. ∠3 = ∠1	3. Vert. ∠'s are =
4. ∠3 = ∠2	4. Trans. law
5. *m ∥ n*	5. If alt. int. ∠'s are = lines are ∥

> **THEOREM 4.4**
>
> If two lines are cut by a transversal and a pair of alternate exterior angles are equal, then the lines are parallel.

The proof of this theorem is similar to that for Theorem 4.3 and is left for you to do as an exercise.

> **THEOREM 4.5**
>
> If two lines are cut by a transversal and two interior angles on the same side of the transversal are supplementary, then the lines are parallel.

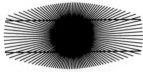

Sometimes our eyes play tricks on us. The two horizontal lines in the figure above, cut by many transversals, are actually parallel. Do they appear this way to you?

> **PRACTICE EXERCISE 1**
>
> Complete the proof of Theorem 4.5.
>
> *Given:* Lines *m* and *n* cut by
>   transversal ℓ
>   ∠1 and ∠2 are supplementary
>
> *Prove: m ∥ n*
>
> *Proof:*
>
> | STATEMENTS | REASONS |
> |---|---|
> | 1. ∠1 and ∠2 are supplementary | 1. _____ |
> | 2. ∠3 and ∠2 are supplementary | 2. _____ |
> | 3. _____ | 3. Supp. of same ∠ are = |
> | 4. *m ∥ n.* | 4. _____ |

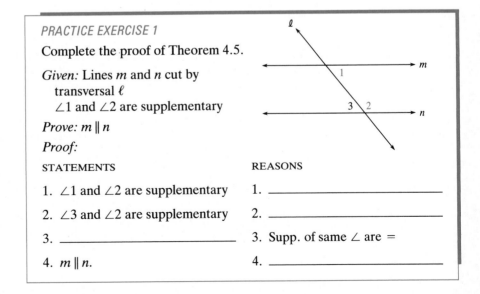

The walls and corners in the rooms of a house are usually not perfectly vertical. A paper hanger uses a plumb line, a string with a weight attached to it, to form a vertical line that is then used to align the first piece of wallpaper. Properties of parallel lines guarantee that the remaining strips of paper will be vertical when seams are aligned.

We now have five ways to prove that two lines m and n are parallel.

1. m and n are both $\perp$ to ℓ. (Theorem 4.1)
2. Alternate interior angles are equal. (Theorem 4.2)
3. Corresponding angles are equal. (Theorem 4.3)
4. Alternate exterior angles are equal. (Theorem 4.4)
5. Interior angles on the same side of a transversal are supplementary. (Theorem 4.5)

The converse of each of these theorems is also true.

THEOREM 4.6   (CONVERSE OF THEOREM 4.1)

If two lines are parallel and a third line is perpendicular to one of them, then it is also perpendicular to the other.

Given: $m \parallel n$
   $\ell \perp m$ (See Figure 4.8.)

Prove: $\ell \perp n$

Construction: Construct line s through P so that $s \perp \ell$. We must show that s and n are the same line.

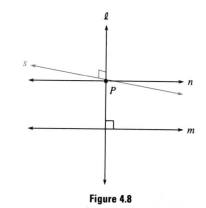

Figure 4.8

Proof:

STATEMENTS	REASONS
1. Construct s through P with $s \perp \ell$	1. Construction 2.4
2. $\ell \perp m$	2. Given
3. $m \parallel s$	3. Lines $\perp$ third line are $\parallel$
4. $m \parallel n$	4. Given
5. s and n are the same line	5. Parallel post.
6. $\ell \perp n$	6. Because $s \perp \ell$ and n and s are the same

THEOREM 4.7   (CONVERSE OF THEOREM 4.2)

If two parallel lines are cut by a transversal, then all pairs of alternate interior angles are equal.

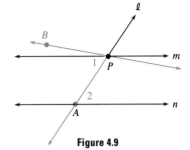

Given: $m \parallel n$ with ℓ a transversal
       (See Figure 4.9.)

Prove: $\angle 1 = \angle 2$

We give an indirect proof by showing
that the assumption $\angle 1 \neq \angle 2$ leads
to a contradiction.

Proof:

Figure 4.9

STATEMENTS	REASONS
1. Assume $\angle 1 \neq \angle 2$	1. Assumption
2. Construct $\angle APB$ such that $\angle APB = \angle 2$	2. Construction 2.2
3. $\overline{BP} \parallel n$	3. Alt. int. $\angle$'s are $=$
4. $m \parallel n$	4. Given

Because $\angle APB = \angle 2$ and $\angle 2 \neq \angle 1$, $\angle APB \neq \angle 1$, so $\overline{BP}$ and m are two distinct lines through P parallel to n, a contradiction of the parallel postulate. Thus, $\angle 1 \neq \angle 2$ leads to a contradiction forcing us to conclude that $\angle 1 = \angle 2$. We could show that the other pair of alternate interior angles are also equal using a similar proof.

THEOREM 4.8   (CONVERSE OF THEOREM 4.3)

If two parallel lines are cut by a transversal, then all pairs of corresponding angles are equal.

PRACTICE EXERCISE 2

Complete the proof of Theorem 4.8.

Given: $m \parallel n$ and m and n are cut
    by transversal ℓ

Prove: $\angle 1 = \angle 2$

Proof:

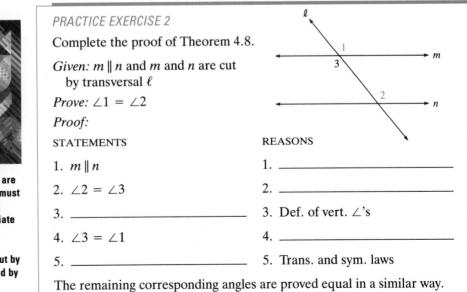

STATEMENTS	REASONS
1. $m \parallel n$	1. _____
2. $\angle 2 = \angle 3$	2. _____
3. _____	3. Def. of vert. $\angle$'s
4. $\angle 3 = \angle 1$	4. _____
5. _____	5. Trans. and sym. laws

The remaining corresponding angles are proved equal in a similar way.

When the pieces in a quilt are sewn together, much care must be taken to maintain equal angles so that the appropriate sides of each quilt block remain parallel. Which theorems involving lines cut by a transversal are suggested by the quilt shown?

The proof of the next theorem is similar to that of Theorem 4.8 and is left for you to do as an exercise.

> **THEOREM 4.9   (CONVERSE OF THEOREM 4.4)**
>
> If two parallel lines are cut by a transversal, then all pairs of alternate exterior angles are equal.

> **THEOREM 4.10   (CONVERSE OF THEOREM 4.5)**
>
> If two parallel lines are cut by a transversal, then all pairs of interior angles on the same side of the transversal are supplementary.

The proof of Theorem 4.10 is left for you to do as an exercise.

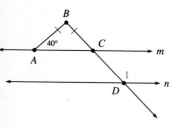

Figure 4.10

E X A M P L E 1   In Figure 4.10, $m \parallel n$ and $AB = BC$. What is the measure of $\angle 1$?

Because $AB = BC$, $\angle BCA = 40°$. Because $\angle BCA$ and $\angle ACD$ are supplementary, $\angle ACD = 140°$. Because $\angle 1$ and $\angle ACD$ are alternate interior angles formed by transversal $\overleftrightarrow{BD}$ cutting parallel lines m and n, $\angle 1 = 140°$.   ◾

ANSWERS TO PRACTICE EXERCISES: **1.** 1. Given   2. Adj. ∠'s whose non-common sides are in a line are supp.   3. $\angle 1 = \angle 3$ 4. If alt. int. ∠'s are = the lines are ∥.   **2.** 1. Given   2. If ∥ lines are cut by a transv. the alt. int. ∠'s are =   3. $\angle 3$ and $\angle 1$ are vert. ∠'s   4. Vert. ∠'s are =   5. $\angle 1 = \angle 2$

4.2 EXERCISES

Exercises 1–12 refer to the figure below in which $m \parallel n$ and ℓ is a transversal.

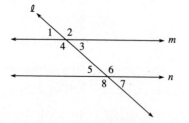

1. List two pairs of alternate interior angles.
2. List two pairs of alternate exterior angles.
3. List four pairs of corresponding angles.

4. List four angles that are supplementary to ∠1.

5. List four angles that are supplementary to ∠2.

6. List three angles that are equal to ∠1.

7. List three angles that are equal to ∠2.

8. If ∠1 = 48°, find the measure of every other angle.

9. If ∠6 = 135°, find the measure of every other angle.

10. Find the measure of each angle if ∠4 is 30° more than twice ∠5.

11. Find the measure of each angle if ∠6 is 15° more than twice ∠1.

12. Find the measure of each angle if ∠2 is 10° more than ∠8.

Exercises 13–20 refer to the figure below.

13. If ∠2 = ∠4, can $\overleftrightarrow{BC}$ and $\overleftrightarrow{DE}$ intersect?

14. If ∠1 = ∠4, can $\overleftrightarrow{BC}$ and $\overleftrightarrow{DE}$ intersect?

15. If ∠2 = ∠5, can $\overleftrightarrow{BC}$ and $\overleftrightarrow{DE}$ intersect?

16. If ∠3 and ∠5 are supplementary, can $\overleftrightarrow{BC}$ and $\overleftrightarrow{DE}$ intersect?

17. If $AB = AC$ and ∠5 = ∠6, can $\overleftrightarrow{BC}$ and $\overleftrightarrow{DE}$ intersect?

18. If $AB = AC$ and ∠4 = ∠6, can $\overleftrightarrow{BC}$ and $\overleftrightarrow{DE}$ intersect?

19. If $AD = AE$ and ∠5 and ∠3 are supplementary, are $\overleftrightarrow{BC}$ and $\overleftrightarrow{DE}$ parallel?

20. If $AD = AE$ and ∠4 = ∠6, are $\overleftrightarrow{BC}$ and $\overleftrightarrow{DE}$ parallel?

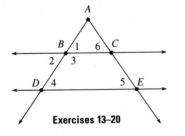

Exercises 13–20

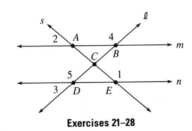

Exercises 21–28

Use the figure above in Exercises 21–28.

21. *Given:* ∠1 is supplementary to ∠2
Prove: m ∥ n

22. *Given:* ∠3 is supplementary to ∠4
Prove: m ∥ n

23. *Given: m ∥ n*
  AB = DE
Prove: △ABC ≅ △CDE

24. *Given: m ∥ n*
  ∠1 = ∠5
Prove: △ABC is isosceles

25. *Given: C* is the midpoint of $\overline{AE}$ and $\overline{BD}$
Prove: m ∥ n

26. *Given: m ∥ n*
  CD = CE
Prove: AC = BC

27. *Given: AC = CE*
  m ∥ n
Prove: DC = CB

28. *Given: AB = DE*
  AD = BE (Draw auxiliary lines.)
Prove: m ∥ n

29. Prove Theorem 4.4.

30. Prove Theorem 4.9.

31. Prove Theorem 4.10.

32. Use the figure below to find the values of x and y that make $\overline{AB} \parallel \overline{CD}$ and $\overline{AD} \parallel \overline{BC}$.

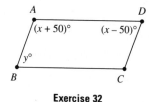

Exercise 32

33. Use the figure below to find the values of x and y that make $\overline{AB} \parallel \overline{CD}$ and $\overline{BC} \parallel \overline{DE}$.

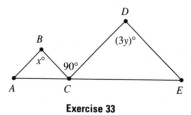

Exercise 33

34. Use the figure below to find the values of x and y that make $m \parallel n$.

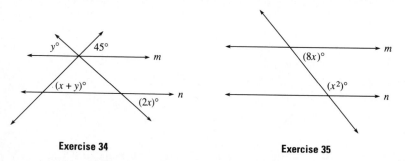

Exercise 34   **Exercise 35**

35. Use the figure above to find the value of x that makes $m \parallel n$.

36. Prove that if a line is drawn through the vertex of an isosceles triangle parallel to the base, then it bisects the exterior angle at the vertex.

4.3 POLYGONS AND ANGLES

A triangle is a special case of a general class of geometric figures called *polygons*. Recall that a triangle is composed of three distinct segments no two of which are on the same line.

> *DEFINITION: POLYGON*
>
> A **polygon** is composed of *n* distinct segments in a plane and possesses the following properties: the segments intersect *only* at their endpoints; exactly two segments contain each endpoint; and no two consecutive segments are on the same line. Each segment is called a **side** of the polygon, and each endpoint is called a **vertex** of the polygon.

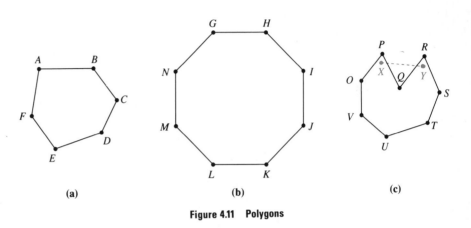

(a)   **(b)**   **(c)**

Figure 4.11 Polygons

Figure 4.11 shows several polygons. The polygon in Figure 4.11(c) possesses a property that we wish to avoid in our work. The polygons that we consider will be assumed to be *convex*.

> *DEFINITION: CONVEX POLYGON*
>
> A polygon is **convex** if, for all pairs of points *X* and *Y* in the interior of the polygon, the segment $\overline{XY}$ is completely in the interior of the polygon.

Notice that the polygon in Figure 4.11(c) is not convex because there are points on $\overline{XY}$ that are not interior to the polygon. The remaining two polygons in Figure 4.11 are convex. When we use the term *polygon* from now on, we will mean *convex polygon*.

What could be more geometrically impressive than the honeycomb of a bee? Note that each opening is in the shape of a convex polygon with six sides.

> *DEFINITION: REGULAR POLYGON*
>
> A polygon is a **regular polygon** if all its sides are equal and all its angles are equal.

Perhaps the most famous polygonal building in the world is the Pentagon in Washington, D.C.

The polygon in Figure 4.11(b) is a regular polygon because $GH = HI = IJ = JK = KL = LM = MN = NG$ and $\angle GHI = \angle HIJ = \angle IJK = \angle JKL = \angle KLM = \angle LMN = \angle MNG = \angle NGH$. Another example of a regular polygon is an equilateral (equiangular) triangle.

Polygons are given special names according to their number of sides as shown in the following table.

Number of Sides	Polygon
3	triangle
4	quadrilateral
5	pentagon
6	hexagon
7	heptagon
8	octagon
9	nonagon
10	decagon
n	n-gon

The polygon in Figure 4.11(a) is a hexagon and the polygon in Figure 4.11(b) is a regular octagon.

The next postulate gives the relationship between the number of sides and the number of angles of a polygon.

POSTULATE 4.2

A polygon has the same number of angles as sides.

DEFINITION: DIAGONAL OF A POLYGON

A **diagonal** of a polygon is a segment that joins two nonadjacent vertices.

In Figure 4.11(a) for example, if we were to construct segment $\overline{AD}$, then $\overline{AD}$ would be a diagonal of the hexagon. Notice in the definition of a diagonal that the term *nonadjacent* is important. Segments joining adjacent vertices are sides, not diagonals.

We have previously defined the perimeter of a triangle. This definition can be extended to all polygons.

DEFINITION: PERIMETER OF A POLYGON

The **perimeter** of a polygon is the sum of the lengths of its sides.

EXAMPLE 1   Consider the polygon in Figure 4.12.

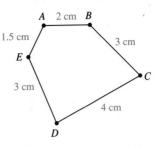

Figure 4.12   Perimeter of a Polygon

(a) What kind of polygon is this? Because it has 5 sides, it is a pentagon.

(b) How many diagonals does it have? Because the diagonals are $\overline{AC}$, $\overline{AD}$, $\overline{BD}$, $\overline{BE}$, and $\overline{CE}$, there are 5 diagonals.

(c) What is the perimeter of the polygon? The perimeter is

$$P = AB + BC + CD + DE + EA.$$
$$= 2 + 3 + 4 + 3 + 1.5 = 13.5 \text{ cm}$$ ▨

To determine properties of the angles of any polygon, we will first prove several theorems about the angles of a triangle and use the results in more general cases. To do this we need to be able to construct the line parallel to a given line and passing through a point not on the given line.

CONSTRUCTION 4.1

Construct the line parallel to a given line that passes through a point not on the given line.

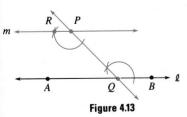

Figure 4.13

Given: Line ℓ and point P (See Figure 4.13.)

To Construct: Line m through P such that $m \parallel \ell$

Construction:

1. Draw any line through P that intersects ℓ at a point we label Q.

2. At P, use Construction 2.2 to construct $\angle RPQ$ such that $\angle RPQ = \angle PQB$.

3. Line $\overleftrightarrow{RP}$, which is equal to m, is the desired line because $m \parallel \ell$ since $\angle RPQ$ and $\angle PQB$ are equal alternate interior angles. By the parallel postulate, m is *the* line satisfying the given conditions.

The followers of Pythagoras (about 584–495 B.C.) are thought to have been the first to prove the next important theorem.

THEOREM 4.11

The sum of the measures of the angles of a triangle is 180°.

Given: $\triangle ABC$

Prove: $\angle A + \angle 1 + \angle B = 180°$

Construction: Construct line $\overleftrightarrow{ED}$ through C parallel to $\overline{AB}$.

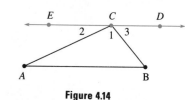

Figure 4.14

Proof:

STATEMENTS	REASONS
1. $\overline{ED} \parallel \overline{AB}$ and C is on $\overline{ED}$	1. Construction 4.1
2. $\angle ECD$ is a straight angle and $\angle ECD = 180°$	2. Def. of st. $\angle$
3. $\angle 2 + \angle 1 + \angle 3 = 180°$	3. Angle add. post.
4. $\angle A = \angle 2$ and $\angle B = \angle 3$	4. Alt. int. $\angle$'s are =
5. $\angle A + \angle 1 + \angle B = 180°$	5. Substitution law

Theorem 4.11 gives rise to a series of corollaries, the proofs of which are requested in the exercises.

COROLLARY 4.12

Any triangle can have at most one right angle or at most one obtuse angle.

COROLLARY 4.13

If two angles of one triangle are equal, respectively, to two angles of another triangle, then the third angles are also equal.

The next corollary gives us another way to prove that two triangles are congruent.

COROLLARY 4.14   AAS = AAS

If two angles and any side of one triangle are equal, respectively, to two angles and the corresponding side of another triangle, then the triangles are congruent.

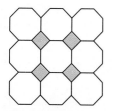

Many floor coverings are made by joining together various polygons without leaving gaps or overlapping the edges. Such coverings are called *tessellations*. The tessellation shown above is made up of squares and regular octagons. Can you find other examples of tessellations?

PRACTICE EXERCISE 1

Complete the proof of Corollary 4.14.

Given: $\triangle ABC$ and $\triangle DEF$
$\quad \angle A = \angle D$ and $\angle B = \angle E$
$\quad CB = FE$

Prove: $\triangle ABC \cong \triangle DEF$

Proof:

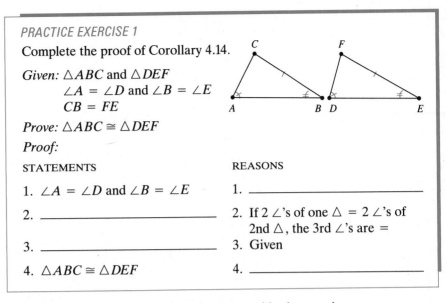

STATEMENTS	REASONS
1. $\angle A = \angle D$ and $\angle B = \angle E$	1. _____
2. _____	2. If 2 $\angle$'s of one $\triangle$ = 2 $\angle$'s of 2nd $\triangle$, the 3rd $\angle$'s are =
3. _____	3. Given
4. $\triangle ABC \cong \triangle DEF$	4. _____

The proof of the next corollary is requested in the exercises.

COROLLARY 4.15

The measure of an exterior angle of a triangle is equal to the sum of the measures of the nonadjacent interior angles.

We now turn our attention to the angles of polygons in general.

THEOREM 4.16

The sum of the measures of the angles of a polygon with n sides is given by the formula $S = (n - 2)180°$.

It is difficult to write a formal two-column proof of this theorem. Instead, we will present a paragraph-style proof using the polygon in Figure 4.15.

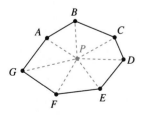

Figure 4.15

PROOF: Select a point P in the interior of the polygon and construct segments joining P to each vertex of the polygon. Because the polygon has n sides, we will obtain n triangles. Because the sum of the angle measures of a triangle is 180°, the sum of the angle measures of these n triangles is $n(180°)$. However, the angles of the triangles with their vertex at P are not parts of the angles of the polygon. Because the sum of these angle measures is 360°, we must subtract this amount from $n(180°)$. Then,

$$n(180°) - 360° = n(180°) - 2(180°)$$
$$= (n - 2)180°.$$

Although the polygon in Figure 4.16 has seven sides, we can see that this polygon was used for illustrative purposes only and that the formula

$$S = (n - 2)180°$$

works for any number of sides n. Notice that when $n = 3$ (a triangle), we obtain $S = (3 - 2)180° = (1)180° = 180°$, the same result we obtained earlier. ■

COROLLARY 4.17

The measure of each angle of a regular polygon with n sides is given by the formula $a = \dfrac{(n - 2)180°}{n}$.

PROOF: Because there are n equal angles in a regular polygon with n sides, if a is the measure of one of these angles, na is the sum of all these measures. By Theorem 4.16,

$$na = (n - 2)180°$$

from which the desired formula results by dividing both sides by n. ■

E X A M P L E 2   Find the sum of the angle measures of a hexagon. Because a hexagon has 6 sides, $n = 6$ in the formula $S = (n - 2)180°$. Thus,

$$S = (6 - 2)180° = (4)180° = 720°. ◪$$

PRACTICE EXERCISE 2

What is the measure of each angle of a regular hexagon?

Sometimes the angles of a polygon are called **interior angles** of the polygon to distinguish them from *exterior angles*. In the polygon in Figure 4.16, $\angle ABF$ is called an **exterior angle** of the polygon.

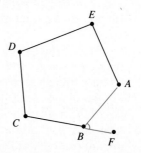

Figure 4.16 Exterior Angle of a Polygon

THEOREM 4.18

The sum of the measures of the exterior angles of a polygon, one at each vertex, is 360°.

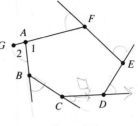

Figure 4.17

PROOF: Consider the polygon in Figure 4.17. Through each vertex we have a straight angle each of which is formed by one interior angle and one exterior angle. For example, straight angle $\angle FAG$ at vertex A is formed by interior angle $\angle 1$ and exterior angle $\angle 2$ because $\angle FAG = \angle 1 + \angle 2$. The total measure of these n straight angles is $(n)180°$, and the sum of the n interior angles of the polygon is $(n - 2)180°$. If we let T be the sum of the n exterior angles, then

$$T + (n - 2)180° = (n)180°.$$

$$
\begin{aligned}
T + (n)(180°) - (2)(180°) &= (n)180° \quad &\text{Distributive law} \\
T - (2)(180°) &= 0 \quad &\text{Subtract } (n)180° \text{ from both sides} \\
T - 360° &= 0 \quad &(2)(180°) = 360° \\
T &= 360° \quad &\text{Add } 360° \text{ to both sides}
\end{aligned}
$$

Thus, the sum of the exterior angles is 360°. ■

COROLLARY 4.19

The measure of each exterior angle of a regular polygon with n sides is determined with the formula $e = \dfrac{360°}{n}$.

The proof of Corollary 4.19 is left for you to do as an exercise.

E X A M P L E   3   What is the measure of each exterior angle of a regular hexagon? Because a hexagon has 6 sides, substitute 6 for n in $e = \dfrac{360°}{n}$.

$$e = \frac{360°}{6} = 60° \quad \blacksquare$$

ANSWERS TO PRACTICE EXERCISES: **1.** 1. Given 2. $\angle C = \angle F$ 3. $CB = FE$ 4. ASA = ASA **2.** 120°

4.3 EXERCISES

In Exercises 1–8, answer each of the following questions for a regular polygon with the given number of sides. **(a)** What is the name of the polygon? **(b)** What is the sum of the angles of the polygon? **(c)** What is the measure of each angle of the polygon? **(d)** What is the sum of the measures of the exterior angles of the polygon? **(e)** What is the measure of each exterior angle of the polygon? **(f)** If each side is 5 cm long, what is the perimeter of the polygon?

1. 3   **2.** 4   **3.** 5   **4.** 6

5. 7   **6.** 8   **7.** 9   **8.** 10

In Exercises 9–12, solve the equation $S = (n - 2)180°$ for n when S is a given value. Find the number of sides of each polygon (if possible) if the given value corresponds to the number of degrees in the sum of the interior angles of a polygon. Remember that n must be a whole number greater than 2, or no such polygon can exist.

9. 1620°   **10.** 2700°   **11.** 2000°   **12.** 3200°

In Exercises 13–16, solve the equation $a = \dfrac{(n - 2)180°}{n}$ for n when a is a given value. Find the number of sides of each polygon (if possible) if the given value is the measure of one interior angle of a regular polygon.

13. 157.5°   **14.** 162°   **15.** 145°   **16.** 105°

17. Two exterior angles of a triangle sum to 200°. What is the measure of the third exterior angle?

18. Three exterior angles of a quadrilateral sum to 300°. What is the measure of the fourth exterior angle?

19. As the number of sides of a regular polygon increases, does each exterior angle increase or decrease?

20. As the number of sides of a regular polygon increases, does an interior angle increase or decrease?

21. What is the smallest angle that any regular polygon can have?

22. What is the largest exterior angle that any regular polygon can have?

23. Find the number of sides of a polygon if the sum of its angles is twice the sum of its exterior angles.

24. If the number of sides of a polygon were doubled, the sum of the angles of the polygon would be increased by 900°. How many sides does the original polygon have?

25. If the sum of the angles of a polygon is equal to the sum of the exterior angles of the polygon, how many sides does the polygon have?

26. By how many degrees is the sum of the angles of a polygon increased when the number of sides is increased by 4?

27. Prove Corollary 4.12.

28. Prove Corollary 4.13.

29. Prove Corollary 4.15.

30. Prove that the exterior angles of a regular polygon are equal.

31. Prove Corollary 4.19.

32. Prove that every point on the bisector of an angle is equidistant from the sides of the angle.

Use the figure below in Exercises 33–34.

33. A city planner wishes to locate the point in Central Park that is equidistant from Elm Street, Washington Avenue, and Park Way. Explain how she can determine this point. [Hint: Use Exercise 32.]

34. Explain how a city planner can find the point in Central Park that is equidistant from a library at point *A*, a monument at point *B*, and a fountain at point *C*.

4.4 PARALLELOGRAMS AND RHOMBUSES

In this and the next section we will study the properties of several quadrilaterals.

> **DEFINITION: PARALLELOGRAM**
> A **parallelogram** is a quadrilateral whose opposite sides are parallel.

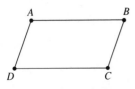

Figure 4.18   □*ABCD*

The symbol □ represents the word *parallelogram*. Figure 4.18 shows □*ABCD* with $\overline{AB} \parallel \overline{CD}$ and $\overline{AD} \parallel \overline{BC}$.

> **THEOREM 4.20**
> Each diagonal divides a parallelogram into two congruent triangles.

Given: □ABCD with diagonal $\overline{BD}$
       (See Figure 4.19.)

Prove: △ABD ≅ △BCD

Proof:

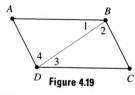
Figure 4.19

STATEMENTS	REASONS
1. □ABCD wth diagonal $\overline{BD}$	1. Given
2. $\overline{AB} \parallel \overline{DC}$ and $\overline{AD} \parallel \overline{BC}$	2. Def. of □
3. ∠1 = ∠3 and ∠2 = ∠4	3. Alt. int. ∠'s are =
4. $BD = BD$	4. Reflexive law
5. △ABD ≅ △BCD	5. ASA = ASA

In a similar manner we could show that diagonal $\overline{AC}$ divides the parallelogram into congruent triangles △ACD and △ABC.

The proof of the next corollary is left for you to do as an exercise.

COROLLARY 4.21

The opposite sides and opposite angles of a parallelogram are equal.

The next theorem provides a way to show that a quadrilateral is a parallelogram.

THEOREM 4.22

If both pairs of opposite sides of a quadrilateral are equal, then the quadrilateral is a parallelogram.

PRACTICE EXERCISE 1

Complete the proof of Theorem 4.22.

Given: $AB = DC$ and $AD = BC$

Prove: ABCD is a □

Construction: Construct diagonal $\overline{BD}$

Proof:

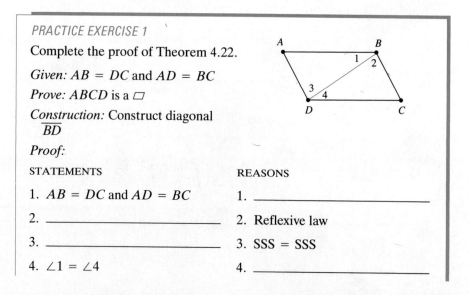

STATEMENTS	REASONS
1. $AB = DC$ and $AD = BC$	1. _____
2. _____	2. Reflexive law
3. _____	3. SSS = SSS
4. ∠1 = ∠4	4. _____

5. $\overline{AB} \parallel \overline{DC}$ 5. _____

6. _____   6. cpoctae

7. _____   7. Alt. int. $\angle$'s are =

8. $ABCD$ is a $\square$ 8. _____

THEOREM 4.23

If both pairs of opposite angles of a quadrilateral are equal, then the quadrilateral is a parallelogram.

The proof of Theorem 4.23 is requested in the exercises.

THEOREM 4.24

Consecutive angles of a parallelogram are supplementary.

The proof of Theorem 4.24 is requested in the exercises.

THEOREM 4.25

If two opposite sides of a quadrilateral are equal and parallel, then the quadrilateral is a parallelogram.

The proof of Theorem 4.25 is requested in the exercises.

THEOREM 4.26

The diagonals of a parallelogram bisect each other.

Given: $\square ABCD$ with diagonals $\overline{AC}$
   and $\overline{BD}$
   (See Figure 4.20.)
Prove: $\overline{AC}$ and $\overline{BD}$ bisect each other
Proof:

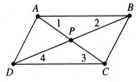

Figure 4.20

STATEMENTS	REASONS
1. $\overline{AC}$ and $\overline{BD}$ are diagonals of $\square ABCD$	1. Given
2. $\angle 1 = \angle 3$ and $\angle 2 = \angle 4$	2. Alt. int. $\angle$'s are =
3. $\overline{AB} = \overline{DC}$	3. Opp. sides of $\square$ are =

4. $\triangle APB \cong \triangle DPC$ 4. ASA = ASA
5. $AP = PC$ and $DP = PB$ 5. cpoctae
6. $\overline{AC}$ and $\overline{BD}$ bisect each other   6. Def. of seg. bisector

The converse of Theorem 4.26 is also true and gives us another way to prove that a quadrilateral is a parallelogram.

THEOREM 4.27

If the diagonals of a quadrilateral bisect each other, then the quadrilateral is a parallelogram.

PRACTICE EXERCISE 2

Complete the proof of Theorem 4.27.

Given: $\overline{AC}$ and $\overline{BD}$ bisect each other

Prove: ABCD is a $\square$

Proof:

STATEMENTS	REASONS
1. $\overline{AC}$ and $\overline{BD}$ bisect each other	1. _____
2. $AP = PC$ and $DP = PB$	2. _____
3. _____	3. Vert. $\angle$'s are =
4. _____	4. SAS = SAS
5. $\angle 3 = \angle 4$	5. _____
6. $\overline{AB} \parallel \overline{DC}$	6. _____
7. $AB = DC$	7. _____
8. _____	8. Two opp. sides are = and $\parallel$

Parallelograms have many applications. For example, in physics when a quantity has both magnitude and direction in a plane, a **vector** is used to describe the quantity. A vector is represented by a line segment, the length of which corresponds to its magnitude. An arrowhead on the segment shows its direction. For example, the velocity of an airplane flying at 400 mph due east can be represented by the vector in Figure 4.21, in which each inch of length corresponds to 100 mph, and the arrow points east in the direction of travel. If the wind is blowing in a northeasterly direction at 50 mph, the vector corresponding to this velocity (of length $\frac{1}{2}$ inch) could be placed with the above vector as shown in Figure 4.22. When the two velocities are combined, the **resultant vector** is found with the **parallelogram law**—by using the diagonal of the parallelogram with the given sides. From the figure we can see that the airplane will fly in the direction of the resultant, and its speed will be a little more than 400 mph, about 425 mph, because the length of the vector is about 4.25 inches.

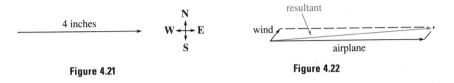

Figure 4.21   Figure 4.22

Using this information we can solve the applied problem given in the chapter introduction.

EXAMPLE 1   The wind is blowing due north at 50 mph. The pilot of an airplane wants to fly due west. Because the wind will take the airplane off course toward the north, the pilot must set a course somewhat to the south of due west to maintain a true westerly resultant direction. If the velocity of the plane is 500 mph, how can a parallelogram be used to assist the pilot in determining the correct course?

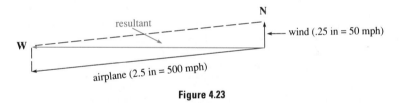

Figure 4.23

According to the parallelogram law, the pilot must set his course in the direction shown in Figure 4.23 so that the resultant of the two forces will be due west. Notice in this case that the actual speed of the plane will be diminished somewhat from the rate of 500 mph because the length of the resultant is a little less than 2.5 inches. ◼

A *rhombus* is a special kind of parallelogram.

RHOMBUS
A rhombus is a parallelogram that has two equal adjacent sides.

THEOREM 4.28
All four sides of a rhombus are equal.

The proof of Theorem 4.28 follows from Corollary 4.21 and is left for you to do as an exercise.

> **THEOREM 4.29**
>
> The diagonals of a rhombus are perpendicular.

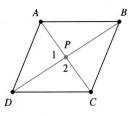

Figure 4.24

Given: □*ABCD* is a rhombus
   (See Figure 4.24.)

Prove: $\overline{AC} \perp \overline{BD}$

Proof:

STATEMENTS	REASONS
1. □*ABCD* is a rhombus	1. Given
2. *AD* = *DC*	2. The sides of a rhombus are =
3. *AP* = *PC*	3. The diag. of □ bisect each other
4. *DP* = *DP*	4. Reflexive law
5. △*APD* ≅ △*DPC*	5. SSS = SSS
6. ∠1 = ∠2	6. cpoctae
7. ∠1 and ∠2 are adjacent angles	7. Def. adj. ∠'s
8. $\overline{AC} \perp \overline{DB}$	8. Def. of ⊥ lines

The converse of Theorem 4.29 is also true, and it provides us with a way to prove that a parallelogram is a rhombus.

> **THEOREM 4.30**
>
> If the diagonals of a parallelogram are perpendicular, then the parallelogram is a rhombus.

> **PRACTICE EXERCISE 3**
>
> Complete the proof of Theorem 4.30.
>
>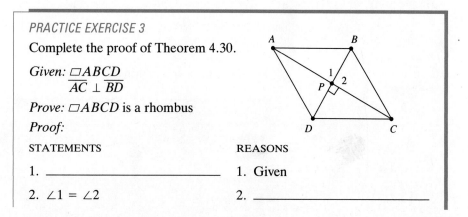
>
> *Given:* □*ABCD*
> $\overline{AC} \perp \overline{BD}$
>
> *Prove:* □*ABCD* is a rhombus
>
> *Proof:*
>
> | STATEMENTS | REASONS |
> |---|---|
> | 1. _____ | 1. Given |
> | 2. ∠1 = ∠2 | 2. _____ |

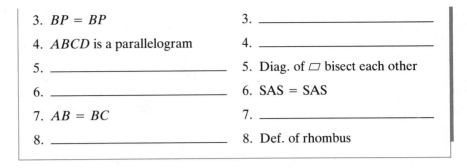

3. $BP = BP$	3. _____
4. $ABCD$ is a parallelogram	4. _____
5. _____	5. Diag. of $\square$ bisect each other
6. _____	6. SAS = SAS
7. $AB = BC$	7. _____
8. _____	8. Def. of rhombus

THEOREM 4.31

The diagonals of a rhombus bisect the angles of the rhombus.

The proof of Theorem 4.31 is requested in the exercises.

ANSWERS TO PRACTICE EXERCISES: **1.** 1. Given   2. $BD = BD$
3. $\triangle ABD \cong \triangle BCD$ 4. cpoctae   5. Alt. int. $\angle$'s are =   6. $\angle 2 = \angle 3$
7. $\overline{BC} \parallel \overline{AD}$ 8. Def. of $\square$ **2.** 1. Given   2. Def. of seg. bisector   3. $\angle 1 = \angle 2$
4. $\triangle APB \cong \triangle DPC$ 5. cpoctae   6. Alt. int. $\angle$'s are =   7. cpoctae   8. $ABCD$
is a parallelogram   **3.** 1. $\overline{AC} \perp \overline{BD}$ 2. Def. of $\perp$ lines   3. Reflexive law
4. Given   5. $AP = PC$ 6. $\triangle APB \cong \triangle BPC$ 7. cpoctae   8. $\square ABCD$ is a
rhombus

4.4 EXERCISES

In Exercises 1–20, refer to the figure below. Answer true or false.

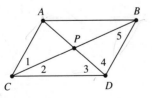

1. If $ABCD$ is a parallelogram, then $\triangle ABC \cong \triangle BCD$.

2. If $ABCD$ is a parallelogram, then $\triangle ABD \cong \triangle ACD$.

3. If $ABCD$ is a parallelogram, then $AB = CD$.

4. If $ABCD$ is a parallelogram, then $CD = DB$.

5. If $ABCD$ is a parallelogram, then $\angle ACD = \angle BAC$.

6. If $ABCD$ is a parallelogram, then $\angle CAB = \angle CDB$.

7. If $AB = CD$ then $ABCD$ is a parallelogram.

8. If $AB = CD$ and $AC = BD$ then $ABCD$ is a parallelogram.

9. If $\angle CAB = \angle BDC$ and $\angle ACD = \angle ABD$ then $ABCD$ is a parallelogram.

10. If $\angle ACD = \angle ABD$ then $ABCD$ is a parallelogram.

11. If $ABCD$ is a parallelogram, then $\angle BAC$ and $\angle ACD$ are supplementary.

12. If $ABCD$ is a parallelogram, then $\angle CAB$ and $\angle CDB$ are supplementary.

13. If $AC = BD$ and $\overline{AC} \parallel \overline{BD}$, then $ABCD$ is a parallelogram.

14. If $AC = BD$ and $\overline{AB} \parallel \overline{CD}$, then $ABCD$ is a parallelogram.

15. If P is the midpoint of $\overline{AD}$ and $\overline{BC}$, then $ABCD$ is a rhombus.

16. If $\overline{AD} \perp \overline{BC}$ then $ABCD$ is a rhombus.

17. If $ABCD$ is a parallelogram and $\angle ACD = 55°$, then $\angle ABD = 55°$.

18. If $ABCD$ is a parallelogram and $\angle BAC = 125°$, then $\angle ACD = 125°$.

19. If $\angle 1 = 25°$ and $\angle 2 = 20°$ in $\square ABCD$, then $\angle 3 + \angle 4 = 135°$.

20. If $\angle 1 = 25°$ in $\square ABCD$, then $\angle 5 = 25°$.

21. Complete the proof of Theorem 4.23.

    Given: ABCD with $\angle A = \angle C$ and $\angle B = \angle D$

    Prove: ABCD is a parallelogram

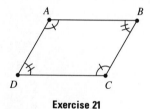

Exercise 21

Proof:

STATEMENTS	REASONS
1. $\angle A + \angle B + \angle C + \angle D = 360°$	1. _____
2. $\angle A = \angle C$ and $\angle B = \angle D$	2. _____
3. $\angle A + \angle B + \angle A + \angle B = 360°$	3. _____
4. $2\angle A + 2\angle B = 360°$	4. Distributive law
5. $\angle A + \angle B = 180°$	5. _____
6. _____	6. Def. of supp. $\angle$'s
7. _____	7. Int. $\angle$'s same side of transv. are supp.

8. $\angle A$ and $\angle D$ are   8. _____
   supplementary

9. _____   9. Int. $\angle$'s same side of transv.
                                                 are supp.

10. _____   10. Def. of $\square$

22. Complete the proof of Theorem 4.25.

   Given: ABCD with $AB = DC$
   and $AB \parallel DC$

   Prove: ABCD is a parallelogram

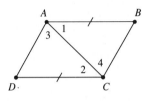

   Exercise 22

   Proof:

STATEMENTS	REASONS
1. Construct diagonal *AB*	1. A line is determined by two points
2. $AB = DC$	2. _____
3. _____	3. Given
4. _____	4. Alt. int. $\angle$'s are $=$
5. $AC = AC$	5. _____
6. _____	6. SAS $=$ SAS
7. _____	7. cpoctae
8. $\overline{AD} \parallel \overline{BC}$	8. _____
9. _____	9. Def. of $\square$

23. Prove Corollary 4.21.   **24.** Prove Theorem 4.24.

25. Prove Theorem 4.28.   **26.** Prove Theorem 4.31.

27. If one angle of a parallelogram is twice the measure of another angle, what is the measure of each angle?

28. If one angle of a parallelogram is $(2x + 20)°$ and a consecutive angle is $(x - 50)°$, find the value of x.

29. *Given: □ABCD is a rhombus*
$\overline{BE} \perp \overline{AD}$ *and* $\overline{DF} \perp \overline{BC}$
Prove: BE = DF

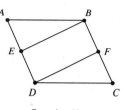

Exercise 29

30. *Given: □ABCD is a rhombus*
Prove: ∠1 and ∠2 are complementary

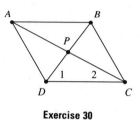

Exercise 30

31. *Given: □ABCD*
$AP = QC$
Prove: PBQD is a parallelogram

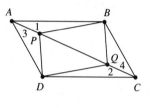

Exercise 31

32. *Given: □ABCD*
$AE = CG$ *and* $BF = DH$
Prove: EFGH is a parallelogram

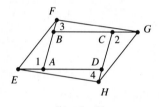

Exercise 32

33. The figure below shows a lawn swing. Explain why the floor of the swing always remains parallel to the ground.

Exercise 33

34. An airplane is flying a course due north at a rate of 400 mph, and the wind is blowing from the west at 75 mph. Make a sketch that illustrates this information and use the parallelogram law to find the resultant of these two velocities. Measure the length of the resultant and estimate the ground speed of the airplane.

35. Two children are dragging a heavy chest by pulling ropes attached to the chest. One is pulling due east with a force of 40 lb, and the other is pulling due south with a force of 60 lb. Use the parallelogram law to find the resultant of these two forces. Measure the length of the resultant to estimate the force required to move the chest in the same manner in the direction of the resultant.

4.5 RECTANGLES, SQUARES, AND TRAPEZOIDS

A *rectangle* is a special type of parallelogram that has many useful applications.

> **DEFINITION: RECTANGLE**
>
> A **rectangle** is a parallelogram with one right angle.

The symbol □ represents the word *rectangle*. Because a rectangle is also a parallelogram, all of the properties of parallelograms are also properties of rectangles. For example, the opposite sides and opposite angles of a rectangle are equal. The next theorem follows from Theorems 4.23 and 4.24 and is left for you to do as an exercise.

> **THEOREM 4.32**
>
> All angles of a rectangle are right angles.

> **THEOREM 4.33**
>
> The diagonals of a rectangle are equal.

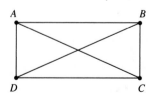

Figure 4.25

Given: □*ABCD* (See Figure 4.25.)

Prove: $AC = BD$

Proof:

STATEMENTS	REASONS
1. □*ABCD*	1. Given
2. $AD = BC$	2. Opp. sides of a □ are =
3. $DC = DC$	3. Reflexive law
4. $\angle ADC$ and $\angle BCD$ are right angles	4. All $\angle$'s of □ are rt. $\angle$'s
5. $\angle ADC = \angle BCD$	5. All rt. $\angle$'s are =
6. $\triangle ADC \cong \triangle BCD$	6. SAS = SAS
7. $AC = BD$	7. cpoctae

The converse of Theorem 4.33 is also true and provides us with another method for proving that a parallelogram is a rectangle.

THEOREM 4.34

If the diagonals of a parallelogram are equal, then the parallelogram is a rectangle.

Theorem 4.34 has many practical applications. To show that a quadrilateral is a rectangle, you must do more than show that the opposite sides are equal (the figure might simply be a parallelogram). Also, it might not be possible to show that the figure has a right angle, but it might be possible to measure the diagonals. For example, a construction worker might measure the diagonals of a foundation for a house to be sure he is laying a rectangular foundation and not simply a foundation with equal opposite sides.

PRACTICE EXERCISE 1

Complete the proof of Theorem 4.34.

Given: □*ABCD*
          AC = BD

Prove: ABCD is a □

Proof:

STATEMENTS	REASONS
1. *ABCD* is a parallelogram	1. _____
2. _____	2. Opp. sides of □ are =
3. *DC = DC*	3. _____
4. _____	4. Given
5. △*ADC* ≅ △*BCD*	5. _____
6. ∠*ADC* = ∠*BCD*	6. _____
7. ∠*ADC* and ∠*BCD* are supplementary	7. _____
8. _____	8. = supp. ∠'s are rt. ∠'s
9. *ABCD* is a rectangle	9. _____

DEFINITION: SQUARE

A square is a rhombus with a right angle.

The symbol □ represents the word *square*. Because a rhombus is a parallelogram, a square is a parallelogram with a right angle, making it a rectangle. Because all sides of a rhombus are equal, a square is a rectangle with all sides equal. Thus, a square has all the properties of both a rhombus and a rectangle.

For example, the diagonals of a square are perpendicular, they bisect each other, and they bisect the angles of the square. Notice that every square is a rhombus, but not every rhombus is a square. Similarly, every square is a rectangle, but not every rectangle is a square.

Because rectangles and squares are perhaps the most important quadrilaterals, we should be able to construct them.

CONSTRUCTION 4.2

Construct a rectangle when two adjacent sides are given.

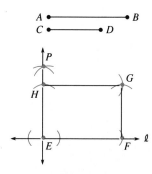

Figure 4.26

Given: Segments $\overline{AB}$ and $\overline{CD}$ (See Figure 4.26.)

To Construct: $\square EFGH$ with $EF = AB$ and $EH = CD$

Construction:

1. Draw line ℓ, identify point E on ℓ, use Construction 2.4 to construct $\overleftrightarrow{EP} \perp \ell$.

2. Use Construction 2.1 to copy $\overline{AB}$ onto line ℓ obtaining $\overline{EF}$, and to copy $\overline{CD}$ onto $\overleftrightarrow{EP}$ obtaining $\overline{EH}$.

3. Set the compass for length AB, place the tip at H, and make an arc. Now set the compass for length CD, place the tip at F, and make an arc that intersects the first arc in point G.

4. Use a straightedge to draw segments $\overline{HG}$ and $\overline{FG}$. Then $EFGH$ is the desired rectangle because it is a parallelogram ($EH = FG$ and $EF = HG$) with right angle $\angle HEF$.

Constructing a square with a given side consists of the same steps as in Construction 4.2 except that $AB = CD$. The remaining construction is left for you to do as an exercise.

CONSTRUCTION 4.3

Construct a square when a side is given.

The following two theorems will be needed for working with the next quadrilateral we will discuss, a *trapezoid*. Recall from Section 2.3 that the distance from a point to a line is the length of a perpendicular segment drawn from the point to the line. We use this fact to prove the following theorem.

THEOREM 4.35

Two parallel lines are always the same distance apart.

Given: $\ell \parallel m$ with A and B arbitrary
points on ℓ, and C and D on m
$\overline{AC} \perp m$ and $\overline{BD} \perp m$
AC is the distance from A to m
BD is the distance from B to m
(See Figure 4.27.)

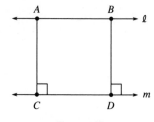

Figure 4.27

Prove: ℓ and m are always the same
distance apart

Proof:

STATEMENTS	REASONS
1. $\ell \parallel m$, $\overline{AC} \perp m$, and $\overline{BD} \perp m$	1. Given
2. $\overline{AC} \parallel \overline{BD}$	2. Lines $\perp$ to third line are $\parallel$
3. $ABCD$ is a parallelogram	3. Opp. sides are $\parallel$
4. $AC = BD$	4. Opp. sides of $\square$ are =
5. Because A and B were arbitrary points on ℓ, ℓ and m are always the same distance apart	5. All points on ℓ are the same distance from m

---

THEOREM 4.36

The segment joining the midpoints of two sides of a triangle is parallel
to the third side and equal to one-half of it.

Given: $\triangle ABC$ with D the midpoint
of $\overline{AB}$ and E the midpoint of $\overline{AC}$
(See Figure 4.28.)

Prove: $\overline{DE} \parallel \overline{BC}$ and $DE = \dfrac{1}{2}BC$

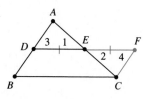

Figure 4.28

Auxiliary lines: Extend $\overline{DE}$ to point
F so that $DE = EF$ by
Construction 2.1
Draw segment $\overline{FC}$

Proof:

STATEMENTS	REASONS
1. D is the midpoint of $\overline{AB}$, and E is the midpoint of $\overline{AC}$ in $\triangle ABC$	1. Given
2. $AE = EC$ and $AD = DB$	2. Def. of midpt.
3. $\angle 1 = \angle 2$	3. Vert. $\angle$'s are =
4. $DE = EF$	4. By construction
5. $\triangle ADE \cong \triangle EFC$	5. SAS = SAS
6. $FC = AD$	6. cpoctae
7. $FC = DB$	7. Transitive law

8. $\angle 3 = \angle 4$	8. cpoctae
9. $\overline{FC} \parallel \overline{DB}$	9. Alt. int. $\angle$'s are =
10. $BCFD$ is a parallelogram	10. Opp. sides are = and $\parallel$
11. $\overline{DE} \parallel \overline{BC}$	11. Opp. sides of $\square$ are $\parallel$
12. $DF = BC$	12. Opp. sides of $\square$ are =
13. $DF = DE + EF$	13. Seg. add. post.
14. $DF = DE + DE$	14. Substitution law
15. $DF = 2DE$	15. Distributive law
16. $2DE = BC$	16. Substitution law
17. $DE = \dfrac{1}{2}BC$	17. Mult.-div. law

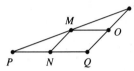

Figure 4.29

E X A M P L E 1   Refer to $\triangle PQR$ in Figure 4.29. If M is the midpoint of $\overline{PR}$, N is the midpoint of $\overline{PQ}$, and $MN = 5$ cm, find the length of $\overline{QR}$. By Theorem 4.36, $MN = \dfrac{1}{2}QR$, so $QR = 10$ cm. ◨

> *PRACTICE EXERCISE 2*
>
> Refer to $\triangle PQR$ in Figure 4.29. Assume $\triangle PQR$ is isosceles with base $\overline{PR}$ and $RQ = 30$ inches, and M and O are the midpoints of $\overline{PR}$ and $\overline{QR}$, respectively. Find the length of $\overline{MO}$.

> *DEFINITION: TRAPEZOID*
>
> A **trapezoid** is a quadrilateral with exactly one pair of parallel sides. The parallel sides are called **bases** and the nonparallel sides are called **legs.** If the legs of a trapezoid are equal, the trapezoid is an **isosceles trapezoid.** A pair of angles of a trapezoid are called **base angles** if they include the same base.

A trapezoid is a figure with properties similar to those of a triangle and a parallelogram. Figure 4.30 shows a trapezoid with bases $\overline{AB}$ and $\overline{CD}$ and sides $\overline{AD}$ and $\overline{BC}$. One pair of base angles is $\angle D$ and $\angle C$, and the other pair is $\angle A$ and $\angle B$. If AD and BC were equal, the trapezoid would be isosceles.

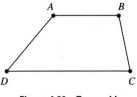

Figure 4.30   Trapezoid

THEOREM 4.37

The base angles of an isosceles trapezoid are equal.

Given: Isosceles trapezoid $ABCD$
with $\overline{AB} \parallel \overline{DC}$ (See Figure 4.31.)

Prove: $\angle D = \angle C$

Auxiliary lines: Construct $\overline{AE}$
through A parallel to $\overline{BC}$ using
Construction 4.1

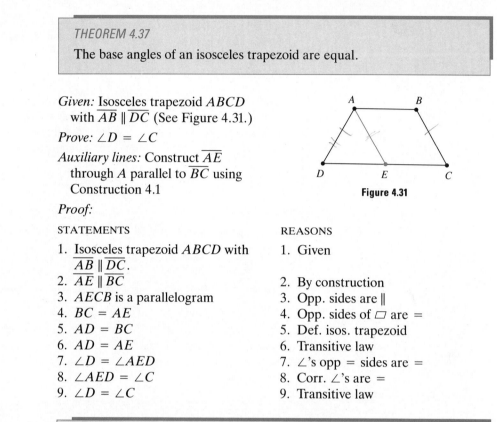

Figure 4.31

Proof:

STATEMENTS	REASONS
1. Isosceles trapezoid $ABCD$ with $\overline{AB} \parallel \overline{DC}$.	1. Given
2. $\overline{AE} \parallel \overline{BC}$	2. By construction
3. $AECB$ is a parallelogram	3. Opp. sides are $\parallel$
4. $BC = AE$	4. Opp. sides of $\square$ are $=$
5. $AD = BC$	5. Def. isos. trapezoid
6. $AD = AE$	6. Transitive law
7. $\angle D = \angle AED$	7. $\angle$'s opp $=$ sides are $=$
8. $\angle AED = \angle C$	8. Corr. $\angle$'s are $=$
9. $\angle D = \angle C$	9. Transitive law

THEOREM 4.38

The diagonals of an isosceles trapezoid are equal.

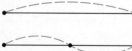

When the string on a guitar is plucked, the vibration creates the sound we hear. Pythagoras knew that the length of the string is related to the pitch of the note. The string in the top figure produces a certain note. If you press the center of the string and pluck it, the note heard is one octave above the first. In general, the most pleasing notes to our ear are formed by dividing the string into an equal number of congruent segments.

The proof of Theorem 4.38 is left for you to do as an exercise.

DEFINITION: MEDIAN OF A TRAPEZOID

The segment joining the midpoints of the legs of a trapezoid is the
median of the trapezoid.

THEOREM 4.39

The median of a trapezoid is parallel to the bases and equal to one-half
their sum.

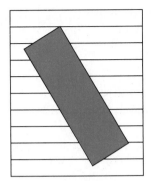

A student wants to cut a rectangular piece of plastic into seven congruent rectangles all with length equal to the width of the plastic rectangle. Assume that a measuring device is not available. Explain how the student can discover where to make the cuts by placing the plastic rectangle on a piece of ruled paper. What theorem are you using?

The proof of Theorem 4.39 is requested in the exercises.

The next theorem provides a way to divide a given line segment into any number of equal parts.

THEOREM 4.40

If three or more parallel lines intercept equal segments on one transversal, then they intercept equal segments on all transversals.

The proof of Theorem 4.40 is requested in the exercises.

CONSTRUCTION 4.4

Divide a given segment into a given number of equal segments.

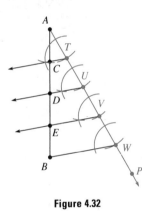

Figure 4.32

Given: Segment $\overline{AB}$ (See Figure 4.32.)

To Construct: Divide $\overline{AB}$ into n equal parts. For purposes of illustration, let $n = 4$. The technique we will use works for any n.

Construction:

1. Choose point P not on $\overleftrightarrow{AB}$ and draw ray $\overrightarrow{AP}$.

2. Set the compass at any length and mark off equal segments $\overline{AT}, \overline{TU}, \overline{UV}$, and $\overline{VW}$, and draw segment $\overline{BW}$.

3. Use Construction 2.2 to construct $\angle ATC$, $\angle TUD$, $\angle UVE$ all of which are equal to $\angle VWB$. The points C, D, and E divide $\overline{AB}$ into equal segments $\overline{AC}, \overline{CD}, \overline{DE}$, and $\overline{EB}$ by Theorem 4.40 because $\overline{AB}$ and $\overline{AP}$ are transversals intercepting parallel lines $\overline{TC}, \overline{UD}$ $\overline{VE}$, and $\overline{WB}$.

ANSWERS TO PRACTICE EXERCISES: **1.** 1. Given   2. $AD = BC$
3. Reflexive law   4. $AC = BD$ 5. SSS = SSS   6. cpoctae   7. Consec. ∠'s of
▱ are supp.   8. $\angle ADC$ and $\angle BCD$ are right angles   9. Def. of ▱   **2.** 15
inches

4.5 EXERCISES

Exercises 1–20 refer to the figure below in which $ABCD$ is a trapezoid with
bases $\overline{AB}$ and $\overline{CD}$, $\overline{EG} \perp \overline{DC}$ and $\overline{FH} \perp \overline{DC}$, E the midpoint of $\overline{AD}$, and F
the midpoint of $\overline{BC}$. Answer true or false.

1. $\overline{AB} \parallel \overline{DC}$.　　　　　　　**2.** $\overline{AD} \parallel \overline{BC}$.　　　　　　　**3.** $\overline{EF} \parallel \overline{AB}$.

4. $\overline{EF} \parallel \overline{DC}$.　　　　　　　**5.** $EG = FH$.　　　　　　　**6.** $AC = BD$.

7. $EFHG$ is a parallelogram.　　　　　**8.** $EFHG$ is a rhombus.

9. $EFHG$ is a rectangle.　　　　　　　**10.** $EFHG$ is a square.

11. $EFHG$ is a trapezoid.　　　　　　　**12.** $EH = FG$.

13. $\overline{EH}$ and $\overline{FG}$ bisect each other.　　　**14.** $\overline{EH} \perp \overline{FG}$.

15. If $AB = 4$ cm and $DC = 6$ cm, then $EF = 5$ cm.

16. If $EF = 10$ inches, then $AB + CD = 30$ inches.

17. If $\angle ADC = 75°$, then $\angle BAD = 105°$.

18. If $\angle ADC = 75°$ and $AD = BC$, then $\angle BCD = 75°$.

19. If $\angle BAD = 110°$ and $AD = BC$, then $\angle ADC = 110°$.

20. If $AD = BC$, then $AC = DB$.

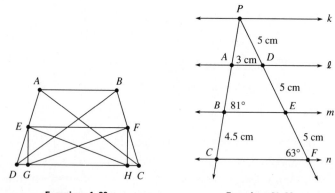

Exercises 1–20　　　　　　　Exercises 21–28

Exercises 21–28 refer to the figure above in which $k \parallel \ell$, $\ell \parallel m$, and $m \parallel n$. Find
the value of each of the following.

21. PA　　　　　**22.** AB　　　　　**23.** $\angle PAD$　　　　　**24.** $\angle BCF$

25. BE　　　　　**26.** CF　　　　　**27.** $\angle DAB$　　　　　**28.** $\angle APD$

29. Complete the proof of Theorem 4.39.

Given: Trapezoid $ABCD$ with $\overline{AB} \parallel \overline{CD}$ and median $\overline{EF}$

Prove: $\overline{EF} \parallel \overline{AB}$, $\overline{EF} \parallel \overline{CD}$, and

$$EF = \frac{1}{2}(AB + CD)$$

Auxiliary lines: Construct $\overleftrightarrow{AF}$. By the parallel postulate $\overleftrightarrow{AF}$ must intersect $\overleftrightarrow{DC}$ at a point, call it G.

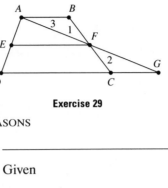

Exercise 29

Proof:

STATEMENTS	REASONS
1. $\overleftrightarrow{AF}$ intersects $\overleftrightarrow{DC}$ at G	1. _____
2. _____	2. Given
3. E is the midpoint of $\overline{AD}$ and F is the midpoint of $\overline{BC}$	3. _____
4. _____	4. Def. of midpoint
5. $\angle 1 = \angle 2$	5. _____
6. _____	6. Given
7. $\angle 3 = \angle G$	7. _____
8. $\triangle ABF \cong \triangle FCG$	8. _____
9. $AF = FG$ and $AB = CG$	9. _____
10. $EF = \frac{1}{2}DG$	10. Theorem 4.36
11. $\overline{DG} = \overline{CG} + \overline{DC}$	11. _____
12. $\overline{DG} = \overline{AB} + \overline{DC}$	12. _____
13. $EF = \frac{1}{2}(AB + DC)$	13. Substitution law
14. $\overline{EF} \parallel \overline{DC}$	14. Theorem 4.36
15. $\overline{EF} \parallel \overline{AB}$	15. _____

30. Complete the proof of Theorem 4.40.

Given: $\ell \parallel m$ and $m \parallel n$
$\qquad AB = BC$

Prove: $DE = EF$

Auxiliary lines: Construct $\overline{AG} \parallel \overline{DE}$ and $\overline{BH} \parallel \overline{EF}$ using Construction 4.1.

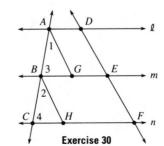

Exercise 30

Proof:

STATEMENTS	REASONS
1. $\overline{AG} \parallel \overline{DE}$ and $\overline{BH} \parallel \overline{EF}$	1. _____

2. $\overline{AG} \parallel \overline{BH}$ 2. _____

3. $\angle 1 = \angle 2$ 3. _____

4. _____   4. Given

5. $\angle 3 = \angle 4$ 5. _____

6. _____   6. ASA = ASA

7. _____   7. cpoctae

8. *ADEG* and *BEFH* are   8. _____
   parallelograms

9. *DE = AG* and *EF = BH*   9. _____

10. _____   10. Sym. and Trans. laws

31. Prove Theorem 4.32.   **32.** Prove Theorem 4.38.

33. Prove that the segments joining the midpoints of the consecutive sides of any quadrilateral form a parallelogram.

34. Prove that the bisectors of two consecutive angles of a parallelogram are perpendicular.

35. Prove that the segments joining the midpoints of the sides of an equilateral triangle form another equilateral triangle.

36. Prove that the segment joining the midpoints of the adjacent sides of a rectangle form a rhombus.

37. Construct a rectangle with sides measuring 3 inches and 2 inches.

38. Construct a square with each side measuring 2 inches.

39. Jeanne wants to enclose a garden with a fence in the shape of a rectangle 15 ft by 20 ft. To be certain she has formed a rectangle, she measures the diagonals and finds they are equal. Does this make the garden rectangular in shape?

40. Galen wishes to find the distance between two points *A* and *B* on opposite sides of a lake. He places a stake at point *C* and determines the midpoints of $\overline{AC}$ and $\overline{BC}$ to be *D* and *E*, respectively (see the figure below). He measures the length of $\overline{DE}$ and finds it to be 56 ft. What is the length of $\overline{AB}$?

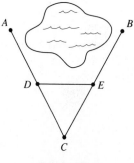

Exercise 40

4.6 AREAS OF POLYGONS

In Section 4.3 we discovered that every polygon has associated with it a measure called its perimeter. Now we consider another measure that is associated with a polygon called its *area*. To be precise, a polygon really has no area, rather it encloses an area. However, for simplicity, and because it is common practice, we shall use the terminology *area of a polygon*.

To measure the length of a segment, we determine how many times it contains a particular unit (such as inch, centimeter, foot, etc.). The same is true for angles in which the common unit of measure is the degree. In a similar manner, we can measure the area of a polygon by determining the number of particular units it contains. For example, suppose we consider the area of a rectangular table top that has sides measuring 3 ft and 5 ft as shown in Figure 4.33.

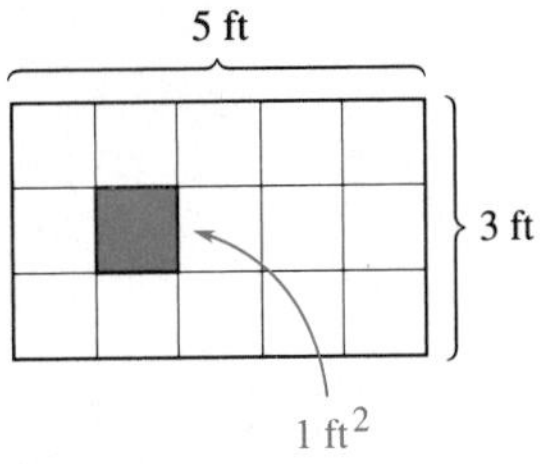

Figure 4.33

The table top can be divided into squares measuring 1 ft on each side. If we say that each of these squares has an area of 1 square foot, we can use this unit of measure to describe the area of the rectangle. In fact, since the rectangle contains 15 square feet in its interior, we say that the rectangle has area 15 square feet. We often abbreviate *square feet* with the symbol ft². Thus, the area of the table top is 15 ft².

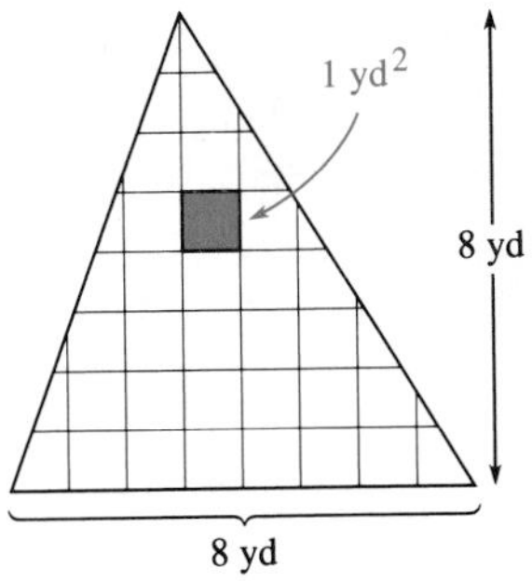

Figure 4.34

EXAMPLE 1 Figure 4.34 shows the sail on a sailboat. Estimate the number of square yards of material in the sail. By counting the number of complete squares in the triangle and approximating the parts of those not totally contained in the triangle, you should come up with about 32 squares. Thus, the area of the sail is about 32 yd². ◪

Clearly, it is difficult to count the number of square units (square inches, square centimeters, square feet) in polygons whose sides "cut off" parts of squares. As a result, it is appropriate for us to search for other ways to find areas. We begin with a rectangle. Notice in Figure 4.33 that the area of the rectangle, 15 ft², can be found by multiplying the lengths of the sides, 3 ft and 5 ft, because $3 \cdot 5 = 15$. This observation leads to our next postulate. For convenience, we'll give names to two parts of a rectangle. Consider $\square ABCD$ in Figure

4.35. The length of $\overline{AB}$, AB, is called the **length** of the rectangle and is denoted by ℓ. The length of $\overline{BC}$, BC, is called the **width** of the rectangle and is denoted by w. Notice that $\ell = AB = DC$ and $w = BC = AD$. It is common practice to call the larger of the two numbers AB and BC the *length* and the smaller of the two the *width* of the rectangle.

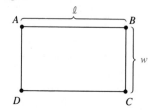

Figure 4.35   Length and Width of a Rectangle

> ### POSTULATE 4.3
>
> The **area of a rectangle** with length ℓ and width w is determined with the formula $A = \ell w$.

E X A M P L E 2   Find the area of $\square ABCD$ if $AB = 7$ ft and $BC = 3$ ft. We are given that $\ell = 7$ ft and $w = 3$ ft, so

$$A = \ell w. \qquad \text{Postulate 4.3}$$
$$= 7 \cdot 3 \qquad \text{Substitute 7 for } \ell \text{ and 3 for } w$$
$$= 21 \qquad \text{Multiply}$$

Thus, the area of the rectangle is 21 ft². ◩

When a given figure can be divided into nonoverlapping parts, the total area of the figure is the sum of the areas of the parts. The next postulate affirms this.

> ### POSTULATE 4.4   ADDITIVE PROPERTY OF AREAS
>
> If lines divide a given area into several smaller nonoverlapping areas, the given area is the sum of the smaller areas.

For example, the area of the trapezoid $ABCD$ in Figure 4.36 is equal to the sum of the areas of $\triangle ADE$, $\triangle BCF$, and $\square ABFE$.

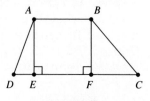

Figure 4.36

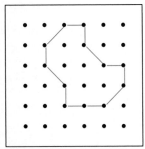

A geoboard is a square piece of wood with nails arranged in rows and columns. If there are 6 rows and 6 columns for a total of 36 nails, the geoboard has dimension 6 × 6. By placing a rubber band around the nails, various polygonal regions can be formed.

We can calculate the area enclosed by the rubber band using Pick's Theorem: If x is the number of nails on the border of the polygon, y is the number of nails on the interior of the polygon, the area of the polygon A is given by $A = \dfrac{x}{2} + y - 1$. What is the area of the polygon shown here?

**A boxing ring is a square
having area 400 ft². What is the
length of each side of the ring?**

Because a square is a rectangle, the proof of the following corollary comes directly from the definition of a square and Postulate 4.3.

> **COROLLARY 4.41**
>
> The **area of a square** with sides of length s is determined with the formula $A = s^2$.

E X A M P L E   3   Find the area of a square with sides of length 5 cm. Because $s = 5$ cm, we have

$$A = s^2$$
$$= 5^2 \qquad 5^2 = 5 \cdot 5 = 25$$
$$= 25 \text{ cm}^2. \qquad \blacksquare$$

Two polygons can have the same area and not be congruent. For example, a square with sides of length 6 inches has area 36 in², the same area as a rectangle with length 9 inches and width 4 inches. Clearly, the square and the rectangle cannot be made to coincide. However, two polygons that are congruent have the same area.

> **POSTULATE 4.5**
>
> Two congruent polygons have the same area.

The following definition aids us in finding the area of a parallelogram.

> **DEFINITION: ALTITUDE AND BASE OF A PARALLELOGRAM**
>
> An **altitude** of a parallelogram is a segment from a vertex of the parallelogram perpendicular to a nonadjacent side (possibly extended). The length of an altitude is called the **height** of the parallelogram and the side to which it is drawn is called the **base** of the parallelogram.

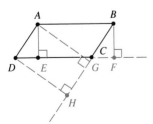

Figure 4.37

In Figure 4.37, $\overline{AE}$ and $\overline{BF}$ are altitudes of $\square ABCD$ with base $\overline{DC}$. The height of $\square ABCD$ is AE (which is equal to BF). Alternatively, if $\overline{BC}$ is extended, we see that $\overline{AG}$ and $\overline{DH}$ are also altitudes of $\square ABCD$ with base $\overline{BC}$. In this case, the height of $\square ABCD$ is AG (which is equal to DH).

> **THEOREM 4.42**
>
> The **area of a parallelogram** with length of base b and height h is determined with the formula $A = bh$.

Given: □*ABCD* with altitude $\overline{BE}$, height $h = BE$, and length of base $b = DC$ (See Figure 4.38.)

Prove: The area of □*ABCD* is determined using the formula $A = bh$

Auxiliary lines: Extend $\overline{DE}$ and construct $\overline{AF} \perp \overline{DE}$

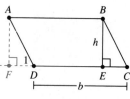

Figure 4.38

Proof:

STATEMENTS	REASONS
1. *ABCD* is a parallelogram	1. Given
2. $AD = BC$	2. Opp. sides of □ are =
3. $\overline{AD} \parallel \overline{BC}$	3. Opp. sides of □ are ∥
4. $\angle C = \angle 1$	4. Corr. ∠'s are =
5. $\overline{AF} \perp \overline{FD}$	5. By construction
6. $\overline{BE} \perp \overline{EC}$	6. Def. of altitude
7. $\angle AFD$ and $\angle BEC$ are right angles	7. ⊥ lines form rt. ∠'s
8. $\angle AFD = \angle BEC$	8. Rt. ∠'s are =
9. $\triangle ADF \cong \triangle BEC$	9. AAS = AAS
10. Area of $\triangle ADF$ = Area of $\triangle BEC$	10. ≅ polygons have = area
11. $AB = b$	11. Opp. sides of □ are =
12. Area of □*ABEF* = bh	12. Area of a □ (Post. 4.3)
13. Area of □*ABEF* = Area of trapezoid *ABED* + Area of $\triangle ADF$	13. Add. prop. of areas
14. Area of □*ABCD* = Area of trapezoid *ABED* + Area of $\triangle BEC$	14. Add. prop. of areas
15. Area of □*ABCD* = Area of trapezoid *ABED* + Area of $\triangle ADF$	15. Substitution law
16. Area of □*ABCD* = Area of □*ABEF*	16. Sym. and trans. laws
17. Area of □*ABCD* = $A = bh$	17. Substitution law

EXAMPLE 4 Find the area of □*ABCD* given in Figure 4.39.

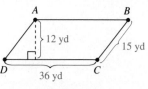

Figure 4.39

The height of the parallelogram is 12 yd and the base is 36 yd. Thus, the area is given by

$$A = bh$$
$$= 36 \cdot 12 \qquad \text{Substitute 36 for } b \text{ and 12 for } h$$
$$= 432 \text{ yd}^2.$$

Notice that the length of side $\overline{BC}$ is 15 yd, but this fact is not used to find the area. It is used to find the perimeter of the parallelogram, 102 yd. ◪

The next theorem provides a way to determine the area of a triangle. We use the term **height** to represent the length of an altitude drawn to a **base** much like that which was done for a parallelogram.

THEOREM 4.43

The **area of a triangle** with length of base b and height h is determined with the formula $A = \dfrac{1}{2}bh$.

Given: $\triangle ABC$ (See Figure 4.40.)

Prove: $A = \dfrac{1}{2}bh$

Auxiliary lines: Construct the line through $A \parallel$ to $\overline{BC}$ and the line through $C \parallel$ to $\overline{AB}$

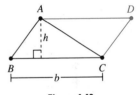

Figure 4.40

Proof:

STATEMENTS	REASONS
1. $\overline{AD} \parallel \overline{BC}$ and $\overline{DC} \parallel \overline{AB}$	1. By construction
2. $ABCD$ is a parallelogram	2. Def. of ▱
3. $\triangle ABC \cong \triangle ADC$	3. Diag. of ▱ form $\cong$ $\triangle$'s
4. Area of $\triangle ABC$ = Area of $\triangle ADC$	4. $\cong$ polygons have = areas
5. Area of ▱$ABCD$ = Area of $\triangle ABC$ + Area of $\triangle ADC$	5. Add. prop. of areas
6. Area of ▱$ABCD$ = Area of $\triangle ABC$ + Area of $\triangle ABC$	6. Substitution law
7. Area of ▱$ABCD$ = 2 (Area of $\triangle ABC$)	7. Distributive law
8. Area of ▱$ABCD$ = bh	8. Formula for area of ▱
9. 2(Area of $\triangle ABC$) = bh	9. Substitution law
10. Area of $\triangle ABC$ = $A = \dfrac{1}{2}bh$	10. Mult.-div. law

EXAMPLE 5   Find the area of a triangle whose base is 4.2 cm and altitude is 6.5 cm.

Substitute 4.2 for b and 6.5 for h in the following formula:

$$A = \frac{1}{2}bh.$$

$$= \frac{1}{2}(4.2)(6.5)$$

$$= \frac{1}{2}(27.3) = 13.65$$

Thus, the area is 13.65 cm². ◨

An **altitude** of a trapezoid is a segment from a vertex of the trapezoid perpendicular to the nonadjacent base. The length of an altitude is called the **height** of the trapezoid.

THEOREM 4.44

The **area of a trapezoid** with length of bases b and b' and height h is determined with the formula $A = \frac{1}{2}(b + b')h$.

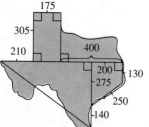

The area of an irregularly shaped region can be approximated by dividing it up into rectangular, trapezoidal, and triangular subregions. Using the information given below approximate the area of the state of Texas.

PRACTICE EXERCISE 1

Complete the proof of Theorem 4.44.

Given: Trapezoid $ABCD$

Prove: $A = \frac{1}{2}(b + b')h$

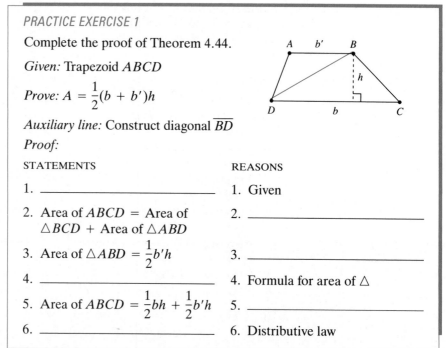

Auxiliary line: Construct diagonal $\overline{BD}$

Proof:

STATEMENTS	REASONS
1. _____	1. Given
2. Area of $ABCD$ = Area of $\triangle BCD$ + Area of $\triangle ABD$	2. _____
3. Area of $\triangle ABD = \frac{1}{2}b'h$	3. _____
4. _____	4. Formula for area of $\triangle$
5. Area of $ABCD = \frac{1}{2}bh + \frac{1}{2}b'h$	5. _____
6. _____	6. Distributive law

E X A M P L E 6   An archway in a building is a rectangle topped by an isosceles trapezoid as shown in Figure 4.41. What is the area of the opening?

To find the total area, we find the area of the rectangle ($\square FCDE$), the area of the trapezoid $(ABCF)$, and add the results using the additive property of areas.

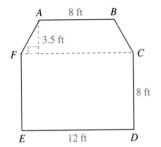

Figure 4.41

$$\text{Area of } FCDE = \ell w = (12)(8) = 96 \text{ ft}^2 \quad \ell = 12 \text{ and } w = 8$$

$$\text{Area of } ABCF = \frac{1}{2}(b + b')h$$

$$= \frac{1}{2}(12 + 8)(3.5) \qquad b = 12, b' = 8, \text{ and } h = 3.5$$

$$= \frac{1}{2}(20)(3.5)$$

$$= (10)(3.5) = 35 \text{ ft}^2$$

Thus, the area of the archway is 96 ft² + 35 ft² = 131 ft². ◪

Because a rhombus is a parallelogram, the area of a rhombus can be determined by using the formula $A = bh$. However, the next theorem provides an alternative way to find the area of a rhombus by using its diagonals. The proof of this theorem is left for you to do as an exercise.

> **THEOREM 4.45**
>
> The **area of a rhombus** with diagonals of length d and d' is determined with the formula $A = \frac{1}{2}dd'$.

ANSWER TO PRACTICE EXERCISE: **1.** 1. *ABCD* is a trapezoid   2. Add. prop. of areas   3. Formula for area of △   4. Area of △*BCD* $= \frac{1}{2}bh$ 5. Substitution law   6. $A = \frac{1}{2}(b + b')h$

4.6 EXERCISES

In Exercises 1–20, find the area and perimeter of each figure.

1.

3 cm

4 cm

2.

4 ft

16 ft

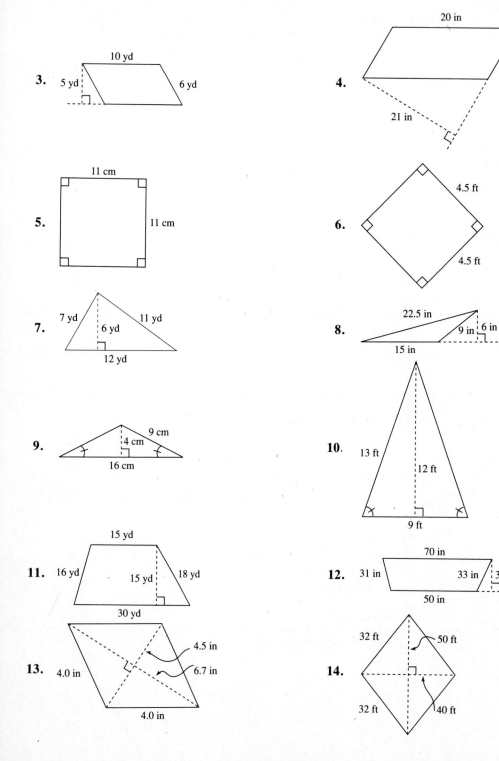

15.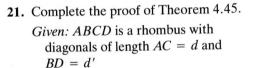

16.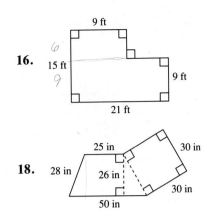

17.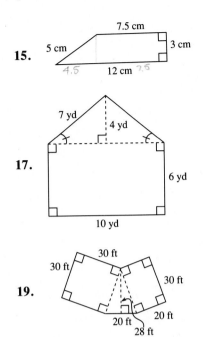

18.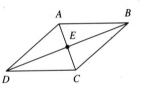

19.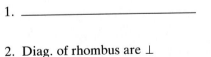

20.

21. Complete the proof of Theorem 4.45.

Given: ABCD is a rhombus with diagonals of length $AC = d$ and $BD = d'$

Prove: $A = \dfrac{1}{2}dd'$

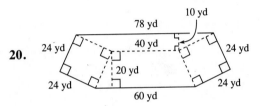

Exercise 21

Proof:

STATEMENTS	REASONS
1. *ABCD* is a rhombus with $AC = d$ and $BD = d'$	1. _____
2. _____	2. Diag. of rhombus are $\perp$
3. Area of *ABCD* = Area of $\triangle ADC$ + Area of $\triangle ABC$	3. _____
4. Area of $\triangle ADC = \dfrac{1}{2}(d)(DE)$	4. _____
5. _____	5. Form. for area of $\triangle$
6. Area of $ABCD = \dfrac{1}{2}(d)(DE) + \dfrac{1}{2}(d)(BE)$	6. _____

7. Area of $ABCD = \frac{1}{2}d(DE + BE)$ 7. _____

8. _____   8. Seg. add. post.

9. _____   9. Substitution law

22. Prove that the median of a triangle divides the triangle into two triangles with the same area. Complete the following proof.

Given: $\triangle ABC$ with median $\overline{AD}$

Prove: Area of $\triangle ABD$ = Area of $\triangle ADC$

Auxiliary line: Construct the altitude from A, $\overline{AE}$

Proof:

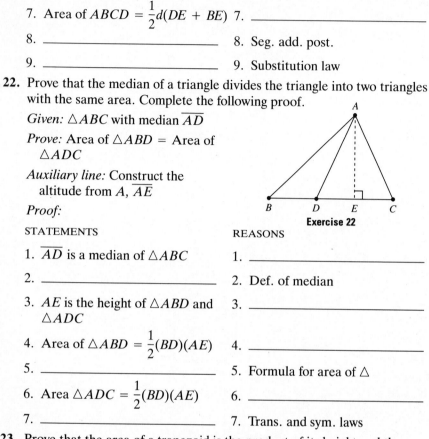

Exercise 22

STATEMENTS	REASONS
1. $\overline{AD}$ is a median of $\triangle ABC$	1. _____
2. _____	2. Def. of median
3. AE is the height of $\triangle ABD$ and $\triangle ADC$	3. _____
4. Area of $\triangle ABD = \frac{1}{2}(BD)(AE)$	4. _____
5. _____	5. Formula for area of $\triangle$
6. Area $\triangle ADC = \frac{1}{2}(BD)(AE)$	6. _____
7. _____	7. Trans. and sym. laws

23. Prove that the area of a trapezoid is the product of its height and the length of its median.

24. Prove that the area of a right triangle is one-half the product of the lengths of its legs.

25. Find the area of a parallelogram with base 18 ft and height 12 ft.

26. Find the area of a rectangle with length 25 cm and width 13 cm.

27. Find the area of a right triangle with legs 3.6 ft and 5.2 ft.

28. Find the area of a triangle with base 4.3 yd and height 6.4 yd.

29. Find the area of a trapezoid with height 7.2 inches and bases 6.3 inches and 11.5 inches.

30. Find the area of a rhombus with diagonals measuring 8.2 ft and 7.1 ft.

31. The length of a rectangle is 24 cm and its area is 168 cm². What is the width of the rectangle?

32. The base of a triangle is 16 ft and its area is 248 ft². What is the height of the triangle?

33. A wall is 41 ft long and 9 ft high. There are two windows in the wall measuring 2 ft by 3 ft. If a single gallon of paint will cover 250 ft², and Randy wants to give the wall two coats of paint, how many gallons of paint will she need?

34. The floor of a family room is rectangular in shape measuring 9 yd by 7 yd. In the center of the room is a firepit measuring 1.5 yd by 1.5 yd. If carpeting costs $19.95 per square yard, neglecting waste, how much will it cost to carpet the room?

Exercises 35–38, refer to the figure below which gives a sketch of the Smith's lot and home.

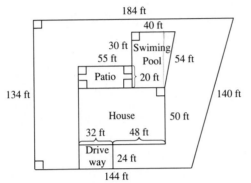

35. The Smiths plan to put outdoor carpeting on their patio. If carpeting costs $15.95 per square yard, neglecting waste, how much will the project cost?

36. How much will it cost for a pool cover made of insulated vinyl costing $12.50 per square yard? Ignore any waste.

37. How much will it cost to put a sealer on the Smith's driveway if one gallon of sealer costs $15.95 and will cover 300 ft²?

38. The Smiths plan to put sod on all areas of their lot not already covered. If the sod costs $0.30 a square foot, how much will the project cost?

39. In the figure below, $ABCD$ is a rectangle with diagonal $\overline{AC}$. Prove that the area of $\square BGFH$ is equal to the area of $\square FJDI$.

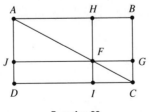

Exercise 39

40. The cross section of an I-beam is shown below. If the width of the upper and lower rectangles is 1.5 inches, find the area of the cross section.

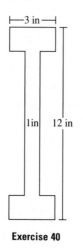

Exercise 40

The areas of certain figures can be found by subtracting areas from larger areas. Use this hint in Exercises 41–42 and find the area of each shaded region.

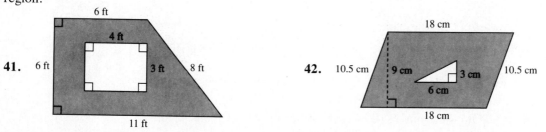

41.

42.

Exercises 43–44 require the ability to solve a quadratic equation using techniques learned in beginning algebra.

43. The area of a rectangle is 120 ft². If the length of the rectangle is 2 ft more than the width, find the length and width.

44. The area of a triangle is 44 cm². If the base of the triangle is 3 cm less than the height, find the base and height.

PROOF TECHNIQUES

To Prove:

Two Lines are the Same

1. Show both are parallel to the same line and both pass through the same point not on the line. (Postulate 4.1, p. 112)

Two Lines Parallel

1. Show they are both perpendicular to a third line. (Theorem 4.1, p. 112)

2. Show a pair of alternate interior angles are equal when the lines are cut by a transversal. (Theorem 4.2, p. 116)

3. Show a pair of corresponding angles are equal when the lines are cut by a transversal. (Theorem 4.3, p. 116)

4. Show a pair of alternate exterior angles are equal when the lines are cut by a transversal. (Theorem 4.4, p. 117)

5. Show two interior angles on the same side of a transversal are supplementary. (Theorem 4.5, p. 117)

Two Lines Perpendicular

1. Show one is parallel to a third line that is perpendicular to the second. (Theorem 4.6, p. 118)
2. Show they are the diagonals of a rhombus. (Theorem 4.29, p. 136)

Two Angles Equal

1. Show they are alternate interior angles formed when parallel lines are cut by a transversal. (Theorem 4.7, p. 118)
2. Show they are corresponding angles formed when parallel lines are cut by a transversal. (Theorem 4.8, p. 119)
3. Show they are alternate exterior angles formed when parallel lines are cut by a transversal. (Theorem 4.9, p. 120)
4. Show they are opposite angles of a parallelogram. (Corollary 4.21, p. 132)
5. Show they are the base angles of an isosceles trapezoid. (Theorem 4.37, p. 146)

Two Angles Supplementary

1. Show they are interior angles on the same side of a transversal that cuts parallel lines. (Theorem 4.10, p. 120)
2. Show they are consecutive angles of a parallelogram. (Theorem 4.24, p. 133)

Two Triangles Congruent

1. Show two angles and any side of one are equal, respectively, to two angles and the corresponding side of the other. (Corollary 4.14, p. 126)
2. Show they are formed when a diagonal divides a parallelogram into triangles. (Theorem 4.20, p. 131)

A Quadrilateral Is a Parallelogram

1. Show both pairs of opposite sides are equal. (Theorem 4.22, p. 132)
2. Show both pairs of opposite angles are equal. (Theorem 4.23, p. 133)
3. Show two opposite sides are equal and parallel. (Theorem 4.25, p. 133)
4. Show the diagonals bisect each other. (Theorem 4.27, p. 134)

Two Segments Equal

1. Show they are diagonals of a rectangle. (Theorem 4.33, p. 141)
2. Show they are diagonals of an isosceles trapezoid. (Theorem 4.38, p. 146)

Section 4.1

1. Discuss the major differences between giving a direct proof and an indirect proof of $P \longrightarrow Q$.

2. Give an indirect proof of the following "theorem."

   Premise 1: If I don't have clean shirts, then I can't go to work.
   Premise 2: If I can't work, then I won't have any money.
   Premise 3: If I don't buy my wife a present, then she'll be unhappy.
   Premise 4: If my wife is unhappy, then she won't wash my shirts.
   Theorem: If I have the money, then I'll buy my wife a present.

3. State the parallel postulate.

4. If $\ell \perp m$, $\ell \perp n$, and m and n are distinct lines, is $m \parallel n$?

5. If $\ell \parallel m$, ℓ, m, and n are distinct lines, P is a point on m, and n is a line through P, is $n \parallel \ell$?

Section 4.2

Exercises 6–12 refer to the figure below in which $\ell \parallel m$.

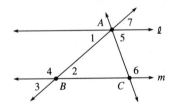

6. Is $\angle 1 = \angle 2$?

7. Is $\angle 5 = \angle 1$?

8. Is $\angle 1 = \angle 3$?

9. Is $\angle 5$ supplementary to $\angle 6$?

10. Give three angles that are equal to $\angle 7$.

11. If $\angle 1 = (x + 20)°$ and $\angle 2 = (3x - 40)°$, find x.

12. If $\angle 4 = (y + 30)°$ and $\angle 7 = (2y - 90)°$, find y.

13. *Given:* $\ell \parallel m$ and $\angle 1 = \angle 2$

    Prove: $\triangle ABC$ is isosceles

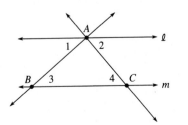

Exercise 13

165

14. Use the figure below and find the value of x and y that make $\ell \parallel m$.

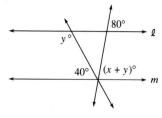

Exercise 14

Section 4.3

15. What is the sum of the angles of a hexagon?

16. What is the measure of each angle of a regular hexagon?

17. What is the sum of the measures of the exterior angles of a hexagon?

18. What is the measure of each exterior angle of a regular hexagon?

19. Give the number of sides of a regular polygon if each interior angle measures 156°.

20. Give the number of sides of a polygon if the sum of its interior angles is 3600°.

21. If three exterior angles of a quadrilateral sum to 325°, what is the measure of the fourth exterior angle?

22. Is an interior angle of a polygon always greater than its corresponding exterior angle?

23. Can a triangle have more than one right angle?

24. What is the relationship between an exterior angle of a triangle and the sum of the measures of the nonadjacent interior angles?

25. If two angles of one triangle are equal to two angles of another triangle, are the third angles equal?

Section 4.4

Exercises 26–35 refer to the figure below in which $ABCD$ is a parallelogram. Answer true or false.

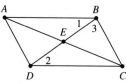

26. $\triangle ABD \cong \triangle BDC$.

27. $AB = DC$.

28. $AD = AB$.

29. $\angle 1 = \angle 2$.

166

30. $\angle BAD = \angle BCD$.

32. $AE = EB$.

34. If $AB = BC$ then $\overline{AC} \perp \overline{BD}$.

31. $\angle BAD$ and $\angle ADC$ are complementary.

33. If $AB = BC$ then $\square ABCD$ is a rhombus.

35. If $\square ABCD$ is a rhombus, then $\angle 1 = \angle 3$.

36. *Given:* $\triangle ABF \cong \triangle EFD$
   F is the midpoint of $\overline{AD}$
   B is the midpoint of $\overline{AC}$

   Prove: $BCDE$ is a parallelogram

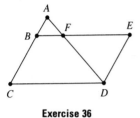

Exercise 36

Section 4.5

37. If the diagonals of a parallelogram are equal, is the parallelogram a rectangle?

38. Is every square also a rhombus?

39. Is every rhombus also a square?

40. Is every trapezoid also a parallelogram?

41. What can be said about the base angles of an isosceles trapezoid?

42. Is the median of a trapezoid perpendicular to the segment joining midpoints of the parallel sides?

43. Draw a segment about 4 inches long and show how to divide it into three equal segments.

44. Construct a square with diagonals the length of a given segment of about 2 inches.

45. *Given:* $\square ABCE$
   $\square BCDE$

   Prove: $\triangle ACD$ is isosceles

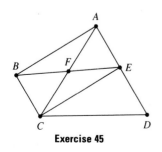

Exercise 45

46. If a trapezoid has bases of 32 ft and 40 ft, what is the length of its median?

Section 4.6

47. If the side of a square is doubled, how does the area change?

48. A room has four rectangular walls each measuring 14 ft by 8 ft. Two of the walls have windows measuring 4 ft by 3 ft. If the walls are to be given two coats of paint and 1 gallon of paint covers 300 ft², how many gallons will be needed for the job?

49. Find the area of a rhombus with diagonals measuring 14 cm and 18 cm.

50. Find the area and perimeter of the figure below.

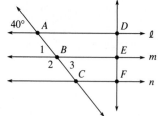

51. Prove that the segments joining the midpoints of adjacent sides of a rhombus form a rectangle.

52. In $\triangle ABC$, $\overline{AE}$ and $\overline{BD}$ are medians intersecting at F. Prove that the Area of $\triangle ABF$ = Area of quadrilateral $DFEC$.

PRACTICE TEST

1. If P is a point not on line ℓ, how many lines through P are parallel to ℓ? Problems 2–5 refer to the figure below. Answer true or false assuming $\ell \parallel m$ and $m \parallel n$.

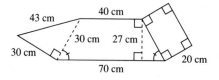

2. $\angle 1 = 40°$.

3. $\angle 2 = 120°$.

4. $\angle 3 = 40°$.

5. If $AB = BC = 3$ cm, and $DE = 2.5$ cm, then $EF = 2.5$ cm.

6. What is the sum of the measures of the angles of an octagon?

7. What is the sum of the measures of the exterior angles of an octagon?

8. What is the measure of each interior angle of a regular octagon?

Problems 9–12 refer to the figure below in which $\overline{AB} \parallel \overline{DC}$ and $\overline{AD} \parallel \overline{BC}$.
Answer true or false.

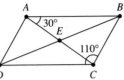

9. $AB = DC$.

10. $\angle ABC = \angle ADC$.

11. $\overline{AC} \perp \overline{BD}$.

12. $\angle ADC = 50°$.

Use the figure below in Problems 13–14. Assume that $ABCD$ is a rectangle and $AE = FC$.

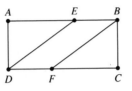

13. Prove that $DE = FB$.

14. Prove that $DEBF$ is a parallelogram.

15. Draw an acute angle and a segment about 2 inches in length. Construct a rhombus with this angle and sides.

16. Find the area of a parallelogram with base 20 ft and height 17 ft.

17. The base of a triangle is 15 yd and the area is 60 yd², what is the height?

18. The figure below shows the family room in a house. The square in the center of the room corresponds to a firepit. If carpeting costs $21.95 per square yard installed, assuming no waste, what will carpeting cost for the entire room?

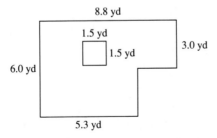

Exercise 18

19. The line segment joining the midpoints of two sides of a triangle measures 44 cm. What is the length of the third side?

20. *Given:* □$ABCD$ with point P between A and B

Prove: Area of $\triangle CDP = \dfrac{1}{2}$ Area of □$ABCD$

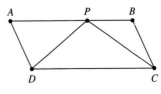

Exercise 20

169

5

RATIO, PROPORTION, AND SIMILARITY

We begin this chapter by reviewing two concepts learned in beginning algebra: ratio and proportion. We'll use these concepts to define the notion of similar polygons and consider applications related to similarity. One application of similar polygons is given below and solved in Example 5 of Section 5.2.

While standing near the Washington Monument in Washington, D.C., Scott discovered that the length of the shadow of the monument was 185 ft at the same time his shadow was 2 ft long. If Scott is 6 ft tall, how did he determine the height of the monument?

5.1 RATIO AND PROPORTION

The topics presented in this section review many of the concepts you learned in beginning algebra. You should also be familiar with basic arithmetic operations on fractions.

DEFINITION: RATIO

The **ratio** of one number a to another number b, $b \neq 0$, is the fraction $\frac{a}{b}$.

The ratio of a to b is sometimes written as $a{:}b$ and read "a is to b."

Ratios compare numbers by using division. They are used in many applications such as rate of speed, gas mileage, and unit cost.

Application	Ratio
32 students for 2 teachers	$\frac{32}{2} = 16$ students per teacher
300 miles in 6 hours	$\frac{300 \text{ mi}}{6 \text{ hr}} = 50$ mph (miles per hour)
100 miles on 5 gallons of gas	$\frac{100 \text{ mi}}{5 \text{ gal}} = 20$ mpg (miles per gallon)
$8 for 4 pounds of meat	$\frac{\$8}{4 \text{ lb}} = \$2$ per lb

E X A M P L E   1   Consider the triangle in Figure 5.1. What is the ratio of the length of the shortest side to the length of the longest side?

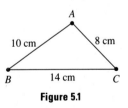

Figure 5.1

Because the shortest side is 8 cm and the longest side is 14 cm, the ratio is $\frac{8}{14}$, which reduces to $\frac{4}{7}$, or 4:7.   ◩

PRACTICE EXERCISE 1

Find the ratio of the length of the longest side to the length of the next longest side in the triangle in Figure 5.1.

Sometimes two pairs of numbers have the same ratio.

DEFINITION: PROPORTION

A statement affirming that two ratios are equal is called a **proportion.**

The following are several examples of proportions:

$$\frac{1}{2} = \frac{2}{4}, \quad \frac{5}{7} = \frac{15}{21}, \quad \text{and} \quad \frac{10}{3} = \frac{40}{12}.$$

The statement "*a* is to *b* as *c* is to *d*" translates to the proportion

$$\frac{a}{b} = \frac{c}{d}$$

assuming $b \neq 0$ and $d \neq 0$. (Remember that division by zero is not defined.) We shall assume from now on that all denominators in ratios are not zero.

DEFINITION: TERMS OF A PROPORTION

Consider the proportion $\frac{a}{b} = \frac{c}{d}$. The numbers *a* and *d* are called **extremes** and *b* and *c* are called **means** of the proportion.

E X A M P L E 2 Name the means and extremes in the proportion

$$\frac{9}{20} = \frac{18}{40}.$$

The means are 20 and 18, and the extremes are 9 and 40. Notice that by the symmetric law

$$\frac{18}{40} = \frac{9}{40}.$$

In this form, the means and extremes become interchanged. ◪

A proportion can be continued to contain several ratios. For example, we might say that *a, b,* and *c* are proportional to *p, q,* and *r*. This means

$$\frac{a}{p} = \frac{b}{q} = \frac{c}{r}.$$

PRACTICE EXERCISE 2

Are 1, 2, 3, and 4 proportional to 5, 10, 15, and 20?

DEFINITION: MEAN PROPORTIONAL

In the proportion $\dfrac{a}{b} = \dfrac{b}{c}$, b is called the **mean proportional** between a and c.

Because $\dfrac{3}{6} = \dfrac{6}{12}$, 6 is a mean proportional between 3 and 12. We know that $\dfrac{3}{6}$ and $\dfrac{6}{12}$ are equal because they both reduce to $\dfrac{1}{2}$. Also, we might observe that the product of the means, $6 \cdot 6 = 36$, is equal to the product of the extremes, $3 \cdot 12 = 36$.

THEOREM 5.1 MEANS-EXTREMES PROPERTY

In any proportion, the product of the means is equal to the product of the extremes. That is,

$$\text{if } \frac{a}{b} = \frac{c}{d}, \text{ then } ad = bc.$$

Given: $\dfrac{a}{b} = \dfrac{c}{d}$

Prove: $ad = bc$

Proof:

STATEMENTS	REASONS
1. $\dfrac{a}{b} = \dfrac{c}{d}$	1. Given
2. $bd = bd$	2. Reflexive law
3. $bd\left(\dfrac{a}{b}\right) = bd\left(\dfrac{c}{d}\right)$	3. Mult.-div. law
4. $ad = bc$	4. Simplify each side

There are many ways to derive different proportions from a given one. The next six theorems show some of these ways.

THEOREM 5.2 RECIPROCAL PROPERTY OF PROPORTIONS

The reciprocals of both sides of a proportion are also proportional. That is,

$$\text{if } \frac{a}{b} = \frac{c}{d}, \text{ then } \frac{b}{a} = \frac{d}{c}.$$

PRACTICE EXERCISE 3

Complete the proof of Theorem 5.2.

Given: $\dfrac{a}{b} = \dfrac{c}{d}$

Prove: $\dfrac{b}{a} = \dfrac{d}{c}$

Proof:

STATEMENTS	REASONS
1. _____	1. Given
2. $ad = bc$	2. _____
3. $ac = ac$	3. _____
4. $\dfrac{ad}{ac} = \dfrac{bc}{ac}$	4. _____
5. _____	5. Simplify the fractions
6. $\dfrac{b}{a} = \dfrac{d}{c}$	6. _____

Notice that because $\dfrac{2}{4} = \dfrac{5}{10}$, by Theorem 5.2 we also have $\dfrac{4}{2} = \dfrac{10}{5}$. If we interchange the means in $\dfrac{2}{4} = \dfrac{5}{10}$, we get $\dfrac{2}{5} = \dfrac{4}{10}$ which is also a proportion. Similarly, by interchanging the extremes, $\dfrac{10}{4} = \dfrac{5}{2}$ is also a proportion. This example introduces the next two theorems, the proofs of which are requested in the exercises.

THEOREM 5.3   MEANS PROPERTY OF PROPORTIONS

If the means are interchanged in a proportion, a new proportion is formed. That is,

$$\text{if } \frac{a}{b} = \frac{c}{d}, \text{ then } \frac{a}{c} = \frac{b}{d}.$$

THEOREM 5.4   EXTREMES PROPERTY OF PROPORTIONS

If the extremes are interchanged in a proportion, a new proportion is formed. That is,

$$\text{if } \frac{a}{b} = \frac{c}{d}, \text{ then } \frac{d}{b} = \frac{c}{a}.$$

Consider the proportion $\frac{2}{8} = \frac{5}{20}$. Suppose we add the denominator to the numerator in each ratio.

$$\frac{2 + 8}{8} = \frac{5 + 20}{20} \quad \text{or} \quad \frac{10}{8} = \frac{25}{20}$$

The results are still proportional. This example introduces the next theorem, the proof of which is requested in the exercises.

> **THEOREM 5.5 ADDITION PROPERTY OF PROPORTIONS**
> If the denominators in a proportion are added to their respective numerators, a new proportion is formed. That is,
> $$\text{if } \frac{a}{b} = \frac{c}{d}, \text{ then } \frac{a + b}{b} = \frac{c + d}{d}.$$

A similar result follows when the denominators are subtracted from the numerators. For example, $\frac{2}{8} = \frac{5}{20}$, and $\frac{2 - 8}{8} = \frac{5 - 20}{20}$ because $\frac{-6}{8} = \frac{-15}{20}$. The proof of the following theorem is requested in the exercises.

> **THEOREM 5.6 SUBTRACTION PROPERTY OF PROPORTIONS**
> If the denominators in a proportion are subtracted from their respective numerators, a new proportion is formed. That is,
> $$\text{if } \frac{a}{b} = \frac{c}{d}, \text{ then } \frac{a - b}{b} = \frac{c - d}{d}.$$

A calendar designer wishes to enlarge a picture for the month of July to fill a space 5 inches high by 8 inches wide. Before the enlargement process, she must crop the picture so that its dimensions are in the same ratio as the dimensions of the space to be filled. How would you crop the picture for the project?

Consider the continued proportion

$$\frac{2}{3} = \frac{4}{6} = \frac{6}{9}.$$

Then consider the following:

$$\frac{2 + 4 + 6}{3 + 6 + 9} = \frac{2}{3}.$$

Because $2 + 4 + 6 = 12$ and $3 + 6 + 9 = 18$, $\frac{12}{18} = \frac{2}{3}$. This example illustrates the following theorem.

THEOREM 5.7

If a, b, c, d, e, and f are numbers satisfying $\frac{a}{b} = \frac{c}{d} = \frac{e}{f}$, then

$$\frac{a + c + e}{b + d + f} = \frac{a}{b}.$$

PRACTICE EXERCISE 4

Complete the proof of Theorem 5.7.

Given: $\dfrac{a}{b} = \dfrac{c}{d} = \dfrac{e}{f}$

Prove: $\dfrac{a + c + e}{b + d + f} = \dfrac{a}{b}$

Proof:

STATEMENTS	REASONS
1. Let $x = \dfrac{a}{b} = \dfrac{c}{d} = \dfrac{e}{f}$	1. _____
2. $a = bx$, $c = dx$, and $e = fx$	2. _____
3. $a + c + e = bx + dx + fx$	3. _____
4. _____	4. Distributive law
5. $\dfrac{a + c + e}{b + d + f} = x$	5. _____
6. _____	6. Substitution law

The process of finding unknown terms in a given proportion, sometimes called **solving the proportion,** applies the means-extremes property. The following examples illustrate.

EXAMPLE 3   Solve each proportion.

(a)
$$\frac{x}{5} = \frac{8}{20}$$

$20x = 5 \cdot 8$	Means-extremes property
$20x = 40$	Simplify
$x = 2$	Divide both sides by 20

(b)
$$\frac{4}{y} = \frac{y}{9}$$

$4 \cdot 9 = y \cdot y$	Means-extremes property
$36 = y^2$	Simplify

Because $6^2 = 36$ and $(-6)^2 = 36$, $y = 6$ or $y = -6$.

(c)

$$\frac{z + 5}{z} = \frac{9}{4}$$

$4(z + 5) = 9z$	Means-extremes property
$4z + 20 = 9z$	Distributive law
$20 = 5z$	Subtract $4z$ from both sides
$4 = z$	Divide both sides by 5 ◨

The proof of the next theorem requires the means-extremes property.

THEOREM 5.8

If three terms of one proportion are equal, respectively, to three terms of another proportion, then the remaining terms are also equal. That is,

$$\text{if } \frac{a}{b} = \frac{c}{x} \text{ and } \frac{a}{b} = \frac{c}{y}, \text{ then } x = y.$$

In Theorem 5.8, note that any of the four terms can be unknown.

E X A M P L E 4   In an election, the winning candidate won by a ratio of 5:4. If she received 600 votes, how many votes did the losing candidate receive?

If a represents the number of votes for the losing candidate, the following proportion describes the problem.

Winning ratio ⟶ $\dfrac{5}{4} = \dfrac{600}{a}$ ⟵ Votes for winner
   ⟵ Votes for loser

$5a = 4 \cdot 600$	Means-extremes property
$5a = 2400$	Simplify
$a = 480$	Divide both sides by 5

Thus, the loser received 480 votes.   ◨

Much of our work in this chapter involves segments that are proportional.

DEFINITION: PROPORTIONAL SEGMENTS

If the lengths of segments are proportional, the segments are called **proportional segments.** That is, segments $\overline{AB}$ and $\overline{CD}$ are proportional to segments $\overline{EF}$ and $\overline{GH}$ when

$$\frac{AB}{CD} = \frac{EF}{GH}.$$

Consider the segments in Figure 5.2. If $AB = 2$ cm, $CD = 3$ cm, $EF = 4$ cm, and $GH = 6$ cm, then because $\frac{2}{3} = \frac{4}{6}$, segments $\overline{AB}$ and $\overline{CD}$ are proportional to segments $\overline{EF}$ and $\overline{GH}$.

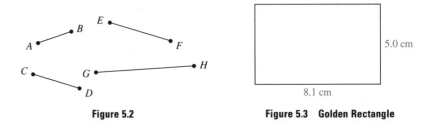

Figure 5.2   **Figure 5.3   Golden Rectangle**

$E\ X\ A\ M\ P\ L\ E\ \ 5$ Consider the rectangle in Figure 5.3. The length ℓ and width w approximately satisfy the proportion $\frac{\ell}{w} = \frac{\ell + w}{\ell}$.

Architects and artists have determined that a rectangle with this property, called a **golden rectangle**, is especially pleasing to the eye. Before the top of the Parthenon was destroyed, the front of the building could be placed into a golden rectangle. The ratio $\frac{\ell}{w}$ is called the **golden ratio** and is approximated by $\frac{8.1}{5.0} = 1.62$. ◪

ANSWERS TO PRACTICE EXERCISES: **1.** $\frac{7}{5}$ or 7:5 **2.** Yes, because $\frac{1}{5} = \frac{2}{10}$ $= \frac{3}{15} = \frac{4}{20}$. **3.** 1. $\frac{a}{b} = \frac{c}{d}$ 2. Means-extremes prop. 3. Reflexive law 4. Mult.-div. law 5. $\frac{d}{c} = \frac{b}{a}$ 6. Sym. law **4.** 1. Given 2. Mult.-div. law 3. Add.-subt. law 4. $a + c + e = (b + d + f)x$ 5. Mult.-div. law 6. $\frac{a + c + e}{b + d + f} = \frac{a}{b}$

5.1 EXERCISES

Write each ratio in Exercises 1–10 as a fraction and simplify.

1. 20 to 35   **2.** 18 to 54   **3.** 8 cm to 32 cm   **4.** 12 ft to 72 ft

5. 200 mi in 4 hr   **6.** 150 mi in 5 hr   **7.** 4 in to 2 ft   **8.** 5 cm to 1 m

9. $\frac{1}{2}$ in to $\frac{3}{8}$ in   **10.** $\frac{1}{4}$ ft to $\frac{3}{2}$ ft

Solve each proportion in Exercises 11–20.

11. $\frac{a}{3} = \frac{14}{21}$ **12.** $\frac{4}{7} = \frac{x}{28}$ **13.** $\frac{40}{35} = \frac{2}{y}$ **14.** $\frac{2}{1} = \frac{1}{a}$

15. $\dfrac{25}{x} = \dfrac{x}{1}$

16. $\dfrac{4}{y} = \dfrac{y}{100}$

17. $\dfrac{a+2}{a} = \dfrac{7}{5}$

18. $\dfrac{x}{x-3} = \dfrac{11}{8}$

19. $\dfrac{y+2}{12} = \dfrac{y-2}{4}$

20. $\dfrac{6}{z-3} = \dfrac{15}{z}$

21. Find the mean proportional between 4 and 36.

22. Find the mean proportional between 36 and 4.

23. Find the mean proportional between 9 and 64.

24. Find the mean proportional between 25 and 144.

25. Representative Wettaw won an election by a ratio of 8 to 5. If he received 10,400 votes, how many votes did his opponent receive?

26. If a wire 70 ft long weighs 84 lb, how much will 110 ft of the same wire weigh?

27. If $\dfrac{1}{2}$ inch on a map represents 10 mi, how many mi are represented by $6\dfrac{1}{2}$ inches?

28. A ranger wants to estimate the number of antelope in a preseve. He catches 58 antelope, tags their ears, and returns them to the preserve. Some time later, he catches 29 antelope and discovers that 7 of them are tagged. Estimate the number of antelope in the preserve.

29. It has been estimated that a family of four will produce 115 lb of garbage in one week. Estimate the number of pounds of garbage produced by 7 such families in one week.

30. A baseball pitcher gave up 60 earned runs in 180 innings. Estimate the number of earned runs he will give up every 9 innings. This number is called his *earned run average*.

31. The mean proportional occurs frequently in nature. For example, a starfish has the shape of a *pentagram* shown below. AB is the mean proportional between BC and AC. That is, $\dfrac{BC}{AB} = \dfrac{AB}{AC}$. Also, the ratios $\dfrac{AD}{AC}, \dfrac{AC}{AB}$, and $\dfrac{AB}{BC}$ all equal the golden ratio. Measure the lengths in the figure and try to confirm these results.

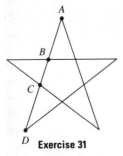

Exercise 31

Window

Door

Fireplace

Exercise 32

32. Proportions are also used to determine actual lengths from scale drawings. For example, shown above is the scale drawing of a family room in which $\frac{1}{8}$ inch corresponds to 1 ft. Use proportions to find the following.

(a) What is the length of the room?

(b) What is the width of the room?

(c) What is the width of the window?

(d) What is the width of the fireplace?

33. Prove Theorem 5.3.

34. Prove Theorem 5.4.

35. Prove Theorem 5.5.

36. Prove Theorem 5.6.

5.2 SIMILAR POLYGONS

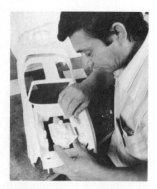

In designing a new automobile, an engineer first makes a scale model. The model and the finished automobile are similar to each other. If the wheel base on the actual car is to be 100 inches in length and the initial model is to be $\frac{1}{20}$ the actual size, what should be the length of the wheel base on the model?

When an artist makes a sketch of a basketball court, or an architect draws a blueprint for a house, or an engineer makes a drawing of a machine part, the result is drawn to scale showing the same objects in reduced sizes. Similarly, a biologist looking through a microscope sees shapes enlarged in size. Two figures with the same shape are called *similar*. In geometry we study *similar polygons*.

> **DEFINITION: SIMILAR POLYGONS**
>
> Two polygons are **similar** if their vertices can be paired in such a way that corresponding angles are equal, and corresponding sides are proportional.

When we refer to similar polygons, we sometimes list the corresponding vertices in the same order. For example, if polygon $ABCDE$ is similar to polygon $PQRST$ in Figure 5.4, we write $ABCDE \sim PQRST$. From the definition of similar polygons,

$$\angle A = \angle P, \angle B = \angle Q, \angle C = \angle R, \angle D = \angle S, \text{ and } \angle E = \angle T,$$

and

$$\frac{AB}{PQ} = \frac{BC}{QR} = \frac{CD}{RS} = \frac{DE}{ST} = \frac{EA}{TP}.$$

When working with similar polygons, it is helpful to label corresponding vertices as A and A' (read A prime), B and B', and so on. The following example illustrates.

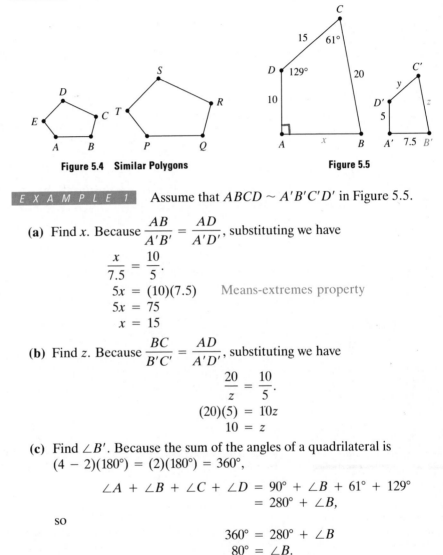

Figure 5.4   Similar Polygons   **Figure 5.5**

E X A M P L E 1   Assume that $ABCD \sim A'B'C'D'$ in Figure 5.5.

(a) Find x. Because $\dfrac{AB}{A'B'} = \dfrac{AD}{A'D'}$, substituting we have

$$\frac{x}{7.5} = \frac{10}{5}.$$
$$5x = (10)(7.5) \quad \text{Means-extremes property}$$
$$5x = 75$$
$$x = 15$$

(b) Find z. Because $\dfrac{BC}{B'C'} = \dfrac{AD}{A'D'}$, substituting we have

$$\frac{20}{z} = \frac{10}{5}.$$
$$(20)(5) = 10z$$
$$10 = z$$

(c) Find $\angle B'$. Because the sum of the angles of a quadrilateral is $(4 - 2)(180°) = (2)(180°) = 360°$,

$$\angle A + \angle B + \angle C + \angle D = 90° + \angle B + 61° + 129°$$
$$= 280° + \angle B,$$

so

$$360° = 280° + \angle B$$
$$80° = \angle B.$$

Because $\angle B' = \angle B$, $\angle B' = 80°$. ◢

PRACTICE EXERCISE 1

Refer to Figure 5.5 in which $ABCD \sim A'B'C'D'$. Find y and $\angle D'$.

Similar triangles, such as $\triangle ABC$ and $\triangle A'B'C'$ in Figure 5.6, have many applications in geometry, and it is appropriate to investigate some of their properties. We can use the definition of similar polygons to prove that two triangles are similar, but the next postulate provides a more useful method.

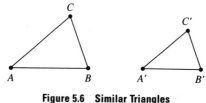

Figure 5.6   Similar Triangles

POSTULATE 5.1   AAA ~ AAA

Two triangles are similar if three angles of one are equal, respectively, to three angles of the other.

Thus, to determine similarity of triangles, we do not need to verify proportionality of corresponding sides, but only equality of corresponding angles. In fact, by Corollary 4.13, we know that if two angles of one triangle are equal, respectively, to two angles of another triangle, then the third angles are equal. So we can show similarity using only two pairs of angles. We have just given an informal proof of the next useful theorem.

THEOREM 5.9   AA ~ AA

Two triangles are similar if two angles of one are equal, respectively, to two angles of the other.

EXAMPLE 2   Refer to Figure 5.7.

Given: $\overline{AB} \perp \overline{BD}$ and $\overline{ED} \perp \overline{BD}$

Prove: $\triangle ABC \sim \triangle CDE$

Proof:

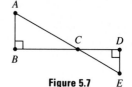

Figure 5.7

STATEMENTS	REASONS
1. $\overline{AB} \perp \overline{BD}$ and $\overline{ED} \perp \overline{BD}$	1. Given
2. $\angle ABC$ and $\angle CDE$ are right angles	2. $\perp$ lines form rt. $\angle$'s
3. $\angle ABC = \angle CDE$	3. Rt. $\angle$'s are $=$
4. $\angle ACB$ and $\angle DCE$ are vertical angles	4. Def. of vert. $\angle$'s
5. $\angle ACB = \angle DCE$	5. Vert. $\angle$'s are $=$
6. $\triangle ABC \sim \triangle CDE$	6. AA ~ AA ▨

> **PRACTICE EXERCISE 2**
> In Figure 5.7 if $AB = 9$ ft, $BC = 12$ ft, $CE = 10$ ft, and $DE = 6$ ft, find AC and CD.

Figure 5.6 clearly shows that similar triangles do not have to be congruent. However, congruent triangles are similar.

THEOREM 5.10

If $\triangle ABC \cong \triangle DEF$, then $\triangle ABC \sim \triangle DEF$

The proof of Theorem 5.10 is left for you to do as an exercise. The next theorem shows that triangles similar to the same triangle are themselves similar.

THEOREM 5.11   TRANSITIVE LAW FOR SIMILAR TRIANGLES

If $\triangle ABC \sim \triangle DEF$ and $\triangle DEF \sim \triangle GHI$, then $\triangle ABC \sim \triangle GHI$.

The proof of Theorem 5.11 is left for you to do as a exercise. The next theorem illustrates how similar triangles are often used to determine certain properties of a given triangle.

THEOREM 5.12   TRIANGLE PROPORTIONALITY THEOREM

A line parallel to one side of a triangle that intersects the other two sides divides the two sides into proportional segments.

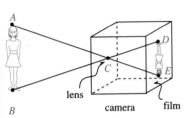

When you take a photograph of a person the image is captured on the film inside the camera as shown. The height of the image on the film will vary in proportion with the distance of the person from the lens. By using the fact that $\triangle$**ABC** $\sim$ $\triangle$**EDC,** and by knowing the height of the person, a photographer can determine the best location of the camera to give any desired image size on the film.

Given: $\triangle ABC$ with $\ell \parallel \overline{BC}$
  (See Figure 5.8.)

Prove: $\dfrac{DB}{AD} = \dfrac{EC}{AE}$

Proof:

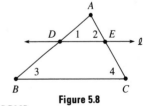

Figure 5.8

STATEMENTS	REASONS
1. $\triangle ABC$ with $\ell \parallel \overline{BC}$	1. Given
2. $\angle 1 = \angle 3$ and $\angle 2 = \angle 4$	2. Corr. $\angle$'s are $=$
3. $\triangle ADE \sim \triangle ABC$	3. AA $\sim$ AA
4. $\dfrac{AB}{AD} = \dfrac{AC}{AE}$	4. Corr. sides of $\sim$ $\triangle$'s are proportional

5. $\dfrac{AB - AD}{AD} = \dfrac{AC - AE}{AE}$ 5. Subtr. prop. of proportions

6. $\overline{AB} - \overline{AD} = \overline{DB}$ and   6. Seg. add. prop.
$\overline{AC} - \overline{AE} = \overline{EC}$

7. $\dfrac{DB}{AD} = \dfrac{EC}{AE}$ 7. Substitution law

E X A M P L E 3   In Figure 5.9, $\overline{DE} \parallel \overline{BC}$, $AC = 8.8$ yd, $AD = 1.3$ yd, and $DB = 3.1$ yd. Find AE and EC.

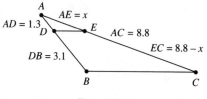

Figure 5.9

Let $x = AE$, then $EC = AC - AE = 8.8 - x$. Using Theorem 5.12,

$$\frac{1.3}{3.1} = \frac{x}{8.8 - x}.$$

$1.3(8.8 - x) = 3.1x$	Means-extremes prop.
$11.44 - 1.3x = 3.1x$	Distributive law
$11.44 = 4.4x$	Add $1.3x$ to both sides
$2.6 = x$	Divide both sides by 4.4

Thus, $AE = 2.6$ yd, and $EC = 8.8 - 2.6 = 6.2$ yd. ◪

Theorem 5.12 offers a way to construct a segment proportional to three given line segments.

CONSTRUCTION 5.1

Construct a segment proportional to three given line segments.

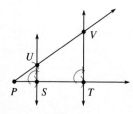

Figure 5.10

Given: Segments $\overline{AB}$, $\overline{CD}$, and $\overline{EF}$ (See Figure 5.10.)

To Construct: $\overline{UV}$ so that $\dfrac{AB}{CD} = \dfrac{EF}{UV}$

Construction:

1. Construct an arbitrary angle with vertex P. On one ray of $\angle P$, construct $\overline{PS}$ such that $PS = AB$ and $\overline{ST}$ such that $ST = CD$.

2. On the other ray of $\angle P$, construct $\overline{PU}$ such that $PU = EF$. Then construct $\overleftrightarrow{SU}$.

3. Use Construction 4.1 to construct the line through T parallel to $\overleftrightarrow{SU}$ intersecting $\overrightarrow{PU}$ at point V. Then $\overline{UV}$ is the desired segment by Theorem 5.12.

THEOREM 5.13   TRIANGLE ANGLE-BISECTOR THEOREM

The bisector of one angle of a triangle divides the opposite side into segments that are proportional to the other two sides.

PRACTICE EXERCISE 3

Complete the proof of Theorem 5.13.

Given: $\overline{AD}$ bisects $\angle A$

Prove: $\dfrac{BD}{DC} = \dfrac{AB}{AC}$

Auxiliary lines: Construct the line through B parallel to $\overline{AD}$ intersecting the extension of $\overline{AC}$ at point E

Proof:

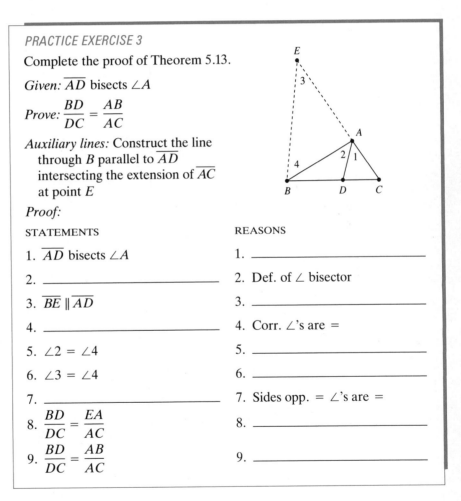

STATEMENTS	REASONS
1. $\overline{AD}$ bisects $\angle A$	1. _____
2. _____	2. Def. of $\angle$ bisector
3. $\overline{BE} \parallel \overline{AD}$	3. _____
4. _____	4. Corr. $\angle$'s are $=$
5. $\angle 2 = \angle 4$	5. _____
6. $\angle 3 = \angle 4$	6. _____
7. _____	7. Sides opp. $= \angle$'s are $=$
8. $\dfrac{BD}{DC} = \dfrac{EA}{AC}$	8. _____
9. $\dfrac{BD}{DC} = \dfrac{AB}{AC}$	9. _____

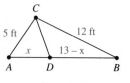

E X A M P L E 4   Refer to Figure 5.11 showing right $\triangle ABC$ with legs 5 ft and 12 ft and hypotenuse 13 ft. If $\overline{CD}$ bisects right $\angle C$, approximate the lengths of $\overline{AD}$ and $\overline{DB}$, correct to the nearest tenth of a foot.

Let $x = AD$, then $DB = AB - AD = 13 - x$. By the triangle angle-bisector theorem,

Figure 5.11

$$\frac{x}{13 - x} = \frac{5}{12}.$$

$$
\begin{aligned}
12x &= 5(13 - x) \qquad &\text{Means-extremes prop.} \\
12x &= 65 - 5x \qquad &\text{Distributive law} \\
17x &= 65 \qquad &\text{Add } 5x \text{ to both sides} \\
x &= \frac{65}{17}
\end{aligned}
$$

Using a calculator, we get $x = 3.8$, to the nearest tenth. Then $13 - x = 9.2$. Thus, $AD = 3.8$ ft and $DC = 9.2$ ft, correct to the nearest tenth of a foot.   ◪

The next example solves the applied problem given in the chapter introduction.

E X A M P L E 5   While standing near the Washington Monument in Washington, D.C., Scott discovered that the length of the shadow of the monument was 185 ft at the same time his shadow was 2 ft long. If Scott is 6 ft tall, how did he determine the height of the monument?

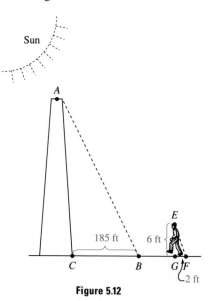

Figure 5.12

Figure 5.12 shows the information given (not drawn to scale). Because the sun is so far away, we will assume that the rays from the sun are parallel, making

$\angle BAC = \angle FEG$. Also, we can assume that $\angle ACB$ and $\angle EGF$ are right angles, hence are equal. Thus, $\triangle ABC \sim \triangle EFG$ by AA $\sim$ AA. If x is the height of the monument (side $\overline{AC}$), then

$$\frac{x}{6} = \frac{185}{2}.$$
$$2x = (6)(185)$$
$$x = 555$$

Thus, Scott concluded that the monument is 555 ft high. ◪

We can use Theorem 5.9 in another way to show how to construct a polygon similar to a given one. We illustrate this with a pentagon, but the process applies to any polygon.

CONSTRUCTION 5.2

Construct a polygon similar to a given polygon.

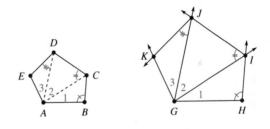

Figure 5.13 Construction of Similar Polygons

Given: Pentagon $ABCDE$ and segment $\overline{GH}$ (See Figure 5.13.)

To Construct: Pentagon $GHIJK$ such that $ABCDE \sim GHIJK$

Construction:

1. Construct diagonal $\overline{AC}$ forming $\angle 1$. Copy $\angle 1$ at G and $\angle B$ at H. Label their point of intersection I. Then $\triangle ABC \sim \triangle GHI$ by AA $\sim$ AA.

2. Now draw diagonal $\overline{AD}$ and repeat the construction in Step 1 obtaining $\triangle GIJ \sim \triangle ACD$.

3. Now construct $\triangle GJK$ in the same manner, so that $\triangle ADE \sim \triangle GJK$.

4. Then $ABCDE \sim GHIJK$.

ANSWERS TO PRACTICE EXERCISES: **1.** 7.5; 129° **2.** $AC = 15$ ft, $CD = 8$ ft **3.** 1. Given 2. $\angle 1 = \angle 2$ 3. By construction 4. $\angle 3 = \angle 1$ 5. Alt. int. $\angle$'s are $=$ 6. Sym. and trans. laws 7. $AE = AB$ 8. $\triangle$ proportionality thm. 9. Substitution law

5.2 EXERCISES

Exercises 1–6 refer to the hexagons in the figure below. Assume that
$ABCDEF \sim A'B'C'D'E'F'$.

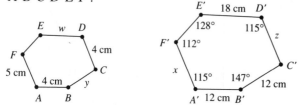

1. Find the value of x.   2. Find the value of y.   3. Find the value of z.   4. Find the value of w.
5. Find $\angle C'$.   6. Find $\angle C$.

In Exercises 7–18, state whether the two polygons are *always, sometimes,* or *never* similar.

7. Two squares.

8. Two rectangles.

9. Two right triangles.

10. Two rhombuses.

11. Two equilateral triangles.

12. Two equiangular triangles.

13. Two isosceles triangles.

14. Two scalene triangles.

15. A right triangle and an acute triangle.

16. An acute triangle and an obtuse triangle.

17. A right triangle and a scalene triangle.

18. A right triangle and an isosceles triangle.

Exercises 19–24 refer to the figure below in which $\overline{DE} \parallel \overline{BC}$.

19. If $AD = 12$ ft, $DB = 6$ ft, and $AE = 20$ ft, find EC.

20. If $BD = 5$ cm, $CE = 9$ cm, and $EA = 27$ cm, find AD.

21. If $AB = 22$ yd, $AE = 8$ yd, and $EC = 3$ yd, find AD and BD.

22. If $AC = 40$ in, $AD = 12$ in, and $DB = 8$ in, find AE and EC.

23. If $AC = 18$ ft, $AE = 12$ ft, and $AB = 15$ ft, find AD and BD.

24. If $AB = 30$ cm, $BD = 5$ cm, and $AC = 42$ cm, find AE and CE.

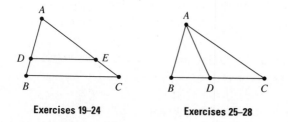

Exercises 19–24   Exercises 25–28

Exercises 25–28 refer to the figure above in which $\overline{AD}$ bisects $\angle A$.

25. If $AB = 6$ ft, $AC = 9$ ft, and $BD = 2$ ft, find DC.

26. If $AB = 20$ cm, $BD = 8$ cm, and $DC = 14$ cm, find AC.

27. If $BC = 45$ in, $AB = 24$ in, and $AC = 36$ in, find BD and DC.

28. If $BC = 28$ yd, $BA = 18$ yd, and $CA = 24$ yd, find CD and DB.

29. Find the height of a tree that casts an 80-foot shadow at the same time that a telephone pole 18 ft tall casts a 12-foot shadow.

30. To measure the distance between points A and B on two islands (see the figure below), a man takes the following measurements on one of the islands: $AC = 5$ yd, $DE = 6$ yd, and $AE = 15$ yd. How far apart are the islands?

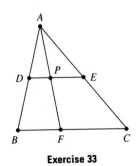

Exercise 30

31. *Given:* $\triangle ABC$ and $\square CDEF$

   Prove: $\dfrac{AD}{DC} = \dfrac{CF}{FB}$

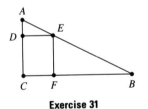

Exercise 31

32. *Given:* $\triangle ABC$ and $\angle 1 = \angle 2$

   Prove: $\dfrac{AD}{AB} = \dfrac{DE}{BC}$

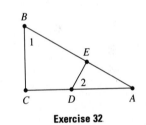

Exercise 32

33. *Given:* $\triangle ABC$ and $\overline{DE} \parallel \overline{BC}$

   Prove: $\dfrac{DP}{BF} = \dfrac{PE}{FC}$

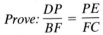

Exercise 33

34. *Given:* $\triangle ABC$, $\overline{AD}$ bisects $\angle BAC$, and $AE = ED$

   Prove: $\dfrac{AE}{AC} = \dfrac{BD}{BC}$

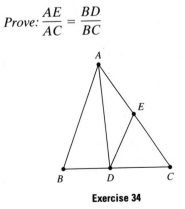

Exercise 34

35. Prove that the ratio of the perimeters of two similar triangles equals the ratio of the lengths of any two corresponding sides.

36. Prove that the ratio of the lengths of the altitudes from corresponding angles in similar triangles equals the ratio of the lengths of any two corresponding sides.

37. Draw segments $\overline{AB}$, $\overline{CD}$, and $\overline{EF}$ such that AB is about 2 inches, CD is about 3 inches, and EF is about 4 inches. Construct $\overline{UV}$ so that
$$\frac{AB}{CD} = \frac{EF}{UV}.$$

38. Construct a quadrilateral similar to a given quadrilateral.

39. Wires are stretched from the top of each of two poles to the bottom of the other as shown in the following figure. If one pole is 4 ft tall and the other 12 ft tall, how far above the ground do the wires cross? [Note: The distance between the poles is not important; that is, x can be found regardless of the values of u and v.]

Exercise 39

40. A *pantograph* is an instrument used for reducing or enlarging a map or drawing. It consists of four bars hinged together at points A, B, C, and D so that $ABCD$ is a parallelogram and P, D, and E lie on the same line. Point P is attached to a drawing table and does not move. To produce a larger triangle II that is similar to triangle I, a stylus that traces triangle I is inserted at D while a pen at E produces triangle II. Although the angles of the parallelogram change, P, D, and E remain on the same line. Assume $PA = 6$ inches and $AB = 12$ inches. Answer the following:

(a) Explain why $\triangle PAD \sim \triangle PBE$.

(b) What is the value of $\dfrac{PA}{PB}$?

(c) What is the value of $\dfrac{PD}{PE}$?

(d) How do the sides of $\triangle$II compare to those of $\triangle$I?

(e) How would this pantograph be used to reduce a drawing?

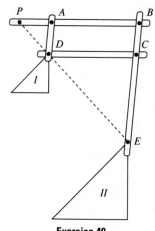

Exercise 40

5.3 MORE THEOREMS ON SIMILAR TRIANGLES

In this section we explore more ways to show that two triangles are similar. The following theorem is the converse of the triangle proportionality theorem.

> **THEOREM 5.14**
>
> If a line intersects and divides two sides of a triangle into proportional segments, then the line is parallel to the third side.

Given: $\triangle ABC$ and line ℓ such that
$$\frac{AD}{DB} = \frac{AE}{EC}$$
(See Figure 5.14.)

Prove: $\ell \parallel \overline{BC}$

Auxiliary line: Construct the line m through B parallel to ℓ and intersecting $\overleftrightarrow{AC}$ in a point, for instance F, forming $\triangle ABF$

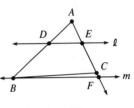

Figure 5.14

Proof:

STATEMENTS	REASONS
1. $m \parallel \ell$	1. By construction
2. $\dfrac{AD}{DB} = \dfrac{AE}{EF}$	2. $\triangle$ proportionality thm.
3. $\dfrac{AD}{DB} = \dfrac{AE}{EC}$	3. Given
4. $EF = EC + CF$	4. Seg. add. post.
5. $EF = EC$	5. If 3 terms of a prop. = 3 terms of another, the fourth terms are =
6. $0 = CF$	6. Add.–subtr. law
7. F and C coincide	7. If $CF = 0$, C and F are the same point
8. Line m coincides with the line $\overleftrightarrow{BC}$	8. Two pts. determine a unique line
9. $\ell \parallel \overline{BC}$	9. Substitution law

> **THEOREM 5.15   SAS ~ SAS**
>
> If an angle of one triangle is equal to an angle of another triangle, and the including sides are proportional, then the triangles are similar.

David Hilbert (1862–1943)

David Hilbert, a German mathematician, played a major role in determining the nature of mathematics as we know it today. In 1899 he published a small but renowned book entitled *Grundlagen der Geometric (Foundations of Geometry)* that had a profound impact on all mathematics in the twentieth century. This work solved many of the logical difficulties found in Euclid's *Elements* and attempted to place geometry in a more formal setting similar to algebra.

Given: $\triangle ABC$, $\triangle DEF$, with $\angle A = \angle D$ and $\dfrac{AB}{DE} = \dfrac{AC}{DF}$
(See Figure 5.15.)

Prove: $\triangle ABC \sim \triangle DEF$

Auxiliary line: Construct E' on $\overline{AB}$ such that $AE' = DE$, and F' on $\overline{AC}$ such that $AF' = DF$. Draw $\overline{E'F'}$.

Proof:

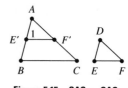

Figure 5.15   SAS ~ SAS

STATEMENTS	REASONS
1. $\angle A = \angle D$	1. Given
2. $AE' = DE$ and $AF' = DF$	2. By construction
3. $\triangle AE'F' \cong \triangle DEF$	3. SAS = SAS
4. $\triangle AE'F' \sim \triangle DEF$	4. Congruent $\triangle$'s are ~
5. $\dfrac{AB}{DE} = \dfrac{AC}{DF}$	5. Given
6. $\dfrac{AB}{AE'} = \dfrac{AC}{AF'}$	6. Substitution law
7. $\dfrac{AB - AE'}{AE'} = \dfrac{AC - AF'}{AF'}$	7. Subtr. law of proportions
8. $E'B = AB - AE'$ and $F'C = AC - AF'$	8. Seg. add. post.
9. $\dfrac{E'B}{AE'} = \dfrac{F'C}{AF'}$	9. Substitution law
10. $\overline{E'F'} \parallel \overline{BC}$	10. Lines div. sides of $\triangle$ in prop. seg. are $\parallel$
11. $\angle 1 = \angle B$	11. Corr. $\angle$'s are =
12. $\angle A = \angle A$	12. Reflexive law
13. $\triangle AE'F' \sim \triangle ABC$	13. AA ~ AA
14. $\triangle ABC \sim \triangle DEF$	14. Trans. law for ~ $\triangle$'s

EXAMPLE 1   Explain why $\triangle ABC$ and $\triangle A'B'C'$, shown in Figure 5.16, are similar.

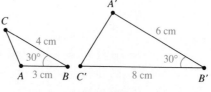

Figure 5.16

Because $\angle B = \angle B'$ and $\dfrac{BC}{BA} = \dfrac{4}{3} = \dfrac{8}{6} = \dfrac{B'C'}{B'A'}$, by Theorem 5.15, $\triangle ABC \sim \triangle A'B'C'$.

> **THEOREM 5.16   SSS ~ SSS**
>
> If three sides of one triangle are proportional to the three corresponding sides of another triangle, then the triangles are similar.

Given: $\triangle ABC$, $\triangle DEF$, with $\dfrac{AB}{DE} = \dfrac{BC}{EF} = \dfrac{AC}{DF}$

(See Figure 5.17.)

Prove: $\triangle ABC \sim \triangle DEF$

Auxiliary line: Construct E' on AB such that $AE' = DE$, and F' on $\overline{AC}$ such that $AF' = DF$. Draw $\overline{E'F'}$

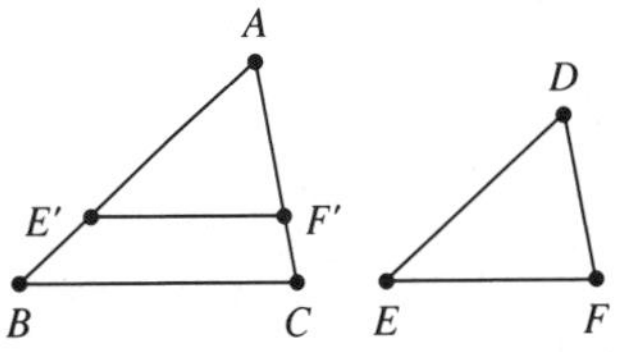

Figure 5.17   SSS ~ SSS

Proof:

STATEMENTS	REASONS
1. $\dfrac{AB}{DE} = \dfrac{AC}{DF}$	1. Given
2. $\angle A = \angle A$	2. Reflexive law
3. $AE' = DE$ and $AF' = DF$	3. By construction
4. $\dfrac{AB}{AE'} = \dfrac{AC}{AF'}$	4. Substitution law
5. $\triangle AE'F' \sim \triangle ABC$	5. SAS ~ SAS
6. $\dfrac{AB}{AE'} = \dfrac{BC}{E'F'}$	6. Corr. sides of ~ $\triangle$'s are proportional
7. $\dfrac{AB}{DE} = \dfrac{BC}{E'F'}$	7. Substitution law
8. $\dfrac{AB}{DE} = \dfrac{BC}{EF}$	8. Given
9. $E'F' = EF$	9. Use Statements 7 and 8 with 3 terms of one prop. = 3 terms of another which makes fourth terms =
10. $\triangle DEF \cong \triangle AE'F'$	10. SSS = SSS
11. $\triangle DEF \sim \triangle AE'F'$	11. Congruent $\triangle$'s are ~
12. $\triangle ABC \sim \triangle DEF$	12. Trans. law for ~ $\triangle$'s

E X A M P L E   2   Is a triangle with sides of 2 cm, 3 cm, and 4 cm similar to a triangle with sides of 10 cm, 15 cm and 20 cm?

Because $\frac{2}{10} = \frac{3}{15} = \frac{4}{20}$ (all fractions reduce to $\frac{1}{5}$), by Theorem 5.16, corresponding sides are proportional making the triangles similar. ◪

The perimeters of two similar triangles have the same ratio as the ratio formed by a pair of corresponding sides (see Exercise 35 in Section 5.2). A similar result holds for the areas of similar triangles.

THEOREM 5.17

The areas of two similar triangles have the same ratio as the squares of the lengths of any two corresponding sides.

PRACTICE EXERCISE 1

Complete the proof of Theorem 5.17.

Given: $\triangle ABC \sim \triangle A'B'C'$

Prove: $\dfrac{\text{Area } \triangle ABC}{\text{Area } \triangle A'B'C'} = \dfrac{(BC)^2}{(B'C')^2}$

Auxiliary lines: Construct altitudes $\overline{AD}$ and $\overline{A'D'}$ from vertices A and A', respectively

Proof:

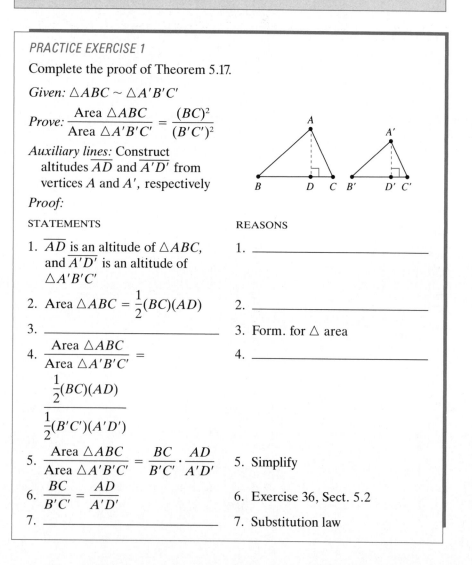

STATEMENTS

1. $\overline{AD}$ is an altitude of $\triangle ABC$, and $\overline{A'D'}$ is an altitude of $\triangle A'B'C'$

2. Area $\triangle ABC = \dfrac{1}{2}(BC)(AD)$

3. _____

4. $\dfrac{\text{Area } \triangle ABC}{\text{Area } \triangle A'B'C'} = \dfrac{\dfrac{1}{2}(BC)(AD)}{\dfrac{1}{2}(B'C')(A'D')}$

5. $\dfrac{\text{Area } \triangle ABC}{\text{Area } \triangle A'B'C'} = \dfrac{BC}{B'C'} \cdot \dfrac{AD}{A'D'}$

6. $\dfrac{BC}{B'C'} = \dfrac{AD}{A'D'}$

7. _____

REASONS

1. _____

2. _____

3. Form. for △ area

4. _____

5. Simplify

6. Exercise 36, Sect. 5.2

7. Substitution law

Right triangles have numerous properties relative to similarity and proportionality. These properties will be discussed in Chapter 6.

ANSWER TO PRACTICE EXERCISE: **1.** 1. By construction 2. Form. for $\triangle$ area 3. Area $\triangle A'B'C' = \frac{1}{2}(B'C')(A'D')$ 4. Mult.-div. law 7. $\dfrac{\text{Area } \triangle ABC}{\text{Area } \triangle A'B'C'}$

$$= \frac{(BC)^2}{(B'C')^2}$$

5.3 EXERCISES

In Exercises 1–10, are the triangles shown below similar based on the given information?

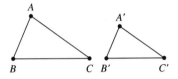

1. $\angle A = 81°$, $\angle A' = 81°$, $A'B' = 6$ cm, $A'C' = 9$ cm, $AB = 12$ cm, and $AC = 18$ cm.

2. $\angle B = 50°$, $\angle B' = 50°$, $BA = 4$ yd, $BC = 10$ yd, $B'A' = 2$ yd, and $B'C' = 5$ yd.

3. $\angle C = 30°$, $\angle C' = 30°$, $AB = 10$ ft, $A'B' = 5$ ft, $BC = 18$ ft, and $B'C' = 9$ ft.

4. $\angle A = 75°$, $\angle A' = 75°$, $BC = 30$ cm, $B'C' = 10$ cm, $AC = 21$ cm, and $A'C' = 7$ cm.

5. $AB = 30$ yd, $AC = 40$ yd, $BC = 50$ yd, $A'B' = 3$ yd, $A'C' = 4$ yd, and $B'C' = 5$ yd.

6. $AB = 25$ ft, $AC = 35$ ft, $BC = 45$ ft, $A'B' = 5$ ft, $A'C' = 7$ ft, and $B'C' = 9$ ft.

7. $AB = 20$ cm, $AC = 30$ cm, $BC = 40$ cm, $A'B' = 10$ cm, $A'C' = 15$ cm, and $B'C' = 25$ cm.

8. $AB = 1.5$ yd, $AC = 2.5$ yd, $BC = 4.0$ yd, $A'B' = 3.0$ yd, $A'C' = 4.5$ yd, and $B'C' = 8.0$ yd.

9. $\angle A = \angle A' = 80°$, $\angle B = 40°$, and $\angle C' = 60°$.

10. $\angle B = \angle B' = 45°$, $\angle A = 78°$, and $\angle C' = 47°$.

11. If $\triangle ABC \sim \triangle A'B'C'$, $AB = 30$ cm and $A'B' = 6$ cm, what is the ratio of the area of $\triangle ABC$ to $\triangle A'B'C'$?

12. If $\triangle ABC \sim \triangle A'B'C'$, $BC = 15$ ft, and $B'C' = 5$ ft, what is the ratio of the area of $\triangle ABC$ to $\triangle A'B'C'$?

13. If $\triangle ABC \sim \triangle A'B'C'$, the area of $\triangle ABC = 49$ yd^2, $AC = 14$ yd, and $A'C' = 20$ yd, what is the area of $\triangle A'B'C'$?

14. If $\triangle ABC \sim \triangle A'B'C'$, the area of $\triangle A'B'C' = 54$ cm^2, $A'C' = 18$ cm, and $AC = 72$ cm, what is the area of $\triangle ABC$?

15. Prove that the line segment joining the midpoints of two sides of a triangle forms a triangle that is similar to the original triangle.

16. Prove that the triangle formed by joining the midpoints of all the sides of a given triangle is similar to the given triangle.

17. Prove that two isosceles triangles with equal vertex angles are similar.

18. Prove that the lengths of corresponding medians in similar triangles are proportional to corresponding sides.

19. Prove that the lengths of corresponding angle bisectors in similar triangles are proportional to corresponding sides.

20. Prove that two triangles are similar if their corresponding sides are parallel.

KEY TERMS AND SYMBOLS

5.1 ratio, , p. 171
proportion, p. 172
terms (of a proportion),
  p. 172
extremes, p. 172
means, p. 172

mean proportional, p. 173
solving proportions, p. 176
proportional segments,
  p. 177

golden rectangle, p. 178
golden ratio, p. 178
5.2 similar polygons (~),
  p. 180

PROOF TECHNIQUES

To Prove:

Two Triangles Similar

1. Show two angles of one are equal to two angles of the other. (Theorem 5.9, p. 182)

2. Show the triangles are congruent. (Theorem 5.10, p. 183)

3. Show an angle of one is equal to an angle of the other and the including sides are proportional. (Theorem 5.15, p. 191)

4. Show the three sides of one are proportional to the three sides of the other. (Theorem 5.16, p. 193)

REVIEW EXERCISES

Section 5.1

In Exercises 1–2, write each ratio as a fraction and simplify.

1. 32 to 40

2. 300 mi in 5 hr

Solve each proportion in Exercises 3–6.

3. $\dfrac{a}{12} = \dfrac{1}{4}$

4. $\dfrac{4}{20} = \dfrac{8}{x}$

5. $\dfrac{y + 5}{y} = \dfrac{21}{6}$

6. $\dfrac{a + 3}{36} = \dfrac{a - 3}{9}$

7. If 6 is to 5 as 24 is to x, find x.

8. Find the mean proportional between 16 and 25.

9. If $\dfrac{1}{4}$ inch on a map represents 20 mi, how many miles are represented by $2\dfrac{1}{4}$ inches?

Answer true or false in Exercises 10–12.

10. If $\dfrac{a}{b} = \dfrac{c}{d}$, then $ac = bd$.

11. If $\dfrac{a}{b} = \dfrac{c}{d}$, then $\dfrac{a + b}{b} = \dfrac{c + d}{d}$.

12. If $\dfrac{a}{b} = \dfrac{c}{d}$, then $\dfrac{a + c}{b + d} = \dfrac{a}{b}$.

Section 5.2

Exercises 13–16 refer to the quadrilaterals in the figure below. Assume that $ABCD \sim A'B'C'D'$.

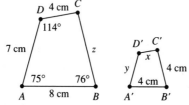

13. Find the value of y.   **14.** Find the value of x.   **15.** Find the value of z.   **16.** Find $\angle C'$.

17. Are two isosceles right triangles always similar?

18. Is a right triangle ever similar to an equilateral triangle?

Exercises 19–22 refer to the figure below in which $\overline{DE} \parallel \overline{BC}$ and $\overline{AF}$ bisects $\angle A$.

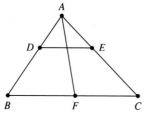

19. If $AD = 20$ ft, $DB = 15$ ft, and $AE = 28$ ft, find EC.

20. If $AC = 50$ ft, $AB = 48$ ft, $DB = 12$ ft, find AE and EC.

21. If $AB = 35$ cm, $AC = 50$ cm, and $BF = 14$ cm, find FC.

22. If $AB = 30$ ft, $AC = 80$ ft, and $BC = 77$ ft, find BF and FC.

23. *Given:* $\triangle ABC$ is isosceles with base $\overline{BC}$, and $\angle C$ is supplementary to $\angle 1$

   Prove: $\dfrac{AD}{DB} = \dfrac{AE}{EC}$

24. *Given:* $\triangle ABC$, $\overline{BD}$ bisects $\angle B$, and $\overline{ED} \parallel \overline{BC}$

   Prove: $\dfrac{AE}{EB} = \dfrac{AB}{BC}$

Exercise 23

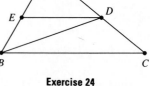

Exercise 24

25. Draw three segments $\overline{AB}$, $\overline{CD}$, and $\overline{EF}$ with AB about 3 cm, CD about 5 cm, and EF about 4 cm. Construct $\overline{UV}$ so that $\dfrac{AB}{CD} = \dfrac{EF}{UV}$.

Section 5.3

In Exercises 26–27 determine whether $\triangle ABC$ is similar to $\triangle A'B'C'$ based on the given information.

26. $\angle A = 35°$, $\angle A' = 35°$, $AB = 12$ ft, $A'B' = 40$ ft, $AC = 3$ ft, and $A'C' = 10$ ft.

27. $AB = 18$ cm, $BC = 12$ cm, $CA = 9$ cm, $A'B' = 9$ cm, $B'C' = 6$ cm, and $C'A' = 5$ cm.

28. If $\triangle ABC \sim \triangle A'B'C'$, the area of $\triangle ABC = 180$ yd², $BC = 30$ yd, and $B'C' = 5$ yd, what is the area of $\triangle A'B'C'$?

29. *Given:* $\dfrac{BC}{CE} = \dfrac{CA}{CD}$

Prove: $\angle B = \angle E$

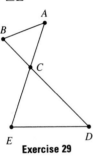

E D

Exercise 29

30. *Given:* $\dfrac{AB}{CD} = \dfrac{AE}{BC} = \dfrac{BE}{BD}$

Prove: $\overline{BC} \parallel \overline{AE}$

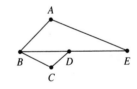

Exercise 30

31. A woman uses a mirror to find the height of a tree by placing the mirror horizontally on the ground and walking backward in the line formed by the tree and mirror until she can see the top of the tree in the mirror. In the figure below, laws of physics tell us that $\angle 1 = \angle 2$. What is the height of the tree if her eyes are 5 ft from the ground, her feet are 8 ft from the mirror, and the mirror is 64 ft from the base of the tree?

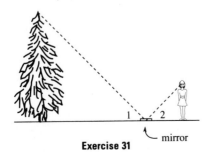

mirror

Exercise 31

1. Write the ratio 400 mi in 10 hr as a fraction and simplify.

2. Solve the proportion. $\dfrac{a}{a-2} = \dfrac{21}{15}$

3. Find the mean proportional between 4 and 36.

4. If 50 ft of wire weighs 65 lb, how much will 70 ft of the same wire weigh?

5. True or false: If $\dfrac{a}{b} = \dfrac{c}{d}$, then $\dfrac{a}{c} = \dfrac{b}{d}$.

6. If $\square ABCD \sim \square A'B'C'D'$, $AB = 18$ ft, $A'B' = 3$ ft, and $BC = 12$ ft, what is the value of $B'C'$?

7. Can two acute triangles be similar?

Problems 8–9 refer to the figure below in which $\overline{DE} \parallel \overline{AB}$ and $\overline{CF}$ bisects $\angle C$.

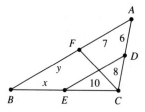

8. Find the value of x.

9. Find the value of y.

10. Is $\triangle ABC \sim \triangle A'B'C'$ if $AB = 30$ yd, $BC = 22$ yd, $CA = 14$ yd, $A'B' = 15$ yd, $B'C' = 11$ yd, and $C'A' = 7$ yd?

11. Is $\triangle ABC \sim \triangle A'B'C'$ if $\angle B = \angle B'$, $BC = 40$ ft, $B'C' = 30$ ft, $BA = 8$ ft, and $B'A' = 5$ ft?

12. How tall is a tower if it casts a shadow 65 ft long at the same time that a building 160 ft tall casts a shadow 130 ft long?

13. In the figure below, $BC = 7$ ft, $AC = 8$ ft, $CD = 21$ ft, and $CE = 24$ ft. Explain why $\overline{AB} \parallel \overline{DE}$.

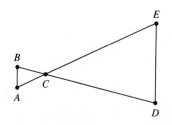

Exercise 13

6

RIGHT TRIANGLES AND THE PYTHAGOREAN THEOREM

In this chapter we study some of the properties of right triangles. Many of the concepts we'll encounter require the use of radicals and often lead to solving quadratic equations, so we begin with a review. Next we investigate several congruence theorems for right triangles and consider properties of altitudes and medians. We conclude with the Pythagorean Theorem and properties of two special right triangles. An application of one of these triangles is presented below and solved in Example 4 of Section 6.3.

A tightrope performer in a circus begins his act by walking up a wire to a platform that is 120 ft high. If the wire makes an angle of 30° with the horizontal, how far does he walk along the wire to reach the platform?

6.1 REVIEW OF RADICALS AND QUADRATIC EQUATIONS (OPTIONAL)

The concepts in this chapter often involve the use of radicals and quadratic equations. This optional review section may be omitted by students with an appropriate background in algebra.

We'll begin by defining square root. If a and x are real numbers such that $x = a^2$ then a is a **square root** of x. For example, because $9 = 3^2$, 3 is a square root of 9, and because $9 = (-3)^2$, -3 is also a square root of 9. Every positive real number x has two square roots, one positive and one negative. The positive square root is called the **principal square root** and is indicated by using a **radical sign** $\sqrt{}$. Thus, the principal square root of 9 is $\sqrt{9}$, which simplifies to 3. We indicate the negative square root of 9 by $-\sqrt{9}$, which simplifies to -3. To represent both square roots of 9 in one expression we write $\pm\sqrt{9}$ or ± 3.

EXAMPLE 1

(a) The principal square root of 25 is $\sqrt{25}$ which simplifies to 5 because $5^2 = 25$.

(b) The negative square root of 25 is $-\sqrt{25}$ which simplifies to -5 because $(-5)^2 = 25$.

(c) The principal square root of -16 is not a real number since no real number squared can be negative, in particular, -16.

(d) The principal square root of 7 is $\sqrt{7}$.   ◼

In Example 1(d), we found the principal square root of 7 to be $\sqrt{7}$, and unlike $\sqrt{25}$, we did not simplify it further. Many square roots, such as $\sqrt{2}$, $\sqrt{3}$, $\sqrt{10}$, and $\sqrt{15}$, cannot be simplified further. In some practical applications it is appropriate to give a decimal approximation for these roots using a calculator with a $\boxed{\sqrt{}}$ key. For example, to approximate $\sqrt{7}$ using a calculator, enter 7 and press the $\boxed{\sqrt{}}$ key. The display will show $\boxed{2.6457513}$. Thus, we might use 2.65 as an approximation of $\sqrt{7}$, correct to the nearest hundredth.

PRACTICE EXERCISE 1

Use a calculator to give an approximation of each square root, correct to the nearest hundredth.
(a) $\sqrt{5}$ (b) $\sqrt{21}$ (c) $\sqrt{102}$ (d) $\sqrt{7735}$

We know $\sqrt{9} = 3$, but how can we determine $\sqrt{18}$? We can use a calculator to approximate $\sqrt{18}$, but instead $\sqrt{18}$ can be simplified using the fact that the square root of a product is the product of the square roots,

$$\sqrt{ab} = \sqrt{a}\,\sqrt{b} \text{ (if } a \geq 0 \text{ and } b \geq 0).$$

Thus,

$$\sqrt{18} = \sqrt{9 \cdot 2} = \sqrt{9}\,\sqrt{2} = 3\sqrt{2}.$$

Early mathematicians commonly used the symbol ℞ to indicate roots. It is a contraction of the Latin word *radix* meaning root. The symbol today is commonly used for medical prescriptions. The symbol √ for square root first appeared in the early 1500s and may have been used because it resembles a lower case r. The radical sign we use today √ was introduced by René Descartes in the early 1600s.

EXAMPLE 2   Simplify each radical.

(a) $\sqrt{27} = \sqrt{9 \cdot 3} = \sqrt{9}\,\sqrt{3} = 3\sqrt{3}$ 9 is a perfect square

(b) $\sqrt{52} = \sqrt{4 \cdot 13} = \sqrt{4}\,\sqrt{13} = 2\sqrt{13}$ 4 is a perfect square

(c) $\sqrt{15} = \sqrt{3 \cdot 5}$ cannot be simplified further because neither 3 nor 5 is a perfect square. ◪

The method of simplifying a radical shown in Example 2 involves factoring the original number and looking for perfect square factors. Radical expressions involving fractions or radicals in a denominator are often simplified using the fact that

$$\sqrt{\frac{a}{b}} = \frac{\sqrt{a}}{\sqrt{b}} \ (\text{if } a \geq 0 \text{ and } b > 0).$$

EXAMPLE 3   Simplify each radical expression.

(a) $\sqrt{\dfrac{1}{3}} = \sqrt{\dfrac{1 \cdot 3}{3 \cdot 3}}$ Multiply numerator and denominator by 3 to obtain a perfect square in the denominator

$ = \sqrt{\dfrac{3}{9}}$

$ = \dfrac{\sqrt{3}}{\sqrt{9}}$

$ = \dfrac{\sqrt{3}}{3}$

(b) $\dfrac{1}{\sqrt{2}} = \dfrac{1 \cdot \sqrt{2}}{\sqrt{2} \cdot \sqrt{2}}$ Multiply numerator and denominator by $\sqrt{2}$

$ = \dfrac{\sqrt{2}}{\sqrt{2 \cdot 2}}$ Use $\sqrt{a}\,\sqrt{b} = \sqrt{ab}$

$ = \dfrac{\sqrt{2}}{\sqrt{4}}$

$ = \dfrac{\sqrt{2}}{2}$ ◪

NOTE: In the past, radicals such as $\sqrt{\dfrac{1}{3}}$ and $\dfrac{1}{\sqrt{2}}$ were simplified to $\dfrac{\sqrt{3}}{3}$ and $\dfrac{\sqrt{2}}{2}$, respectively, because it was easier to compute approximations for these numbers from simplified forms. Now that we have calculators, this simplification isn't necessary. However, it is easier to compare answers when expressions are simplified in this way because it is not obvious at a glance that $\frac{1}{\sqrt{2}}$ and $\frac{\sqrt{2}}{2}$ represent the same number.   □

An equation of the form $ax^2 + bx + c = 0$ where a, b, and c are real numbers, $a \neq 0$, is a **quadratic equation.** When $b = 0$, we can solve the quadratic equation by solving for x^2 and taking the square root of both sides as shown in the next example.

E X A M P L E   4   Solve each quadratic equation.

(a) $x^2 - 16 = 0$

$$x^2 = 16 \qquad \text{Add 16 to both sides}$$
$$x = \pm\sqrt{16} \qquad \text{Take square root of both sides and use } \pm \text{ to indicate both roots.}$$
$$x = \pm 4 \qquad \sqrt{16} = 4$$

Thus, the solutions are 4 and -4.

(b) $2x^2 - 48 = 0$

$$2x^2 = 48 \qquad \text{Add 48 to both sides}$$
$$x^2 = 24 \qquad \text{Divide both sides by 2}$$
$$x = \pm\sqrt{24} \qquad \text{Take square root of both sides}$$
$$x = \pm\sqrt{4 \cdot 6} \qquad \text{Factor 24 and identify perfect square 4}$$
$$x = \pm\sqrt{4}\sqrt{6}$$
$$x = \pm 2\sqrt{6}$$

Thus, the solutions are $2\sqrt{6}$ and $-2\sqrt{6}$.   ◪

PRACTICE EXERCISE 2

Find the mean proportional between 18 and 45.

NOTE: In Chapter 5 when we found the mean proportional between two numbers their product was always a perfect square so that the answer could be determined by inspection. Practice Exercise 2 takes the challenge a step further.   □

Some quadratic equations can be solved by first factoring and then using the next theorem.

THEOREM 6.1   ZERO-PRODUCT RULE

If a and b are real numbers such that $ab = 0$, then $a = 0$ or $b = 0$.

PROOF: We must show that $a = 0$ or $b = 0$. If $a = 0$, then we are finished. So assume $a \neq 0$, then we must show that $b = 0$. If $a \neq 0$, then $\frac{1}{a}$ exists, and if we multiply both sides of $ab = 0$ by $\frac{1}{a}$ we have

$$\frac{1}{a}(ab) = \frac{1}{a}(0).$$
$$\left(\frac{1}{a}a\right)b = 0$$
$$1 \cdot b = 0$$
$$b = 0$$

Thus, we have $a = 0$ or $b = 0$.   ■

The next example shows how Theorem 6.1 is used to solve a quadratic equation.

E X A M P L E 5   Solve $2x^2 - 5x - 3 = 0$.

$$(2x + 1)(x - 3) = 0 \qquad \text{Factor left side}$$
$$2x + 1 = 0 \text{ or } x - 3 = 0 \qquad \text{Use Theorem 6.1}$$
$$2x = -1 \qquad\qquad x = 3$$
$$x = -\frac{1}{2}$$

Thus, the solutions are $-\frac{1}{2}$ and 3.   ◪

PRACTICE EXERCISE 3

Solve $5x^2 + 6x - 8 = 0$.

Many quadratic equations that cannot be solved by factoring and using the zero-product rule are solved using the **quadratic formula.**

THEOREM 6.2   QUADRATIC FORMULA

The solutions to $ax^2 + bx + c = 0$, $a \neq 0$, are given by the formula
$$x = \frac{-b \pm \sqrt{b^2 - 4ac}}{2a}.$$

The proof of Theorem 6.2 is requested in the exercises.

EXAMPLE 6   Solve $2x^2 - 4x + 1 = 0$.

First we identify $a = 2$, $b = -4$, and $c = 1$, and then substitute these values into the quadratic formula.

$$x = \frac{-b \pm \sqrt{b^2 - 4ac}}{2a}$$

$$= \frac{-(-4) \pm \sqrt{(-4)^2 - 4(2)(1)}}{2(2)}$$

$$= \frac{4 \pm \sqrt{16 - 8}}{4}$$

$$= \frac{4 \pm \sqrt{8}}{4}$$

$$= \frac{4 \pm 2\sqrt{2}}{4}$$

$$= \frac{2(2 \pm \sqrt{2})}{4}$$

$$= \frac{2 \pm \sqrt{2}}{2}$$

Thus, the solutions are $\dfrac{2 + \sqrt{2}}{2}$ and $\dfrac{2 - \sqrt{2}}{2}$. ◪

ANSWERS TO PRACTICE EXERCISES: **1. (a)** 2.24 **(b)** 4.58 **(c)** 10.10 **(d)** 87.95 **2.** $9\sqrt{10}$ or $-9\sqrt{10}$ **3.** $\dfrac{4}{5}$, -2

6.1 EXERCISES

In Exercises 1–8, simplify each radical without using a calculator.

1. $\sqrt{4}$ **2.** $-\sqrt{4}$ **3.** $-\sqrt{81}$ **4.** $\sqrt{81}$

5. $\sqrt{121}$ **6.** $-\sqrt{169}$ **7.** $-\sqrt{225}$ **8.** $\sqrt{400}$

In Exercises 9–12, use a calculator to approximate each radical, correct to the nearest hundredth.

9. $\sqrt{39}$ **10.** $\sqrt{87}$ **11.** $\sqrt{431}$ **12.** $\sqrt{6219}$

Simplify each radical in Exercises 13–20 by removing perfect square factors. Do not use a calculator.

13. $\sqrt{12}$ **14.** $\sqrt{20}$ **15.** $\sqrt{45}$ **16.** $\sqrt{63}$

17. $\sqrt{50}$ **18.** $\sqrt{72}$ **19.** $\sqrt{300}$ **20.** $\sqrt{288}$

Simplify each radical expression in Exercises 21–28 by writing the expression with no fractions under the radical or with no radical in the denominator.

21. $\sqrt{\dfrac{1}{5}}$ 　　　 **22.** $\sqrt{\dfrac{2}{3}}$ 　　　 **23.** $\dfrac{1}{\sqrt{5}}$ 　　　 **24.** $\dfrac{2}{\sqrt{3}}$

25. $3\sqrt{\dfrac{2}{5}}$ 　　 **26.** $2\sqrt{\dfrac{3}{7}}$ 　　 **27.** $\dfrac{2}{\sqrt{10}}$ 　　 **28.** $\dfrac{5}{\sqrt{15}}$

Solve each quadratic equation in Exercises 29–40.

29. $x^2 - 81 = 0$ 　　　　　　　　**30.** $x^2 - 121 = 0$

31. $x^2 - 8 = 0$ 　　　　　　　　**32.** $2x^2 - 24 = 0$

33. $x^2 + x - 6 = 0$ 　　　　　　**34.** $x^2 - 3x - 4 = 0$

35. $2x^2 + 5x - 3 = 0$ 　　　　　**36.** $3x^2 - 5x - 2 = 0$

37. $x^2 + 2x - 1 = 0$ 　　　　　　**38.** $x^2 + 3x - 2 = 0$

39. $3x^2 - x - 3 = 0$ 　　　　　　**40.** $5x^2 + 2x - 1 = 0$

Solve each proportion in Exercises 41–46.

41. $\dfrac{x}{3} = \dfrac{27}{x}$ 　　　　　　　　**42.** $\dfrac{2}{a} = \dfrac{a}{32}$

43. $\dfrac{4}{y} = \dfrac{y}{11}$ 　　　　　　　　**44.** $\dfrac{x}{5} = \dfrac{10}{x}$

45. $\dfrac{a + 1}{4} = \dfrac{5}{a}$ 　　　　　　**46.** $\dfrac{y - 2}{3} = \dfrac{2}{y + 3}$

47. Find the mean proportional between 30 and 21.

48. Find the mean proportional between 10 and 30.

49. Give reasons in the proof of the theorem: If $x^2 = a^2$, then $x = \pm a$.

　　Proof:

STATEMENTS	REASONS
1. $x^2 = a^2$	1. _____
2. $x^2 - a^2 = 0$	2. _____
3. $(x - a)(x + a) = 0$	3. _____
4. $x - a = 0$ or $x + a = 0$	4. _____
5. $x = a$ or $x = -a$	5. _____
6. $x = \pm a$	6. _____

$$\dfrac{-b \pm \sqrt{b^2 - 4ac}}{2a}$$

50. Give reasons in the proof of Theorem 6.2.

Proof:

STATEMENTS	REASONS
1. $ax^2 + bx + c = 0, a \neq 0$	1. _____
2. $x^2 + \dfrac{b}{a}x + \dfrac{c}{a} = 0$	2. _____
3. $x^2 + \dfrac{b}{a}x = -\dfrac{c}{a}$	3. _____
4. $x^2 + \dfrac{b}{a}x + \dfrac{b^2}{4a^2} = -\dfrac{c}{a} + \dfrac{b^2}{4a^2}$	4. _____
5. $\left(x + \dfrac{b}{2a}\right)^2 = \dfrac{b^2 - 4ac}{4a^2}$	5. _____
6. $x + \dfrac{b}{2a} = \pm\sqrt{\dfrac{b^2 - 4ac}{4a^2}}$	6. _____
7. $x + \dfrac{b}{2a} = \dfrac{\pm\sqrt{b^2 - 4ac}}{2a}$	7. _____
8. $x = \dfrac{-b \pm\sqrt{b^2 - 4ac}}{2a}$	8. _____

Another formula for finding the area of a triangle when the lengths of the three sides a, b, and c, are known was developed by Heron of Alexandria (about 125 B.C.), a Greek mathematician also known as Hero. If $s = \frac{1}{2}(a + b + c)$, s is one-half the perimeter of the triangle. Hero's Formula gives the area of the triangle as

$$A = \sqrt{s(s - a)(s - b)(s - c)}.$$

Use this formula in Exercises 51–52 to find the area of each triangle with the given sides, correct to the nearest tenth of a unit.

51. $a = 8$ cm, $b = 11$ cm, and $c = 15$ cm   **52.** $a = 4.3$ ft, $b = 3.7$ ft, and $c = 5.2$ ft

---

6.2 PROPERTIES OF RIGHT TRIANGLES

Recall that a right triangle is a triangle that contains a right angle. The hypotenuse of a right triangle is the side opposite the right angle, and the remaining two sides are called legs. In this section we prove congruence and similarity theorems that apply to right triangles. We'll present three theorems using the terminology of right triangles with A representing *acute angle*, L representing *leg,* and *H* representing *hypotenuse.* The first theorem is a special case of Postulate 3.2, ASA = ASA, or Corollary 4.14, AAS = AAS, for arbitrary triangles.

THEOREM 6.3   LA = LA

If a leg and acute angle of one right triangle are equal, respectively, to a leg and the corresponding acute angle of another right triangle, then the two right triangles are congruent.

The next theorem is a special case of Corollary 4.14, AAS = AAS.

THEOREM 6.4   HA = HA

If the hypotenuse and an acute angle of one right triangle are equal, respectively, to the hypotenuse and an acute angle of another triangle, then the two right triangles are congruent.

The next theorem is a special case of Postulate 3.1, SAS = SAS.

THEOREM 6.5   LL = LL

If the two legs of one right triangle are equal, respectively, to the two legs of another right triangle, then the two right triangles are congruent.

You may have noticed that we had no theorem of the form SSA = SSA for arbitrary triangles. For example, the triangles in Figure 6.1 have this property but they are clearly not congruent. The next theorem, however, provides the counterpart for right triangles giving us a totally new congruence theorem.

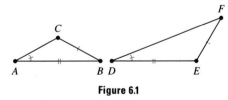

Figure 6.1

THEOREM 6.6   HL = HL

If the hypotenuse and a leg of one right triangle are equal, respectively, to the hypotenuse and a leg of another right triangle, then the two right triangles are congruent.

Given: Right triangles $\triangle ABC$ and $\triangle DEF$ with $\angle C$ and $\angle F$ right angles, $AB = DE$, and $AC = DF$ (See Figure 6.2.)

Prove: $\triangle ABC \cong \triangle DEF$

Auxiliary lines: Extend $\overleftrightarrow{FE}$, construct $\overline{FG}$ such that $FG = BC$, and construct $\overline{DG}$

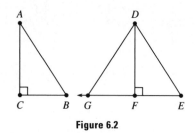

Figure 6.2

Proof:

STATEMENTS	REASONS
1. $\triangle ABC$ and $\triangle DEF$ are right triangles with $\angle C$ and $\angle F$ right angles	1. Given
2. $AB = DE$ and $AC = DF$	2. Given
3. $\angle DFG$ is a right angle making $\triangle DFG$ a right triangle	3. Supp. of rt $\angle$ is also a rt $\angle$
4. $BC = FG$	4. By construction
5. $\triangle ABC \cong \triangle DFG$	5. LL = LL
6. $AB = DG$	6. cpoctae
7. $DE = DG$	7. Sym. and trans. laws
8. $\angle DGF = \angle DEF$	8. $\angle$'s opp = sides of $\triangle$ are =
9. $\triangle DFG \cong \triangle DEF$	9. HA = HA
10. $\triangle ABC \cong \triangle DEF$	10. Transitive law for $\cong$

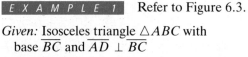 **Refer to Figure 6.3.**

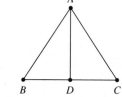

Given: Isosceles triangle $\triangle ABC$ with base $\overline{BC}$ and $\overline{AD} \perp \overline{BC}$

Prove: $BD = DC$

Proof:

Figure 6.3

STATEMENTS	REASONS
1. $\triangle ABC$ is isosceles with base $\overline{BC}$	1. Given
2. $AB = AC$	2. Def. of isos. $\triangle$
3. $\overline{AD} \perp \overline{BC}$	3. Given
4. $\angle ADB$ and $\angle ADC$ are right angles	4. $\perp$ lines form rt $\angle$'s
5. $\triangle ADB$ and $\triangle ADC$ are right triangles	5. Def. of rt. $\triangle$
6. $AD = AD$	6. Reflexive law
7. $\triangle ADB \cong \triangle ADC$	7. HL = HL
8. $BD = DC$	8. cpoctae ◪

The next theorem provides a useful property of the median from the right angle in a right triangle.

THEOREM 6.7

The median from the right angle in a right triangle is one-half the length of the hypotenuse.

PRACTICE EXERCISE 1

Complete the proof of Theorem 6.7.

Given: $\triangle ABC$ is a right triangle with median $\overline{CD}$

Prove: $CD = \dfrac{1}{2}BA$

Auxiliary line: Construct ℓ through D parallel to AC

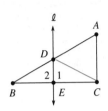

Proof:

STATEMENTS	REASONS
1. $\triangle ABC$ is a right triangle with median $\overline{CD}$	1. _____
2. _____	2. Def. of median
3. $\dfrac{AD}{DB} = \dfrac{CE}{EB}$	3. _____
4. $CE = EB$	4. _____
5. $\overline{DE} \perp \overline{BC}$	5. _____
6. _____	6. $\perp$ lines form right $\angle$'s
7. $DE = DE$	7. _____
8. _____	8. LL = LL
9. $BD = CD$	9. _____
10. $BD + DA = BA$	10. _____
11. $BD + BD = BA$	11. _____
12. $2BD = BA$	12. _____
13. _____	13. Substitution law
14. $CD = \dfrac{1}{2}BA$	14. _____

A surveyor uses an instrument called a transit to measure angles. Angle measurements are important in the construction of highways, bridges, tunnels, and shopping malls, where right triangles are used extensively.

The altitude from the right angle in a right triangle also has an important property of similarity.

THEOREM 6.8

The altitude from the right angle to the hypotenuse in a right triangle forms two right triangles that are similar to each other and to the original triangle.

Given: Right triangle $\triangle ABC$ with altitude $\overline{CD}$ from right angle $\angle C$ (See Figure 6.4.)

Prove: $\triangle ACD \sim \triangle BCD$, $\triangle ACD \sim \triangle ABC$, and $\triangle BCD \sim \triangle ABC$

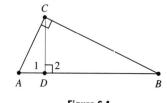

Figure 6.4

Proof:

STATEMENTS	REASONS
1. $\overline{CD}$ is an altitude of right triangle $\triangle ABC$ drawn from right $\angle C$	1. Given
2. $\overline{CD} \perp \overline{AB}$	2. Def. of altitude
3. $\angle 1$ and $\angle 2$ are right angles	3. $\perp$ lines form rt $\angle$'s
4. $\angle 1 = \angle ACB$	4. Rt. $\angle$'s are $=$
5. $\angle A = \angle A$	5. Reflexive law
6. $\triangle ACD \sim \triangle ABC$	6. AA $\sim$ AA
7. $\angle 2 = \angle ACB$	7. Rt. $\angle$'s are $=$
8. $\angle B = \angle B$	8. Reflexive law
9. $\triangle BCD \sim \triangle ABC$	9. AA $\sim$ AA
10. $\triangle ACD \sim \triangle BCD$	10. Trans. law for $\sim$

Theorem 6.8 has two important corollaries that often lead to working with quadratic equations.

COROLLARY 6.9

The altitude from the right angle to the hypotenuse in a right triangle is the mean proportional between the segments of the hypotenuse.

Given: Right triangle $\triangle ABC$ with altitude $\overline{CD}$ from right angle $\angle C$ (See Figure 6.5.)

Prove: $\dfrac{AD}{CD} = \dfrac{CD}{DB}$

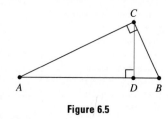

Figure 6.5

Proof:

STATEMENTS	REASONS
1. $\triangle ABC$ is a right triangle with altitude $\overline{CD}$ from right angle $\angle C$	1. Given

2. $\triangle ACD \sim \triangle BCD$ 2. Theorem 6.8

3. $\dfrac{AD}{CD} = \dfrac{CD}{DB}$ 3. Corr. sides of $\sim \triangle$'s are
proportional

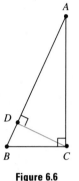

Figure 6.6

$\boxed{E\ X\ A\ M\ P\ L\ E\ \ 2}$ Find the length of $\overline{CD}$ in Figure 6.6 if $DB = 5$ cm and $AD = 16$ cm. Also give the approximate length of $\overline{CD}$ to the nearest hundredth of a centimeter.

By Corollary 6.9 we have

$$\frac{AD}{CD} = \frac{CD}{DB}.$$

Let $x = CD$, substitute 16 for AD and 5 for DB, and solve for x.

$$\frac{16}{x} = \frac{x}{5}$$

$x^2 = 16 \cdot 5$ Means-extremes property

$x = \pm\sqrt{16 \cdot 5}$ Take square root of both sides

$x = \pm 4\sqrt{5}$ Simplify

Because CD is a distance and cannot be negative, we discard $-4\sqrt{5}$. Thus, $CD = 4\sqrt{5}$ cm. Using a calculator to approximate $\sqrt{5}$, we find that CD is about 8.94 cm. ◪

COROLLARY 6.10

If the altitude is drawn from the right angle to the hypotenuse in a right triangle, then each leg is the mean proportional between the hypotenuse and the segment of the hypotenuse adjacent to the leg.

Given: Refer to Figure 6.6 showing
right triangle $\triangle ABC$ with altitude
$\overline{CD}$ from the right angle

Prove: $\dfrac{AB}{BC} = \dfrac{BC}{BD}$ and $\dfrac{AB}{AC} = \dfrac{AC}{AD}$

Proof:

STATEMENTS	REASONS
1. $\triangle ABC$ is a right triangle with altitude $\overline{CD}$ from right angle $\angle C$	1. Given
2. $\triangle ABC \sim \triangle BCD$	2. Theorem 6.8
3. $\dfrac{AB}{BC} = \dfrac{BC}{BD}$	3. Corr. sides of $\sim \triangle$'s are proportional
4. $\triangle ABC \sim \triangle ACD$	4. Theorem 6.8
5. $\dfrac{AB}{AC} = \dfrac{AC}{AD}$	5. Corr. sides of $\sim \triangle$'s are proportional

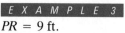 Find the length of $\overline{PS}$ in Figure 6.7 if $QS = 12$ ft and $PR = 9$ ft.

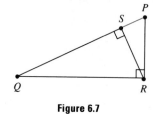

Figure 6.7

By Corollary 6.10 we have

$$\frac{PQ}{PR} = \frac{PR}{PS}.$$

Because $PQ = QS + PS$, $PQ = 12 + PS$. Let $x = PS$, and substitute 9 for PR and $12 + x$ for PQ.

$$\frac{12 + x}{9} = \frac{9}{x}$$

$$x(12 + x) = 81 \qquad \text{Means-extremes property}$$

$$12x + x^2 = 81 \qquad \text{Distributive law}$$

$$x^2 + 12x - 81 = 0$$

Using the quadratic formula, we determine $x = -6 \pm 3\sqrt{13}$. Because the length of a segment cannot be negative, we discard $-6 - 3\sqrt{13}$. Thus, $PS = (-6 + 3\sqrt{13})$ ft. In an applied problem, we would approximate this value. Using a calculator, we find $PS = 4.82$ ft, correct to the nearest hundredth of a foot. ▧

ANSWER TO PRACTICE EXERCISE: **1.** 1. Given 2. $AD = DB$ 3. A line ∥ to one side and intersecting two sides of a △ divides the sides into proportional segments 4. From Statements 2 and 3 5. A line ⊥ to one of two ∥ lines is ⊥ to other 6. ∠1 and ∠2 are right angles 7. Reflexive law 8. $\triangle DBE \cong \triangle DEC$ 9. cpoctae 10. Seg. add. post. 11. Substitution law 12. Distributive law 13. $2CD = BA$ 14. Mult.-div. law

6.2 *EXERCISES*

Exercises 1–6 refer to $\triangle ABC$ and $\triangle DEF$ given below. State why $\triangle ABC \cong \triangle DEF$ under the given conditions.

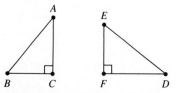

1. BC = 5 cm, EF = 5 cm, $\angle A$ = 35°, $\angle D$ = 35°
2. AB = 15 ft, ED = 15 ft, $\angle B$ = 40°, $\angle E$ = 40°
3. BC = 7 yd, EF = 7 yd, AC = 10 yd, DF = 10 yd
4. AB = 12 cm, BC = 5 cm, ED = 12 cm, EF = 5 cm
5. ED = 4.3 yd, FE = 3.1 yd, AB = 4.3 yd, BC = 3.1 yd
6. AC = 9.2 ft, $\angle A$ = 32°, FD = 9.2 ft, $\angle D$ = 32°

Exercises 7–22 refer to the figure below in which $\triangle ABC$, $\triangle ACD$, and $\triangle BCD$ are all right triangles, and E is the midpoint of $\overline{AB}$. When appropriate, give an approximate answer correct to the nearest hundredth of a unit.

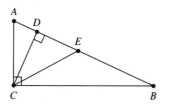

7. If AB = 20 cm, find CE.
9. If AD = 4 cm and BD = 9 cm, find CD.
11. If CE = 5 ft, find AB.
13. If AD = 3 cm and CD = 9 cm, find BD.
15. If AB = 32 yd and AD = 2 yd, find AC.
17. If AB = 10 ft and AD = 3 ft, find AC.
19. If BD = 8 cm and AC = 7 cm, find AD.
21. If CE = 5 yd and BD = 8 yd, find BC.

8. If AB = 68 yd, find CE.
10. If AD = 18 cm and BD = 50 cm, find CD.
12. If CE = 13 yd, find AB.
14. If BD = 18 ft and CD = 6 ft, find AD.
16. If AB = 50 cm and BD = 32 cm, find BC.
18. If AB = 14 yd and BD = 7 yd, find BC.
20. If AD = 2 ft and BC = 11 ft, find BD.
22. If CE = 7 cm and AD = 2 cm, find AC.

Exercises 23–24 refer to the figure below.

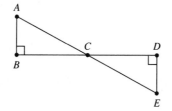

23. *Given: C is the midpoint of $\overline{AE}$ and of $\overline{BD}$, $\overline{AB} \perp \overline{BD}$, and $\overline{DE} \perp \overline{BD}$*
    Prove: $\triangle ABC \cong \triangle CDE$

24. *Given: $AB = DE$, $\overline{AB} \perp \overline{BD}$, and $\overline{DE} \perp \overline{BD}$*
    Prove: $AC = CE$

Exercises 25–26 refer to the figure below.

25. *Given:* $\overline{BD} \perp \overline{AC}, \overline{CE} \perp \overline{AB}$, and $CD = BE$
      Prove: $\triangle ABC$ is isosceles

26. *Given:* $\overline{BD} \perp \overline{AC}, \overline{CE} \perp \overline{AB}$, and
         $BE = CE$
       Prove: $AE = AD$

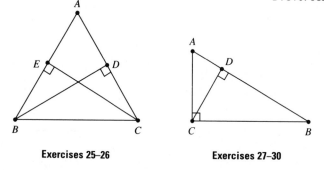

Exercises 25–26   Exercises 27–30

Exercises 27–30 refer to the figure above in which $\overline{CD}$ is an altitude of right triangle $\triangle ABC$.

27. If $AD = 4$ cm and $BD = 16$ cm, find the area of $\triangle ABC$.

28. If $AD = 9$ ft and $BD = 25$ ft, find the area of $\triangle ABC$.

29. If $AD = 4$ cm and $BD = 16$ cm, find the area of $\triangle ADC$.

30. If $AD = 9$ ft and $BD = 25$ ft, find the area of $\triangle ADC$.

31. A right triangle has hypotenuse 13 cm and one leg 12 cm. Find the area of the triangle.

32. A right triangle has hypotenuse 15 ft and one leg 9 ft. Find the area of the triangle.

6.3 THE PYTHAGOREAN THEOREM

The Pythagorean Theorem is perhaps the most famous and most useful of all theorems in geometry. It has numerous applications in algebra, trigonometry, and calculus as well as many practical applications in everyday life. There is evidence that a special case of the theorem was known to the Egyptians as long ago as 2000 B.C. And it is believed that Pythagoras, in about 525 B.C., gave the first deductive proof of the theorem, probably the one presented by Euclid in his book *Elements*. More than 250 different proofs have since been given, including one by President James A. Garfield in 1876, most of which involve finding areas of various figures. The proof we present now uses Corollary 6.10 and is perhaps the simplest of all the proofs of this theorem.

THEOREM 6.11   THE PYTHAGOREAN THEOREM

In a right triangle, the square of the length of the hypotenuse is equal to the sum of the squares of the lengths of the legs.

Given: Right triangle $\triangle ABC$ with
$c = AB$, $b = AC$, and $a = BC$
(See Figure 6.8.)

Prove: $a^2 + b^2 = c^2$

Auxiliary line: Construct altitude $\overline{CD}$ from vertex C to hypotenuse $\overline{AB}$

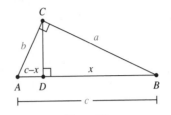

Figure 6.8

Proof:

STATEMENTS	REASONS
1. $\triangle ABC$ is a right triangle	1. Given
2. $\overline{CD}$ is the altitude from vertex C to hypotenuse $\overline{AB}$	2. By construction
3. $\dfrac{c}{a} = \dfrac{a}{x}$ and $\dfrac{c}{b} = \dfrac{b}{c-x}$	3. Corollary 6.10
4. $a^2 = cx$ and $b^2 = c(c - x)$	4. Means-extremes prop.
5. $a^2 + b^2 = cx + c(c - x)$	5. Add.-subtr. prop.
6. $a^2 + b^2 = c^2$	6. Distributive law and simplify result

NOTE: Often we abbreviate the statement of the Pythagorean Theorem to say "In a right triangle, the sum of the squares of the legs equals the square of the hypotenuse." Also, for convenience, we usually designate the right angle in the right triangle C so we can use $a^2 + b^2 = c^2$ for the Pythagorean Theorem, with a the length of the side opposite $\angle A$, b the length of the side opposite $\angle B$, and c the length of the hypotenuse opposite $\angle C$.   □

The Pooles own a compact station wagon that has a rectangle-shaped rear door opening 28 inches high and 44 inches wide. If they purchase a square piece of wood paneling that is 48 inches on a side, will they be able to transport the paneling in their wagon?

E X A M P L E   1   In right triangle $\triangle ABC$ (C is the right angle), $a = 12$ cm and $b = 7$ cm. Find the length of the hypotenuse, c.

Substituting into the Pythagorean Theorem we have

$$a^2 + b^2 = c^2$$
$$12^2 + 7^2 = c^2$$
$$144 + 49 = c^2$$
$$193 = c^2$$
$$\sqrt{193} = c$$

We use only the principal square root in this case because the length of a hypotenuse must be positive. We can also approximate $\sqrt{193}$ using a calculator to find $c = 13.9$ cm, correct to the nearest tenth of a centimeter.   ◪

PRACTICE EXERCISE 1

In right triangle $\triangle ABC$, $c = 32$ ft and $a = 18$ ft, find b.

Many applied problems can be solved using the Pythagorean Theorem.

E X A M P L E 2 A 100-foot tower is to be supported by four guy wires attached to the top of the tower and to points on the ground that are 35 ft from the base of the tower. Assume that each wire will require an extra 2 ft for attaching to the tower and to the points on the ground. How much wire will be needed for this project?

We can make a sketch showing one of these wires and the tower as shown in Figure 6.9. The wire forms the hypotenuse of a right triangle with legs measuring 35 ft and 100 ft. Let x be the length of the wire as shown. Then by the Pythagorean Theorem,

$$x^2 = 35^2 + 100^2$$
$$= 1225 + 10{,}000$$
$$= 11{,}255.$$
$$x = \sqrt{11{,}225} \approx 105.95 \text{ ft}$$

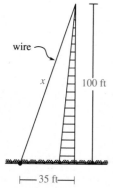

wire

x 100 ft

35 ft

Figure 6.9

Add the 2 feet for attachment.

$$105.95 + 2 = 107.95 \text{ ft}$$

With four such wires, we would have

$$4(107.95) = 431.8 \text{ ft}.$$

Thus, about 432 ft of wire are required to secure the tower. ◪

NOTE: In Example 2 we used the symbol $\approx$ which stands for the phrase "is approximately equal to." This is often used in applied problems to indicate approximate or rounded values. □

The converse of the Pythagorean Theorem is also true and can be used to prove that a given triangle is a right triangle when its sides are given.

THEOREM 6.12   CONVERSE OF THE PYTHAGOREAN THEOREM

If the sides of a triangle have lengths a, b, and c, and $a^2 + b^2 = c^2$, then the triangle is a right triangle.

Given: $\triangle ABC$ with $a^2 + b^2 = c^2$
  (See Figure 6.10.)

Prove: $\triangle ABC$ is a right triangle

Construction: Construct a right
  triangle $\triangle EFG$ with $e = a$, $f = b$,
  and $\angle G$ a right angle. (If we show
  $\triangle ABC \cong \triangle EFG$, then $\angle C$ is also
  a right angle making $\triangle ABC$ a
  right triangle.)

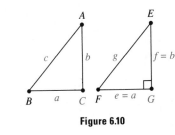

Figure 6.10

Proof:

STATEMENTS	REASONS
1. $\triangle ABC$ with $a^2 + b^2 = c^2$	1. Given
2. $\triangle EFG$ with $e = a$, $f = b$, and $\angle G$ a right angle	2. By construction
3. $e^2 + f^2 = g^2$	3. Pythagorean Theorem
4. $a^2 + b^2 = g^2$	4. Substitution Law
5. $g^2 = c^2$	5. Trans. and Sym. Laws
6. $g = c$	6. The principal square roots of equal numbers are equal
7. $\triangle ABC \cong \triangle EFG$	7. SSS = SSS
8. $\angle C = \angle G$	8. cpoctae
9. $\angle C$ is a right angle	9. From Statement 2 because $\angle G$ is a right angle
10. $\triangle ABC$ is a right triangle	10. Def. of rt. $\triangle$

Pythagoras (About 584–495 B.C.)

Pythagoras was one of the most remarkable mathematicians of all time, giving us many geometry proofs, including one of the theorem bearing his name. He formed a society of mathematicians and philosophers, called the Pythagoreans, that was made up of two groups—listeners and mathematicians. To become a member of the group of mathematicians one had to first prove to be a good listener. The Pythagoreans are credited with discovering the relationship between musical harmony and the length of the strings of a musical instrument, and with developing the concept of irrational numbers.

E X A M P L E   3   A triangle has sides of 10 cm, 24 cm, and 26 cm. Determine if the triangle is a right triangle.

We know by the preceding theorem that a triangle is a right triangle if the sum of the squares of two sides is equal to the square of the other side. Because

$$10^2 = 100, 24^2 = 576, \text{ and } 26^2 = 676$$

and

$$10^2 + 24^2 = 100 + 576 = 676 = 26^2,$$

the triangle is a right triangle.   ▨

PRACTICE EXERCISE 2

Is a triangle with sides measuring 4 ft, 8 ft, and 9 ft a right triangle?

Certain right triangles with acute angles of 45° and 45° (an isosceles right triangle), and of 30° and 60°, play an important role in the study of trigonometry.

These triangles are often referred to as a 45°-45°-right triangle and a 30°-60°-right triangle, respectively. The next two theorems present properties of the sides of these special triangles.

THEOREM 6.13 45°-45°-RIGHT TRIANGLE THEOREM

In a 45°-45°-right triangle, the hypotenuse is $\sqrt{2}$ times as long as each (equal) leg.

A baseball diamond is a square with sides 90 ft in length. Explain how you would use properties of a 45°-45°-right triangle to find the distance from third base to first base.

The proof of Theorem 6.13 is left for you to do as an exercise.

THEOREM 6.14 30°-60°-RIGHT TRIANGLE THEOREM

In a 30°-60°-right triangle, the leg opposite the 30°-angle is one-half as long as the hypotenuse, and the leg opposite the 60°-angle is $\sqrt{3}$ times as long as the leg opposite the 30°-angle and $\dfrac{\sqrt{3}}{2}$ times as long as the hypotenuse.

Given: Right triangle $\triangle ABC$ with
$\angle A = 60°$, $\angle B = 30°$ and
$\angle C = 90°$
(See Figure 6.11.)

Prove: $b = \dfrac{1}{2}c$, $a = \sqrt{3}b$, and $a = \dfrac{\sqrt{3}}{2}c$.

Construction: Construct the
extension of $\overline{AC}$ and $\angle CBD = 30°$
(copy $\angle ABC$) with side
intersecting $\overleftrightarrow{AC}$ at D forming
$\angle BCD = 90°$.

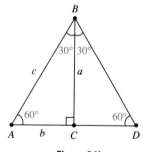

Figure 6.11

Proof:

STATEMENTS	REASONS
1. Right triangle $\triangle ABC$ with $\angle A = 60°$, $\angle B = 30°$, and $\angle C = 90°$	1. Given
2. $\triangle BCD$ with $\angle CBD = 30°$ and $\angle BCD = 90°$	2. By construction
3. $\angle D = 60°$	3. Sum of $\angle$'s of $\triangle = 180°$
4. $AB = BD$	4. Sides opp. $= \angle$'s are $=$
5. $\triangle ABC \cong \triangle BCD$	5. HA = HA
6. $AC = CD$	6. cpoctae
7. $AD = AC + CD$	7. Seg. Add. post.

8. $AD = AC + AC = 2AC$ 8. Substitution and distributive laws

9. $\angle ABC = 30° + 30° = 60°$ 9. Angle Add. Post.

10. $AD = AB$ 10. Sides opp $= \angle$'s are $=$

11. $2AC = AB$ 11. Sym. and trans. laws

12. $AC = \dfrac{1}{2}AB$ 12. Mult.-div. law

13. $b = \dfrac{1}{2}c$ 13. Substitution law

14. $c^2 = a^2 + b^2$ 14. Pythagorean Theorem

15. $c^2 = a^2 + \left(\dfrac{1}{2}c\right)^2$ 15. Substitution law

16. $c^2 = a^2 + \dfrac{1}{4}c^2$ 16. Simplify

17. $a^2 = \dfrac{3}{4}c^2$ 17. Simplify

18. $a = \dfrac{\sqrt{3}}{2}c$ 18. Take principal square root of both sides

19. $a = \dfrac{\sqrt{3}}{2}(2b)$ 19. Substitute $2b$ for c because $b = \dfrac{1}{2}c$

20. $a = \sqrt{3}b$ 20. Simplify

We can use Theorem 6.14 to solve the applied problem from the chapter introduction.

E X A M P L E 4   A tightrope performer in a circus begins his act by walking up a wire to a platform that is 120 ft high. If the wire makes an angle of 30° with the horizontal, how far does he walk along the wire to reach the platform?

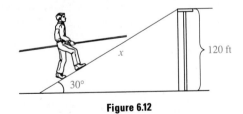

Figure 6.12

Figure 6.12 shows a sketch of the information given. We must find x. By Theorem 6.14, the side opposite the 30°-angle, with length 120 ft, is one-half the hypotenuse, with length x. Thus, we have

$$120 = \frac{1}{2}x$$
$$240 = x.$$

Thus, the tightrope walker walks a distance of 240 ft to reach the platform. ◢

PRACTICE EXERCISE 3

Two airplanes leave the same airport at the same time, one flying due north and the other flying due east. If each is flying at a rate of 450 mph, use Theorem 6.13 to find the distance between the two after 2 hours, correct to the nearest tenth of a mile.

ANSWERS TO PRACTICE EXERCISES: **1.** $b = 10\sqrt{7}$ ft, which is approximately 26.5 ft **2.** No, because $4^2 + 8^2 \neq 9^2$. **3.** 1272.8 mi

---

6.3 EXERCISES

In Exercises 1–8, use the Pythagorean Theorem to find the length of the missing side in right triangle $\triangle ABC$ with right angle C.

1. If $a = 3$ cm and $b = 4$ cm, find c.

2. If $a = 12$ yd and $b = 5$ yd, find c.

3. If $b = 25$ ft and $c = 65$ ft, find a.

4. If $a = 12$ cm and $c = 20$ cm, find b.

5. If $a = 6$ yd and $c = 11$ yd, find b.

6. If $b = 14$ ft and $c = 23$ ft, find a.

7. If $c = 2\sqrt{97}$ cm and $a = 8$ cm, find b.

8. If $c = 2\sqrt{130}$ cm and $b = 22$ cm, find a.

In Exercises 9–14, is the triangle with sides of the given lengths a right triangle?

9. 15 cm, 20 cm, 25 cm

10. 15 ft, 36 ft, 39 ft

11. 3 yd, 7 yd, $\sqrt{58}$ yd

12. $3\sqrt{3}$ cm, 6 cm, 3 cm

13. $\sqrt{7}$ ft, $\sqrt{2}$ ft, 9 ft

14. $\sqrt{11}$ yd, $\sqrt{5}$ yd, 16 yd

Exercises 15–26 refer to the 45°-45°-right triangle shown below.

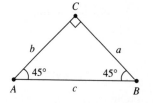

15. If $a = 10$ ft, find c.

16. If $b = 15$ cm, find c.

17. If $a = 3\sqrt{2}$ yd, find b.

18. If $b = 7\sqrt{2}$ ft, find a.

19. If $a = 3\sqrt{2}$ cm, find c.

20. If $b = 7\sqrt{2}$ yd, find c.

21. If $b = 3\sqrt{3}$ ft, find c.

22. If $a = 4\sqrt{5}$ cm, find c.

23. If $c = 6$ yd, find a.

24. If $c = 10$ ft, find b.

25. If $c = \dfrac{\sqrt{2}}{2}$ cm, find b.

26. If $c = \dfrac{\sqrt{3}}{3}$ yd, find a.

Exercises 27–38 refer to the 30°-60°-right triangle shown below.

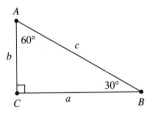

27. If $b = 10$ ft, find c.

28. If $b = 60$ cm, find c.

29. If $c = 16$ yd, find b.

30. If $c = 34$ ft, find b.

31. If $b = 7$ cm, find a.

32. If $b = 13$ yd, find a.

33. If $a = 2\sqrt{3}$ ft, find b.

34. If $a = 7\sqrt{3}$ cm, find b.

35. If $c = \sqrt{3}$ yd, find a.

36. If $c = 8\sqrt{3}$ ft, find a.

37. If $a = \sqrt{3}$ cm, find c.

38. If $a = 2\sqrt{6}$ yd, find c.

39. If the sides of a square are 4 inches long, what is the length of a diagonal?

40. If a rectangle has sides of 14 ft and 5 ft, what is the length of a diagonal?

41. A ladder 18 ft long is placed against the side of a building with the base of the ladder 6 ft from the building. To the nearest tenth of a foot, how far up the building will the ladder reach?

42. A telephone pole 35 ft tall has a guy wire attached to it 5 ft from the top and tied to a ring on the ground 15 ft from the base of the pole. Assume that an extra 2 feet of wire are needed to attach the wire to the ring and the pole. What length of wire is needed for the job? Give an answer to the nearest tenth of a foot.

43. A 400-foot tower has a guy wire attached to it that makes a 60°-angle with level ground. How far from the base of the tower is the wire anchored? Give an answer correct to the nearest tenth of a foot.

44. Two hikers leave their camp at the same time. When Dick is 6.5 mi due east of the camp, Vickie is due north of Dick and northeast of the camp. How far from the camp is Vickie? Give an answer correct to the nearest tenth of a mile.

45. Find the length of an altitude of an equilateral triangle with sides measuring 10 ft.

46. Find the area of an equilateral triangle with sides measuring 10 ft.

47. Prove Theorem 6.13.

48. Prove that the area of an isosceles right triangle is one-fourth the square of the length of the hypotenuse.

49. Find the length d of a diagonal of a cube with sides of length x. See the figure below.

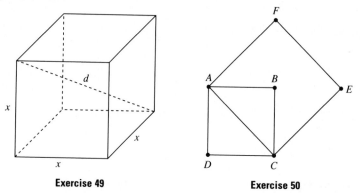

Exercise 49   Exercise 50

50. Prove that the area of square $ABCD$ is half the area of square $ACEF$ in the figure above.

51. Draw a line segment about 1 inch in length. Construct an isosceles right triangle with legs equal in length to this segment. Then the hypotenuse is $\sqrt{2}$ times as long as the given segment. Using this hypotenuse as one leg, construct another right triangle with the second leg equal in length to the original segment. What is the length of the new hypotenuse? Can you continue this process to find a segment with length $\sqrt{4}$ times the length of the original segment? with $\sqrt{5}$ times the length of the original segment?

52. One proof of the Pythagorean Theorem involves expressing algebraically the areas of the two squares given below and equating the results. Notice that each contains copies of a right triangle with legs measuring a and b and hypotenuse measuring c. Note that both squares have sides of length $a + b$ making both areas $(a + b)^2$. Using this information, show that $a^2 + b^2 = c^2$.

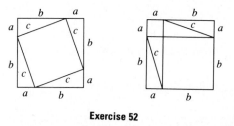

Exercise 52

53. In the left figure for Exercise 52, you probably assumed that the "inside" quadrilateral with sides of length c was a square. Prove that this is indeed the case.

54. Use the figure below to explain why the original statement of the Pythagorean Theorem was: "In a right triangle, the square *on* the hypotenuse is equal to the sum of the squares *on* the legs."

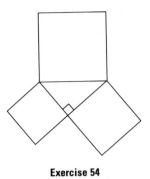

Exercise 54

55. Another proof of the Pythagorean Theorem, the one written by President Garfield, uses the figure given below. Refer to the figure to answer the following:

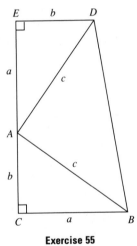

Exercise 55

(a) Show that $\triangle ABD$ is a right triangle. [Hint: Show $\angle DAB$ is a right angle.]
(b) Write three expressions for the areas of the three triangles.
(c) Show that quadrilateral $BCED$ is a trapezoid.
(d) Write an expression for the area of trapezoid $BCED$.
(e) Equate an expression for the area of the trapezoid with an expression formed by adding the areas of the three triangles and simplify the result to obtain $a^2 + b^2 = c^2$.

KEY TERMS AND SYMBOLS

6.1 square root, p. 202
principal square root,
p. 202
radical, p. 202

quadratic equation, p. 204
quadratic formula, p. 205
6.3 Pythagorean Theorem,
p. 217

approximately equal to
(≈), p. 218

PROOF TECHNIQUES

To Prove:

Right Triangles Congruent

1. Show a leg and an acute angle of one triangle equal to a leg and corresponding acute angle of the other triangle. (Theorem 6.3, p. 209)
2. Show the hypotenuse and an acute angle of one triangle equal to the hypotenuse and acute angle of the other triangle. (Theorem 6.4, p. 209)
3. Show that the legs of one triangle are equal to the legs of the other triangle. (Theorem 6.5, p. 209)
4. Show that the hypotenuse and a leg of one triangle are equal to the hypotenuse and a leg of the other triangle. (Theorem 6.6, p. 209)

A Triangle is a Right Triangle

1. Show that it contains a right angle.
2. Show that it satisfies $a^2 + b^2 = c^2$. (Theorem 6.11, p. 217)

REVIEW EXERCISES

Section 6.1

Simplify each radical in Exercises 1–3 without using a calculator.

1. $\sqrt{25}$ **2.** $-\sqrt{25}$ **3.** $\sqrt{64}$

Use a calculator in Exercises 4–6 and approximate each radical, correct to the nearest hundredth.

4. $\sqrt{83}$ **5.** $\sqrt{159}$ **6.** $\sqrt{9542}$

Simplify each radical in Exercises 7–9 by removing perfect square factors. Do not use a calculator.

7. $\sqrt{28}$ **8.** $\sqrt{75}$ **9.** $\sqrt{98}$

Simplify each radical expression in Exercises 10–12 by writing the expression with no fractions under the radical and with no radical in the denominator.

10. $\sqrt{\dfrac{1}{7}}$

11. $\dfrac{1}{\sqrt{6}}$

12. $\dfrac{5}{\sqrt{10}}$

Solve each quadratic equation in Exercises 13–16.

13. $x^2 - 64 = 0$

14. $x^2 + 6x - 7 = 0$

15. $5x^2 - 9x - 2 = 0$

16. $2x^2 + 7x + 4 = 0$

Solve each proportion in Exercises 17–18.

17. $\dfrac{x}{7} = \dfrac{14}{x}$

18. $\dfrac{y + 2}{20} = \dfrac{4}{y}$

19. Find the mean proportional between 50 and 60.

Section 6.2

Exercises 20–22 refer to $\triangle ABC$ and $\triangle PQR$ given below. State why $\triangle ABC \cong \triangle PQR$ under the given conditions.

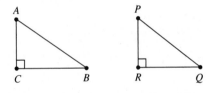

20. $AB = 17$ cm, $PQ = 17$ cm, $\angle A = 38°$, $\angle P = 38°$

21. $AC = 10.5$ yd, $PR = 10.5$ yd, $AB = 17.3$ yd, $PQ = 17.3$ yd

22. $BC = 9.2$ ft, $QR = 9.2$ ft, $AC = 8.7$ ft, $PR = 8.7$ ft

Exercises 23–27 refer to the figure below in which E is the midpoint of $\overline{AB}$, $\overline{CD} \perp \overline{AB}$, and $\angle C$ is a right angle in $\triangle ABC$.

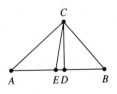

23. If $AB = 44$ cm, find CE.

24. If $AD = 25$ ft and $DB = 16$ ft, find CD.

25. If $AB = 28$ cm and $AD = 16$ cm, find AC.

26. If $BD = 5$ yd and $AC = 6$ yd, find AD.

27. If $AD = 36$ ft and $DB = 25$ ft, find the area of $\triangle ABC$.

Exercises 28–29 refer to the figure below in which $ABCD$ is a rhombus with diagonals $\overline{AC}$ and $\overline{BD}$.

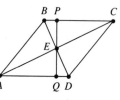

28. *Given:* $\overline{PQ} \perp \overline{BC}$
   Prove: $(PE)^2 = (BP)(PC)$

29. *Given:* $\angle EQA$ is a right angle
   Prove: $(AE)^2 = (AQ)(AD)$

30. A pasture is in the shape of a right triangle with hypotenuse 100 yd and one leg 80 yd. What is the area of the pasture?

Section 6.3

In Exercises 31–32, use the Pythagorean Theorem to find the length of the missing side in right triangle $\triangle ABC$ with right angle C.

31. $a = 11$ cm and $b = 8$ cm, find c.

32. $c = 2\sqrt{170}$ ft and $a = 14$ ft, find b.

33. Is the triangle with sides 6 yd, $\sqrt{11}$ yd, and 7 yd a right triangle?

Exercises 34–42 refer to the figure below.

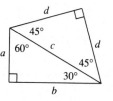

34. If $a = 30$ cm, find c.

35. If $a = 20$ yd, find b.

36. If $c = 8$ ft, find a.

37. If $c = 12$ cm, find b.

38. If $b = 3\sqrt{3}$ yd, find a.

39. If $b = 7$ ft, find c.

40. If $c = \dfrac{\sqrt{2}}{5}$ yd, find d.

41. If $d = 5\sqrt{2}$ cm, find c.

42. If $a = 3$ ft, find d.

43. Find the length of a diagonal of a rectangle with sides 18 ft and 7 ft. Give an answer correct to the nearest tenth of a foot.

44. A mountain road is inclined 30° with the horizontal. If a pickup truck drives 2 mi on this road, what change in altitude has been achieved?

45. Prove that the area of an equilateral triangle with side x is $\dfrac{\sqrt{3}}{4}x^2$.

1. Simplify $\sqrt{49}$ without using a calculator.
2. Approximate $\sqrt{431}$ correct to the nearest hundredth.
3. Simplify $\sqrt{80}$ by removing perfect square factors.
4. Simplify $\sqrt{\dfrac{3}{11}}$ and write without a radical in the denominator.
5. Solve $2x^2 + x - 3 = 0$.
6. Find the mean proportional between 15 and 20.
7. *Given:* $\overline{BA} \perp \overline{AD}$, $\overline{BC} \perp \overline{CD}$, and $AB = BC$
   Prove: $\angle 1 = \angle 2$

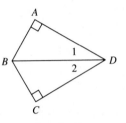

Exercise 7

Exercises 8–12 refer to the figure below in which $\overline{BD} \perp \overline{AC}$ and $\overline{AB} \perp \overline{BC}$.

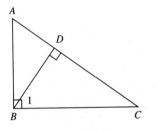

8. If $CD = 12$ cm and $AD = 5$ cm, find BD.
9. If $BC = 18$ yd and $AC = 24$ yd, find CD.
10. If $BD = 8$ ft and $AC = 20$ ft, find AD.
11. If $\angle 1 = 45°$ and $BC = 6$ cm, find BD.
12. If $\angle A = 60°$ and $AB = 14$ yd, find AD.
13. A garden is in the shape of a square 12 yd on a side. Mary wishes to place a picket fence across one diagonal. To the nearest tenth of a yard, how much fencing will she need?
14. In right triangle $\triangle ABC$, $\angle C$ is the right angle, $a = \sqrt{11}$ cm, and $c = 6$ cm, find b.
15. Find the area of an isosceles triangle with equal sides of length x and base angles 30°.

7

CIRCLES

In this chapter we study properties of a circle and the arcs, lines, and angles associated with a circle. We'll also consider regular polygons as they are inscribed in and circumscribed around circles.

The circle is frequently used in architecture and design because of its symmetric properties. Other applications of circles are found in science and engineering, including the one given below which is solved in Section 7.3, Example 1.

Assume that a cross section of the earth is a circle with radius 4000 mi. If a communications satellite is in orbit 110 mi above the surface of the earth, what is the approximate distance from the satellite to the horizon, the farthest point that can be seen on the surface of the earth?

7.1 CIRCLES AND ARCS

One of the most familiar of all geometric figures is the *circle*. We'll begin by reviewing a few familiar terms and introducing some new ones.

> **DEFINITION: CIRCLE**
>
> A **circle** is the set of all points in a plane that are located a fixed distance from a fixed point called its **center.** A line segment joining the center of a circle to one of its points is called the **radius** of the circle.

Figure 7.1 shows a circle with center O and radius $\overline{OP}$. Although the radius of a circle is a segment, it is common practice to call the radius the length of the segment and denote the radius by r. For example, if $OP = 5$ cm in Figure 7.1, we might say that the radius of the circle is $r = 5$ cm. The segment $\overline{QR}$ in Figure 7.1 passing through center O is called a **diameter** of the circle. We often use d to represent the diameter of a circle. In this case $d = QR$, and it follows that $QR = QO + OR = r + r = 2r$, which proves the following theorem.

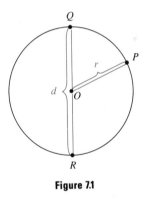

Figure 7.1

> **THEOREM 7.1**
>
> The diameter d of a circle is twice the radius r of the circle. That is, $d = 2r$.

The **circumference** of a circle is the distance around the circle (similar to the perimeter of a polygon). The ratio of the circumference C of any circle to its diameter d is the constant irrational number **pi**, denoted by π. Thus,

$$\pi = \frac{C}{d}.$$

We often approximate π using 3.14, but the actual value is the unending non-repeating decimal 3.14159265358. . . .

POSTULATE 7.1   CIRCUMFERENCE OF A CIRCLE

The circumference C of any circle with radius r and diameter d is determined with the formula $C = 2\pi r = \pi d$.

EXAMPLE 1   Find the circumference of a circle with radius 3.50 cm. Give an approximate value of the circumference, correct to the nearest hundredth of a centimeter, using $\pi \approx 3.14$.

Because $C = 2\pi r$, substituting 3.5 for r, we have $C = 2\pi(3.50) = 7\pi$ cm as the exact value of the circumference. Substituting 3.14 for π we find $C = 7\pi \approx 7(3.14) = 21.98$ cm.   ▨

NOTE: If you have a scientific calculator and use the $\boxed{\pi}$ key in your calculations, the answers shown on the display will not always be exactly the same as those for which you use 3.14 for π.   □

PRACTICE EXERCISE 1

What is the radius of a circle, correct to the nearest tenth of a foot, if the circumference is 23.0 ft?

The next postulate gives the formula for the area of a circle. Figure 7.2 helps to show why the formula is true.

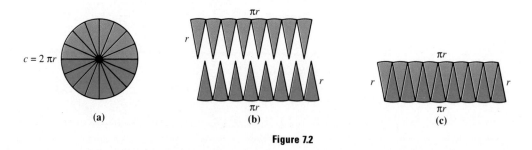

(a)   (b)   (c)

Figure 7.2

Suppose we cut the circle in Figure 7.2(a) along all the radii (plural of radius). When we place the top half of the circle above the bottom half as in Figure 7.2(b) and then slide them together as in Figure 7.2(c), the figure formed is approximately a parallelogram with height r. The length of the base of the "parallelogram" is πr because $2\pi r$ is the circumference, and half the circle is on the top and half is on the bottom. Because the area of a parallelogram is determined with the formula $A = bh$, the area of the circle is

$$A = bh = (\pi r)(r) = \pi r^2.$$

POSTULATE 7.2   AREA OF A CIRCLE

The area of a circle with radius r is determined with the formula
$A = \pi r^2$.

E X A M P L E   2   In the machine part shown in Figure 7.3, each circular hole has radius 3 cm. Using 3.14 for π, find the area of the remaining metal.

First find the area of the trapezoid.

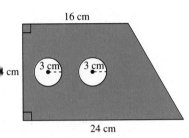

16 cm

cm

3 cm   3 cm

24 cm

Figure 7.3

$$A = \frac{1}{2}(b + b')h$$

$$= \frac{1}{2}(16 + 24)(14)$$

$$= \frac{1}{2}(40)(14)$$

$$= (20)(14) = 280 \text{ cm}^2$$

Now find the area of each circular hole.

$$A = \pi r^2 \approx 3.14(3)^2 = (3.14)(9) = 28.3 \text{ cm}^2$$

Thus, the two circles have a combined area of

$$2(28.3) = 56.6 \text{ cm}^2.$$

The area of the remaining metal is the area of the trapezoid minus the area of the circles, $280 - 56.6 = 223.4$. Thus, the remaining area is 223.4 cm². ◻

PRACTICE EXERCISE 2

A triangle with base 10 inches and height 5 inches has three holes drilled through it, each with diameter 2 inches. What is the remaining area of the triangle after the holes are drilled? Use 3.14 for π.

Recall that two geometric figures are congruent if they can be made to coincide.

POSTULATE 7.3   CONGRUENT CIRCLES

If two circles are congruent, then their radii and diameters are equal. Conversely, if the radii or diameters are equal, then two circles are congruent.

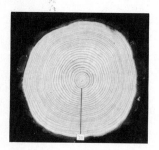

All circles are similar in shape, and thus there is no need for a formal definition of similarity. Circles that lie in the same plane and have a common center are called *concentric circles*. The concentric circles of tree rings are used by foresters to study the climate and the ecology of the region in which the tree grew.

NOTE: As we've seen in postulates and definitions, many times both the direct statement and its converse are true. By stating "if and only if," both statements can be made at the same time. For example, Postulate 7.3 could be written as follows: Two circles are congruent if and only if their radii or their diameters are equal.   □

The study of a circle's continuous parts reveals its many properties.

DEFINITION: ARCS AND SEMICIRCLES

An **arc** of a circle forms a continuous part of the circle. An arc of a circle whose endpoints are the endpoints of a diameter of the circle is called a **semicircle.** An arc that is longer than a semicircle is called a **major arc** of the circle, and an arc that is shorter than a semicircle is called a **minor arc** of the circle.

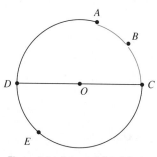

Figure 7.4   Arcs and Semicircles

Arcs of a circle can be named using three points on the arc. In Figure 7.4, minor arc ABC (shown in color), denoted by $\overset{\frown}{ABC}$ has endpoints A and C. Major arc $\overset{\frown}{ADC}$ has the same endpoints as minor arc $\overset{\frown}{ABC}$ (shown in black). If $\overline{DC}$ is a diameter of the circle, then $\overset{\frown}{DAC}$ and $\overset{\frown}{DEC}$ are both semicircles. If we use two letters to name an arc, such as $\overset{\frown}{AC}$, we always mean the minor arc with endpoints A and C. That is, $\overset{\frown}{AC} = \overset{\frown}{ABC}$ in Figure 7.4.

The next postulate is similar to Postulate 1.13 for segments.

POSTULATE 7.4   ARC ADDITION POSTULATE

Let A, B, and C be three points on the same circle with B between A and C. Then $\overset{\frown}{AC} = \overset{\frown}{AB} + \overset{\frown}{BC}$, $\overset{\frown}{BC} = \overset{\frown}{AC} - \overset{\frown}{AB}$, and $\overset{\frown}{AB} = \overset{\frown}{AC} - \overset{\frown}{BC}$.

Contrary to what we might expect, arcs are *not* measured using their length. Instead, the measure of an arc is given using an angle in a circle.

> **DEFINITION: CENTRAL ANGLE**
>
> An angle with sides that are radii of a circle and vertex the center of the circle is called a **central angle.**

Figure 7.5 shows central angle $\angle AOB$ that **intercepts** minor arc $\overarc{AB}$.

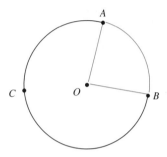

Figure 7.5   Central Angle $\angle AOB$

> **DEFINITION: MEASURE OF AN ARC**
>
> The **measure of an arc** is the number of degrees in the central angle that intercepts the arc.

If $\angle AOB = 84°$ in Figure 7.5, then the measure of $\overarc{AB}$ is 84°. We often abbreviate this statement and simply say that $\overarc{AB} = 84°$. Because $360° - 84° = 276°$, the measure of $\overarc{ACB}$ is 276°. Notice that every minor arc has a measure less than 180°, every major arc has a measure greater than 180°, and every semicircle has a measure equal to 180°. Also, equal central angles intercept equal arcs.

> **DEFINITION: INSCRIBED ANGLE**
>
> An angle whose vertex is on a circle and whose sides intersect the circle in two other points is called an **inscribed angle.**

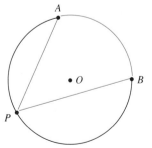

Figure 7.6   Inscribed Angle $\angle APB$

In Figure 7.6, $\angle APB$ is an inscribed angle in the circle with center O. We say that $\angle APB$ **intercepts** $\overarc{AB}$ and is **inscribed** in $\overarc{APB}$. The next theorem presents a property involving the measure of an inscribed angle.

> ### THEOREM 7.2
>
> The measure of an inscribed angle is one-half the measure of its intercepted arc.

The complete proof of Theorem 7.2 can be accomplished by considering three cases.

Case 1: The center of the circle is on one of the sides of the inscribed angle.

Case 2: The center of the circle is in the interior of the inscribed angle.

Case 3: The center of the circle is in the exterior of the inscribed angle.

We will prove the theorem for Case 1. Proofs of Cases 2 and 3 can be accomplished by constructing a line segment through the center of the circle and following the proof of Case 1. Proof of Case 1:

Given: Inscribed angle $\angle APB$ with O on $\overline{PB}$ (See Figure 7.7.)

Prove: $\angle APB = \dfrac{1}{2}\widehat{AB}$

Auxiliary line: Construct Segment $\overline{AO}$

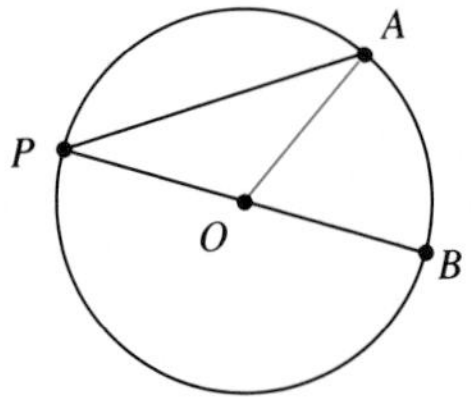

Proof:

Figure 7.7

STATEMENTS	REASONS
1. O is on side $\overline{PB}$ of inscribed angle $\angle APB$	1. Given
2. $\angle AOB = \widehat{AB}$	2. Def. of measure of arc
3. $PO = AO$	3. Radii are equal
4. $\angle APB = \angle A$	4. $\angle$'s opp. $=$ sides are $=$
5. $\angle APB + \angle A = \angle AOB$	5. Ext. $\angle$ = sum of nonadj. int. $\angle$'s
6. $\angle APB + \angle APB = \angle AOB$	6. Substitution law
7. $2\angle APB = \angle AOB$	7. Distributive law
8. $\angle APB = \dfrac{1}{2}\angle AOB$	8. Mult.-div. law
9. $\angle APB = \dfrac{1}{2}\widehat{AB}$	9. Substitution law

EXAMPLE 3 In Figure 7.8, assume that $\angle AOB = 116°$. What is the measure of $\angle ACB$?

Because central angle $\angle AOB = 116°$, $\widehat{AB} = 116°$. With inscribed angle $\angle ACB$ intercepting $\widehat{AB}$, its measure is one-half the measure of $\widehat{AB}$. Thus,

$$\angle ACB = \frac{1}{2}\widehat{AB} = \frac{1}{2}(116°) = 58°. \quad \blacksquare$$

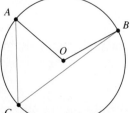

Figure 7.8

To find the center of a given circle, a student places a sheet of paper with one corner on the circle at point *P*, locates points *Q* and *R*, and draws chord *QR*.

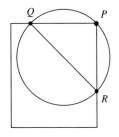

She then places the paper so that the corner is on another point *S* on the circle, locates points *T* and *V*, and draws chord *TV*.

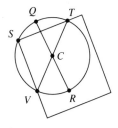

The point of intersection of *QR* and *TV, C,* is the center of the circle. Can you explain why this procedure works?

PRACTICE EXERCISE 3

Assume that $\angle ACB = 60°$ in Figure 7.8. What is the measure of $\overset{\frown}{AB}$? What is the measure of $\overset{\frown}{ACB}$?

Theorem 7.2 has two useful corollaries.

COROLLARY 7.3

Inscribed angles that intercept the same or equal arcs are equal.

COROLLARY 7.4

Every angle inscribed in a semicircle is a right angle.

E X A M P L E 4 Use Figure 7.9 to answer the following.

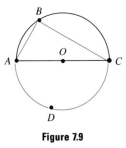

Figure 7.9

(a) What is the measure of $\overset{\frown}{ADC}$?
Because $\overset{\frown}{ADC}$ is a semicircle, $\overset{\frown}{ADC} = 180°$

(b) What is the measure of $\angle ABC$?
Because $\angle ABC$ is inscribed in semicircle $\overset{\frown}{ABC}$, $\angle ABC$ is a right angle so, $\angle ABC = 90°$. ◻

ANSWERS TO PRACTICE EXERCISES: **1.** 3.7 ft **2.** 15.6 in² **3.** $\overset{\frown}{AB} = 120°$; $\overset{\frown}{ACB} = 240°$

7.1 EXERCISES

In Exercises 1–4, find the diameter of each circle with the given radius.

1. $r = 11$ in **2.** $r = 5.8$ cm **3.** $r = \dfrac{3}{4}$ ft **4.** $r = 13.25$ yd

In Exercises 5–8, find the radius of each circle with the given diameter.

5. $d = 16$ in **6.** $d = 4.8$ cm **7.** $d = \dfrac{2}{3}$ ft **8.** $d = 22.42$ yd

In Exercises 9–12, find the circumference and area of each circle with the given radius or diameter. Leave the answer in terms of π.

9. $r = 7$ yd **10.** $r = 6.2$ mi **11.** $d = \dfrac{4}{3}$ cm **12.** $d = 12.78$ ft

In Exercises 13–16, find the approximate circumference and area of each circle with the given radius or diameter by using 3.14 for π.

13. $r = 4.10$ ft **14.** $r = \dfrac{3}{4}$ cm **15.** $d = 12.00$ mi **16.** $d = 18.36$ yd

17. How much more cross-sectional area is there for water to pass through in a $\dfrac{3}{4}$-inch diameter water hose than there is in a $\dfrac{1}{2}$-inch diameter water hose? Use 3.14 for π.

18. A circular garden has radius 9 m. If a 1-meter-wide circular walk surrounds it, what is the area of the walk? Use 3.14 for π. [Hint: Find the area of a circle with 10-meter radius and subtract the area of the garden.]

19. A 12-inch diameter pizza costs $4.50. A 16-inch diameter pizza costs $7.50. Which pizza costs less per square inch? Use 3.14 for π.

20. A patio is in the shape of a trapezoid with bases 8.1 yd and 6.7 yd and height 5.8 yd. A circular dining area in the center of the patio has diameter 3.2 yd and is covered with Mexican tile. Assuming no waste, how much will it cost to the nearest dollar, to cover the remainder of the patio with outdoor carpeting that costs $18.50 per square yard? Use 3.14 for π.

In Exercises 21–24, find the area of metal remaining on each machine part with circular holes drilled in it. Use 3.14 for π.

21.

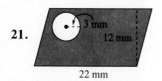

22.

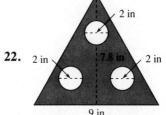

23.

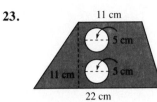

24.

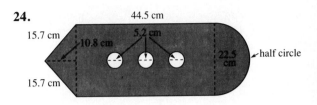

25. Use the figure below to answer each question.
  (a) What is $\angle AOC$ called with respect to the circle?
  (b) What is $\angle ABC$ called with respect to the circle?
  (c) What is the measure of $\overset{\frown}{AC}$?
  (d) What is the measure of $\overset{\frown}{ABC}$?
  (e) What is the measure of $\angle ABC$?
  (f) What is the measure of the arc intercepted by $\angle ABC$?

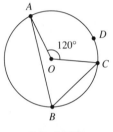

Exercise 25

26. Use the figure below to answer each question.
  (a) What is $\angle AOC$ called with respect to the circle?
  (b) What is $\angle ABC$ called with respect to the circle?
  (c) What is the measure of $\angle AOC$?
  (d) What is the measure of $\overset{\frown}{AC}$?
  (e) What is the measure of $\overset{\frown}{ABC}$?
  (f) What is the measure of the arc intercepted by $\angle ABC$?

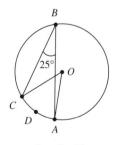

Exercise 26

27. Use the figure below to answer each question.
    (a) What is the measure of $\overset{\frown}{ADC}$?
    (b) What is the measure of $\overset{\frown}{ABC}$?
    (c) What is the measure of $\angle AOC$?
    (d) What is the measure of $\angle AEC$?

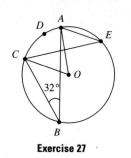

Exercise 27

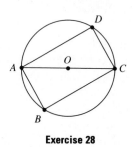

Exercise 28

28. Use the figure above to answer each question.
    (a) What is the measure of $\angle AOC$?
    (b) What is the measure of $\overset{\frown}{ABC}$?
    (c) What is the measure of $\angle ABC$?
    (d) What is the measure of $\angle ADC$?

29. *Given:* $\overline{AB} \perp \overline{BC}$
    Prove: $\overset{\frown}{ADC} = 180°$

30. *Given:* $\overline{AB} \parallel \overline{CD}$
    Prove: $\overset{\frown}{AC} = \overset{\frown}{BD}$
        [Hint: Draw auxiliary segment $\overline{AD}$.]

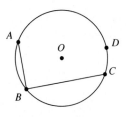

Exercise 29

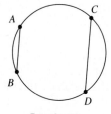

Exercise 30

31. *Given:* $\overline{AB}$ is a diameter
        $\overline{AC} \parallel \overline{OD}$
    Prove: $\overset{\frown}{BD} = \overset{\frown}{DC}$

32. *Given:* $\overline{AB}$ is a diameter
        $\overline{CD} \perp \overline{AB}$
    Prove: $(CD)^2 = (AD)(DB)$

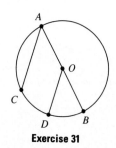

Exercise 31

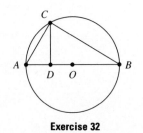

Exercise 32

In Exercises 33–34, assume that the earth is 93,000,000 miles from the sun and that the orbit of the earth is a circle. Use 3.14 for π.

33. What distance does the earth travel during one year?

34. What distance does the earth travel during one day? [*Note:* Use 365 days in a year.]

35. What distance does the earth travel during one minute?

7.2 CHORDS AND SECANTS

In Section 7.1 we studied some of the properties of radii and diameters of a circle. Now we'll examine several properties of other segments and lines relative to a circle.

> **DEFINITION: CHORD**
>
> A line segment joining two distinct points on a circle is called a **chord** of the circle.

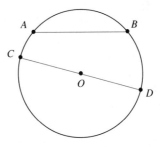

Figure 7.10 Chords of a Circle

In Figure 7.10, $\overline{AB}$ is a chord of the circle centered at point O. Notice that diameter $\overline{CD}$ is a special chord that passes through the center O. We say that minor arc $\overset{\frown}{AB}$ is the arc formed by chord $\overline{AB}$.

> **THEOREM 7.5**
>
> When two chords of a circle intersect, each angle formed is equal to one-half the sum of its intercepted arc and the arc intercepted by its vertical angle.

The Greek astronomer
Eratosthenes (about 276–192
B.C.) was first to calculate the
circumference of the earth. In
the third century B.C., he
accomplished this amazing feat
by assuming that the earth was
a sphere and that the sun's rays
were parallel lines. His
approximation for the earth's
circumference was extremely
close to the accepted value
determined by sophisticated
modern technology.

Given: Chords $\overline{AB}$ and $\overline{CD}$ that intersect at point P (See Figure 7.11.)

Prove: $\angle 1 = \dfrac{1}{2}(\widehat{BC} + \widehat{AD})$

Auxiliary line: Construct segment $\overline{AC}$

Proof:

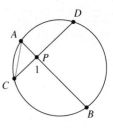

Figure 7.11

STATEMENTS	REASONS
1. Chords $\overline{AB}$ and $\overline{CD}$ intersect at point P	1. Given
2. $\angle CAB = \dfrac{1}{2}\widehat{BC}$ and $\angle ACD = \dfrac{1}{2}\widehat{AD}$	2. Inscribed $\angle$'s have meas. = to $\dfrac{1}{2}$ intercepted arc
3. $\angle 1 = \angle CAB + \angle ACD$	3. Ext. $\angle$ of $\triangle$ = sum nonadj. int. $\angle$'s
4. $\angle 1 = \dfrac{1}{2}\widehat{BC} + \dfrac{1}{2}\widehat{AD}$	4. Substitution law
5. $\angle 1 = \dfrac{1}{2}(\widehat{BC} + \widehat{AD})$	5. Distributive law

EXAMPLE 1   In Figure 7.12, assume that $\widehat{AC} = 30°$ and $\widehat{DB} = 52°$. Find the measure of $\angle 1$.

By Theorem 7.5,

$$\angle 1 = \frac{1}{2}(\widehat{AC} + \widehat{BD})$$

$$= \frac{1}{2}(30° + 52°)$$

$$= \frac{1}{2}(82°) = 41°. \quad \blacksquare$$

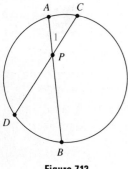

Figure 7.12

PRACTICE EXERCISE 1

In Figure 7.12, assume that $\widehat{AD} = 112°$ and $\widehat{BC} = 176°$. Find the measure of $\angle 1$.

THEOREM 7.6

In the same circle, the arcs formed by equal chords are equal.

Given: $\overline{AB}$ and $\overline{CD}$ are chords with
  $AB = CD$ (See Figure 7.13.)

Prove: $\overset{\frown}{AB} = \overset{\frown}{CD}$

Auxiliary lines: Construct radii $\overline{AO}$,
  $\overline{BO}$, $\overline{CO}$ and $\overline{DO}$

Proof:

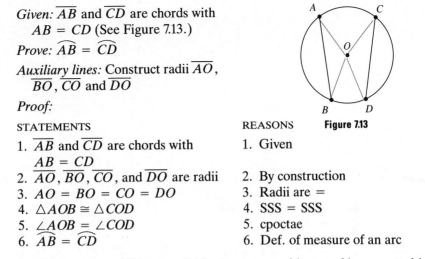

Figure 7.13

STATEMENTS	REASONS
1. $\overline{AB}$ and $\overline{CD}$ are chords with $AB = CD$	1. Given
2. $\overline{AO}, \overline{BO}, \overline{CO}$, and $\overline{DO}$ are radii	2. By construction
3. $AO = BO = CO = DO$	3. Radii are =
4. $\triangle AOB \cong \triangle COD$	4. SSS = SSS
5. $\angle AOB = \angle COD$	5. cpoctae
6. $\overset{\frown}{AB} = \overset{\frown}{CD}$	6. Def. of measure of an arc

The converse of Theorem 7.6 is also true, and its proof is requested in the exercises.

THEOREM 7.7

In the same circle, the chords formed by equal arcs are equal.

DEFINITION: BISECTOR OF AN ARC

A line that divides an arc into two arcs with the same measure is called a **bisector** of the arc.

THEOREM 7.8

A line drawn from the center of a circle perpendicular to a chord bisects the chord and the arc formed by the chord.

Given: Chord $\overline{AB}$ with $\overline{OD} \perp \overline{AB}$
  (See Figure 7.14.)

Prove: $AD = DB$ and $\overset{\frown}{AC} = \overset{\frown}{CB}$

Auxiliary lines: Construct radii $\overline{OA}$
  and $\overline{OB}$

Proof:

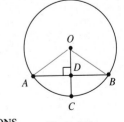

Figure 7.14

STATEMENTS	REASONS
1. $\overline{AB}$ is a chord with $\overline{OD} \perp \overline{AB}$	1. Given
2. $OA = OB$	2. Radii are =

3. $OD = OD$ 3. Reflexive law
4. $\triangle ADO \cong \triangle BDO$ 4. HL = HL
5. $AD = DB$ 5. cpoctae
6. $\angle AOC = \angle BOC$ 6. cpoctae
7. $\overset{\frown}{AC} = \overset{\frown}{CB}$ 7. Def. of measure of an arc

The converse of Theorem 7.8 is also true.

THEOREM 7.9

A line drawn from the center of a circle to the midpoint of a chord (not a diameter) or to the midpoint of the arc formed by the chord is perpendicular to the chord.

The proof of Theorem 7.9 is requested in the exercises.

E X A M P L E 2   In Figure 7.15, $AB = 14$ cm, $ED = 7$ cm, and $\overset{\frown}{AB} = 84°$. Find CE and $\overset{\frown}{FD}$.

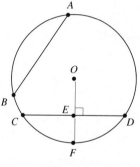

Figure 7.15

When an object is seen from two different positions, it seems to change location. The change is called *parallax*. Astronomers calculate the distance to a planet or star by using information resulting from the parallax. When a planet is viewed from two points on earth, from *A* it appears in line with star *S*, but from *B* it is not. By measuring various angles in this figure, an astronomer can approximate the distance to the planet.

Because $\overline{OE} \perp \overline{CD}$, $\overline{OE}$ bisects chord $\overline{CD}$ and arc $\overset{\frown}{CD}$ by Theorem 7.8. Then $CE = ED$, so $CE = 7$ cm. Because chords $\overline{AB}$ and $\overline{CD}$ are both 14 cm in length, by Theorem 7.6 $\overset{\frown}{AB} = \overset{\frown}{CD} = 84°$. Then $\overset{\frown}{FD}$ is one-half of $\overset{\frown}{CD}$, so $\overset{\frown}{FD} = 42°$. ◻

THEOREM 7.10

In the same circle, equal chords are equidistant from the center of the circle.

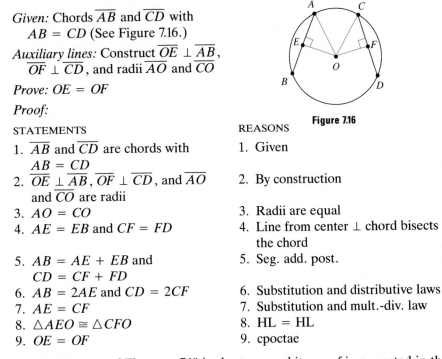

Given: Chords $\overline{AB}$ and $\overline{CD}$ with
   $AB = CD$ (See Figure 7.16.)

Auxiliary lines: Construct $\overline{OE} \perp \overline{AB}$,
   $\overline{OF} \perp \overline{CD}$, and radii $\overline{AO}$ and $\overline{CO}$

Prove: $OE = OF$

Proof:

Figure 7.16

STATEMENTS	REASONS
1. $\overline{AB}$ and $\overline{CD}$ are chords with $AB = CD$	1. Given
2. $\overline{OE} \perp \overline{AB}$, $\overline{OF} \perp \overline{CD}$, and $\overline{AO}$ and $\overline{CO}$ are radii	2. By construction
3. $AO = CO$	3. Radii are equal
4. $AE = EB$ and $CF = FD$	4. Line from center $\perp$ chord bisects the chord
5. $AB = AE + EB$ and $CD = CF + FD$	5. Seg. add. post.
6. $AB = 2AE$ and $CD = 2CF$	6. Substitution and distributive laws
7. $AE = CF$	7. Substitution and mult.-div. law
8. $\triangle AEO \cong \triangle CFO$	8. HL = HL
9. $OE = OF$	9. cpoctae

The converse of Theorem 7.10 is also true, and its proof is requested in the exercises.

THEOREM 7.11

In the same circle, chords equidistant from the center of the circle are equal.

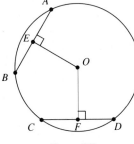

Figure 7.17

E X A M P L E   3   In Figure 7.17, $CF = 5$ ft, $OE = 7$ ft, and $AB = 10$ ft. Find OF.

   Because $\overline{OF} \perp \overline{CD}$, F is the midpoint of $\overline{CD}$. Thus $CF = FD = 5$ ft making $CD = 10$ ft. Because $AB = 10$ ft, $CD = AB$, and by Theorem 7.10, $OE = OF$. Thus, $OF = 7$ ft.   ▨

THEOREM 7.12

The perpendicular bisector of a chord passes through the center of the circle.

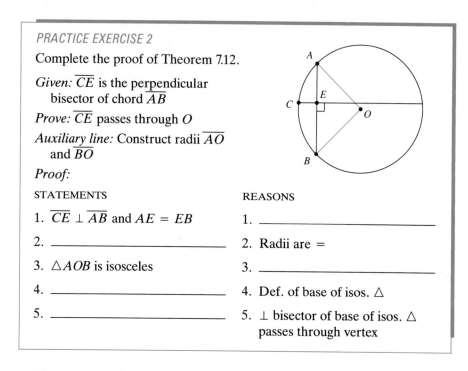

PRACTICE EXERCISE 2

Complete the proof of Theorem 7.12.

Given: $\overline{CE}$ is the perpendicular bisector of chord $\overline{AB}$

Prove: $\overline{CE}$ passes through O

Auxiliary line: Construct radii $\overline{AO}$ and $\overline{BO}$

Proof:

STATEMENTS	REASONS
1. $\overline{CE} \perp \overline{AB}$ and $AE = EB$	1. _____
2. _____	2. Radii are $=$
3. $\triangle AOB$ is isosceles	3. _____
4. _____	4. Def. of base of isos. $\triangle$
5. _____	5. $\perp$ bisector of base of isos. $\triangle$ passes through vertex

The next example shows how Theorem 7.12 can be used to determine the center of a given arc or circle.

E X A M P L E 4   A curve in a highway has the shape of an arc of a circle as shown in Figure 7.18. Explain how to find the center of this arc (the center of the circle that contains the arc).

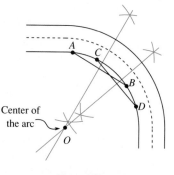

Figure 7.18

Select two points A and B on the arc and construct the perpendicular bisector of $\overline{AB}$. Now select two other points C and D on the arc and construct the perpendicular bisector of $\overline{CD}$. Because both of these bisectors pass through the center of the circle by Theorem 7.12, the point at which they intersect must be the center of the circle, hence the center of the arc.   ◼

NOTE: All theorems considered thus far in this section are true for congruent circles as well as for the same circle.   ☐

We can use similar triangles to prove a property of chords that intersect inside a circle.

> ### THEOREM 7.13
>
> If two chords intersect inside a circle, the product of the lengths of the segments of one chord is equal to the product of the lengths of the segments of the other.

Given: Chords $\overline{AB}$ and $\overline{CD}$ that
  intersect at point P
  (See Figure 7.19.)

Prove: $(AP)(PB) = (CP)(PD)$

Auxiliary lines: Construct $\overline{AC}$ and
  $\overline{BD}$

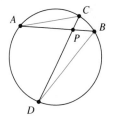

Figure 7.19

Proof:

STATEMENTS	REASONS
1. Chords $\overline{AB}$ and $\overline{CD}$ intersect at P	1. Given
2. $\overline{AC}$ and $\overline{BD}$ are chords	2. By construction
3. $\angle A = \angle D$ and $\angle C = \angle B$	3. Inscribed angles intercepting = arcs are =
4. $\triangle APC \sim \triangle BPD$	4. AA ~ AA
5. $\dfrac{AP}{PD} = \dfrac{CP}{PB}$	5. Corr. sides of ~ $\triangle$'s are proportional
6. $(AP)(PB) = (CP)(PD)$	6. Means-extremes prop.

> ### PRACTICE EXERCISE 3
>
> Refer to Figure 7.19 in which $AP = 3$ cm, $CP = 4$ cm, and $PD = 6$ cm. Find PB.

A line and a circle in the same plane can intersect in one or two points or none at all. Lines that intersect a circle in two points have several important properties.

> ### DEFINITION: SECANT
>
> If a line intersects a circle in two points, the line is called a **secant**.

Line ℓ in Figure 7.20 is a secant that intersects the circle in points A and B. Notice that this secant determines chord $\overline{AB}$. All secants determine a chord. The next theorem gives a relationship between two secants and arcs on the circle.

Figure 7.20   Secant

THEOREM 7.14

If two secants intersect forming an angle outside the circle, then the measure of this angle is one-half the difference of the intercepted arcs.

Given: Secants $\overleftrightarrow{PA}$ and $\overleftrightarrow{PB}$ forming exterior angle $\angle APB$
(See Figure 7.21.)

Prove: $\angle APB = \dfrac{1}{2}(\widehat{AB} - \widehat{CD})$

Auxiliary line: Construct $\overline{BC}$

Proof:

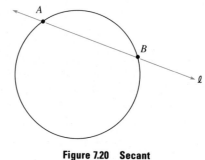

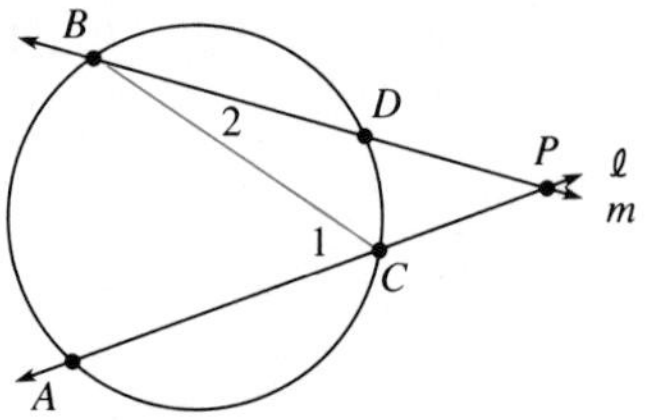

Figure 7.21

STATEMENTS	REASONS
1. Secants $\overleftrightarrow{PA}$ and $\overleftrightarrow{PB}$ forming $\angle APB$	1. Given
2. $\angle 1 = \angle 2 + \angle APB$	2. Ext $\angle$ of $\triangle$ = sum of remote int. $\angle$'s
3. $\angle APB = \angle 1 - \angle 2$	3. Add.-subtr. law
4. $\angle 1 = \dfrac{1}{2}\widehat{AB}$ and $\angle 2 = \dfrac{1}{2}\widehat{CD}$	4. Inscribed $\angle = \dfrac{1}{2}$ meas. of its intercepted arc
5. $\angle APB = \dfrac{1}{2}\widehat{AB} - \dfrac{1}{2}\widehat{CD}$	5. Substitution law
6. $\angle APB = \dfrac{1}{2}(\widehat{AB} - \widehat{CD})$	6. Distributive law

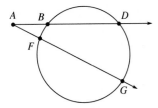

Figure 7.22

$E\ X\ A\ M\ P\ L\ E\ 5$ Refer to Figure 7.22. Assume that $\overset{\frown}{DG} = 110°$ and $\overset{\frown}{BF} = 40°$. Find $\angle A$.

By Theorem 7.14,

$$\angle A = \frac{1}{2}(\overset{\frown}{DG} - \overset{\frown}{BF})$$

$$= \frac{1}{2}(110° - 40°)$$

$$= \frac{1}{2}(70°) = 35° \quad \blacksquare$$

The next theorem is similar to Theorem 7.13.

THEOREM 7.15

If two secants are drawn to a circle from an external point, the product of the lengths of one secant segment and its external segment is equal to the product of the lengths of the other secant segment and its external segment.

Given: Secants $\overleftrightarrow{PA}$ and $\overrightarrow{PB}$
 PA and PB are the lengths of the two secant segments
 (See Figure 7.23.)

Prove: $(PA)(PC) = (PB)(PD)$

Auxiliary lines: Construct segments $\overline{AD}$ and $\overline{BC}$

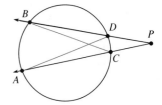

Figure 7.23

Proof:

STATEMENTS	REASONS
1. PA and PB are the lengths of the secant segments formed by secants $\overleftrightarrow{PA}$ and $\overrightarrow{PB}$	1. Given
2. $\angle DBC = \angle DAC$	2. Both inscribed $\angle$'s intercept same arc
3. $\angle P = \angle P$	3. Reflexive law
4. $\triangle APD \sim \triangle BPC$	4. AA ~ AA
5. $\dfrac{PA}{PB} = \dfrac{PD}{PC}$	5. Corr. sides of ~ $\triangle$'s are proportional
6. $(PA)(PC) = (PB)(PD)$	6. Means-extremes prop.

E X A M P L E 6   In Figure 7.23 on page 249, assume that $BD = 8$ cm, $PD = 7$ cm, and $PC = 6$ cm. Find AC.

Let $x = AC$, then $PA = AC + PC = x + 6$. Also, $PB = BD + PD = 8 + 7 = 15$. Use Theorem 7.15 and substitute.

$$(PA)(PC) = (PB)(PD)$$
$$(x + 6)(6) = (15)(7)$$
$$6x + 36 = 105$$
$$6x = 69$$
$$x = 11.5$$

Thus, $AC = 11.5$ cm. ◨

ANSWERS TO PRACTICE EXERCISES: **1.** 36° **2.** 1. Given 2. $AO = BO$ 3. Def. of isos. △ 4. $\overline{AB}$ is the base of △AOB 5. $\overline{CE}$ passes through O **3.** 8 cm

---

7.2 EXERCISES

Exercises 1–10 refer to the figure below.

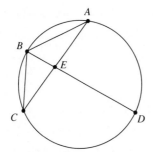

1. If $\overset{\frown}{AB} = 58°$ and $\overset{\frown}{CD} = 130°$, find $\angle AEB$.   **2.** If $\overset{\frown}{BC} = 64°$ and $\overset{\frown}{AD} = 120°$, find $\angle AED$.

3. If $AB = BC$ and $\overset{\frown}{BC} = 60°$, find $\overset{\frown}{AB}$.   **4.** If $\overset{\frown}{AB} = \overset{\frown}{BC}$ and $AB = 8$ cm, find BC.

5. If $\overline{BD}$ passes through the center of the circle and is perpendicular to $\overline{AC}$, and $AE = 12$ inches, find CE.

6. If $\overline{BD}$ passes through the center of the circle and is perpendicular to $\overline{AC}$, and $\overset{\frown}{AB} = 55°$, find $\overset{\frown}{BC}$.

7. If $\overline{BD}$ passes through the center of the circle, $AE = 5$ ft, and $CE = 5$ ft, find $\angle AED$.

8. If $\overline{BD}$ passes through the center of the circle, $\overset{\frown}{AB} = 63°$, and $\overset{\frown}{BC} = 63°$, find $\angle DEC$.

9. If $AE = 5$ cm, $EC = 6$ cm, and $BE = 3$ cm, find ED.

10. If $BE = 4$ ft, $ED = 6$ ft, and $AE = 3$ ft, find CE.

Exercises 11–16 refer to the figure below in which O is the center of the circle.

11. If $AB = 14$ ft, $CD = 14$ ft, and $OF = 11$ ft, find OE.

12. If $OE = 7$ cm, $OF = 7$ cm, and $AB = 9$ cm, find CD.

13. If $AE = 4$ cm, $OE = 5$ cm, and $OF = 5$ cm, find DF.

14. If $OE = 6$ ft, $OF = 6$ ft, and $\overset{\frown}{AGB} = 60°$, find $\overset{\frown}{CHD}$.

15. If $\overset{\frown}{AGB} = \overset{\frown}{DHC}$ and $OE = 15$ cm, find OF.

16. If $\overset{\frown}{AG} = \overset{\frown}{HC}$ and $OF = 9$ inches, find OE.

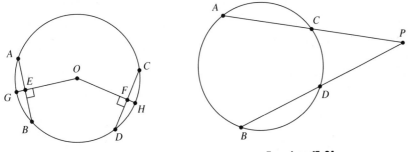

Exercises 11–16   **Exercises 17–24**

Exercises 17–24 refer to the figure above.

17. If $\overset{\frown}{AB} = 130°$ and $\overset{\frown}{CD} = 50°$, find $\angle P$.   **18.** If $\overset{\frown}{AB} = 125°$ and $\overset{\frown}{CD} = 47°$, find $\angle P$.

19. If $\overset{\frown}{AB} = 120°$ and $\angle P = 45°$, find $\overset{\frown}{CD}$.   **20.** If $\overset{\frown}{CD} = 56°$ and $\angle P = 42°$, find $\overset{\frown}{AB}$.

21. If $AP = 12$ ft, $CP = 5$ ft, and $PB = 15$ ft, find PD.

22. If $BP = 25$ yd, $AP = 15$ yd, and $CP = 5$ yd, find PD.

23. If $AC = 5$ cm, $CP = 4$ cm, and $DP = 3$ cm, find BD.

24. If $BD = 3$ ft, $DP = 5$ ft, and $AC = 6$ ft, find CP.

Exercises 25–26 refer to the figure below.

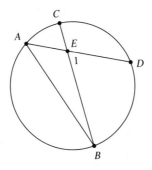

25. If $\angle 1 = 40°$ and $\angle DAB = 30°$, find $\overset{\frown}{AC}$ and $\angle ABC$.

26. If $\angle 1 = 50°$ and $\angle DAB = 35°$, find $\overset{\frown}{AC}$ and $\angle ABC$.

27. *Given:* $\overset{\frown}{BC} + \overset{\frown}{AD} = 180°$

    Prove: $\overline{AC} \perp \overline{BD}$

28. *Given:* $AB = BC$

    Prove: $\angle P = \dfrac{1}{2}(\overset{\frown}{AB} - \overset{\frown}{AD})$

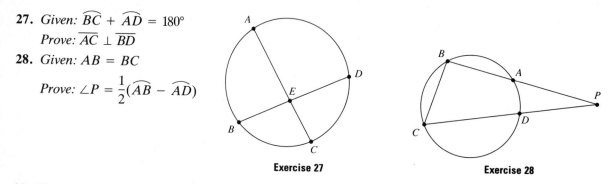

Exercise 27   Exercise 28

29. If two equal chords of a circle intersect, prove that the segments of one are equal, respectively, to the segments of the other.

30. If two equal chords $\overline{AB}$ and $\overline{CD}$ meet at point P when extended, prove that the secant segments $\overline{AP}$ and $\overline{CP}$ are equal.

31. Prove Theorem 7.7.

32. Prove Theorem 7.9.

33. Prove Theorem 7.11.

34. How far is a 12-centimeter chord from the center of a circle with diameter 20 cm?

35. What is the radius of a circle in which a chord 10 inches long is 1 inch from the midpoint of the arc it forms?

7.3 TANGENTS

In Section 7.2 we studied secants, lines that intersect a circle in two distinct points. We'll now consider lines that intersect a circle in exactly one point.

> **DEFINITION: TANGENT**
>
> If a line intersects a circle in one and only one point, the line is called a **tangent** to the circle. The point of intersection is called a **point of tangency**.

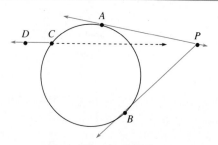

Figure 7.24 Tangent Lines

In Figure 7.24, $\overleftrightarrow{AP}$ is a tangent to the circle with A the point of tangency. We might say that ray $\overrightarrow{PB}$ and segment $\overline{PB}$ are tangent to the circle because they are contained in the tangent line $\overleftrightarrow{BP}$. $\overleftrightarrow{AP}$ and $\overleftrightarrow{BP}$ are said to be tangents to the circle from the same common external point P. Notice that $\overline{DC}$ is not a tangent because when $\overline{DC}$ is extended, it will intersect the circle in two points. Actually, $\overleftrightarrow{DC}$ is a secant.

The next two properties of a tangent can be proved using properties of right triangles that will be presented in Chapter 8; however, here we'll consider them as postulates.

POSTULATE 7.5

If a line is perpendicular to a radius of a circle and passes through the point where the radius intersects the circle, then the line is a tangent.

POSTULATE 7.6

A radius drawn to the point of tangency of a tangent is perpendicular to the tangent.

Postulate 7.5 provides a convenient way to construct a tangent to a circle at a point on the circle.

CONSTRUCTION 7.1

Construct a tangent to a circle at a given point on the circle.

Given: P is a point on a circle with center O (See Figure 7.25.)

To Construct: $\overleftrightarrow{AP}$ tangent to the circle

Construction:

1. Construct radius $\overline{OP}$ and extend it forming $\overleftrightarrow{OP}$.
2. Use Construction 2.4 to construct $\overleftrightarrow{AP}$ perpendicular to $\overleftrightarrow{OP}$ passing through P. By Postulate 7.5, $\overleftrightarrow{AP}$ is the desired tangent.

Postulate 7.5 together with Corollary 7.4 give us a way to construct a tangent to a circle from a point outside the circle.

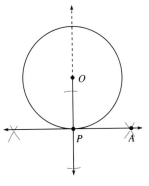

Figure 7.25

CONSTRUCTION 7.2

Construct a tangent to a circle from a point outside the circle.

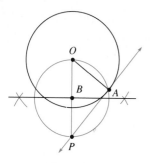

Figure 7.26

Given: P is a point outside a circle with center O (See Figure 7.26.)

To Construct: $\overleftrightarrow{PA}$ tangent to the circle

Construction:

1. Construct $\overline{OP}$ and use Construction 2.3 to locate the midpoint B of $\overline{OP}$.
2. Construct the circle with B as the center and PB as the radius.
3. Let A be a point of intersection of the two circles.
4. Construct $\overleftrightarrow{PA}$ and $\overline{OA}$. Then by Corollary 7.4, $\angle OAP$ is a right angle because it is inscribed in a semicircle. Thus, $\overleftrightarrow{PA} \perp \overline{OA}$, and by Postulate 7.5, $\overleftrightarrow{PA}$ is a desired tangent.

The next example solves the applied problem given in the chapter introduction.

E X A M P L E   1   Assume that a cross-section of the earth is a circle with radius 4000 mi. If a communications satellite is in orbit 110 mi above the surface of the earth, what is the approximate distance from the satellite to the horizon, the farthest point that can be seen on the surface of the earth?

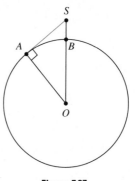

Figure 7.27

In Figure 7.27, S corresponds to the satellite, $\overline{OA}$ the radius of the earth, and SB the height of the satellite above the surface of the earth. The distance from the satellite to the horizon is SA, the length of segment $\overline{SA}$, which is included in tangent $\overleftrightarrow{SA}$ to the earth's surface from S. Because $\triangle AOS$ is a right triangle, we can use the Pythagorean Theorem to obtain

$$(OS)^2 = (SA)^2 + (OA)^2.$$

Thus,

$$
\begin{aligned}
(SA)^2 &= (OS)^2 - (OA)^2 \\
&= [OB + SB]^2 - (OA)^2 \\
&= [4000 + 110]^2 - (4000)^2 \\
&= (4110)^2 - (4000)^2 \\
&= 892{,}100.
\end{aligned}
$$

Figuring the square root with a calculator we have
$$SA \approx 945.$$

Thus, the distance to the horizon is about 945 mi. ◪

PRACTICE EXERCISE 1

Repeat Example 1 for a space station located 95 mi above the surface of the earth.

We now consider two properties of angles, one formed by a tangent and a chord and the other formed by a tangent and a secant.

THEOREM 7.16

The angle formed by a tangent and a chord has a measure one-half its intercepted arc.

Given: $\angle APB$ formed by tangent $\overleftrightarrow{PB}$ and chord $\overline{PA}$
(See Figure 7.28.)

Prove: $\angle APB = \frac{1}{2}\overset{\frown}{AP}$

Auxiliary line: Construct diameter $\overline{CP}$

Proof:

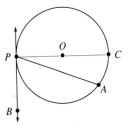

Figure 7.28

STATEMENTS	REASONS
1. $\angle APB$ is formed by tangent $\overleftrightarrow{PB}$ and chord $\overline{PA}$	1. Given
2. $\overline{CP}$ is a diameter forming semicircle $\overset{\frown}{CAP}$	2. By construction
3. $\angle APB = \angle CPB - \angle CPA$	3. Angle add. post.
4. $\angle CPB = 90°$	4. Radius drawn to pt. of tangency is $\perp$ to tan.
5. $\angle CPA = \frac{1}{2}\overset{\frown}{AC}$	5. Inscribed $\angle = \frac{1}{2}$ intercepted arc
6. $\angle APB = 90° - \frac{1}{2}\overset{\frown}{AC}$	6. Substitution law
7. $2\angle APB = 180° - \overset{\frown}{AC}$	7. Mult.-div. law
8. $\overset{\frown}{CAP} = 180°$	8. Semicircle measures 180°
9. $2\angle APB = \overset{\frown}{CAP} - \overset{\frown}{AC}$	9. Substitution law
10. $\overset{\frown}{CAP} - \overset{\frown}{AC} = \overset{\frown}{AP}$	10. Arc add. post.
11. $2\angle APB = \overset{\frown}{AP}$	11. Transitive law
12. $\angle APB = \frac{1}{2}\overset{\frown}{AP}$	12. Mult.-div. law

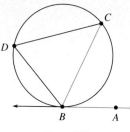

Figure 7.29

E X A M P L E 2 In Figure 7.29, $\overleftrightarrow{AB}$ is a tangent and $\angle CDB = 64°$. Find $\angle CBA$.

Because $\angle CDB$ is an inscribed angle intercepting arc $\overparen{BC}$, and $\angle CDB = 64°$, $\overparen{BC} = 128°$. Thus, by Theorem 7.16, $\angle CBA = \frac{1}{2}(128°) = 64°$. ◪

THEOREM 7.17

The angle formed by the intersection of a tangent and a secant has a measure one-half the difference of the intercepted arcs.

Given: Tangent $\overleftrightarrow{AP}$ and secant $\overleftrightarrow{PC}$
  (See Figure 7.30.)

Prove: $\angle P = \frac{1}{2}(\overparen{AC} - \overparen{AB})$

Auxiliary line: Construct chord $\overline{AC}$

Proof:

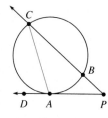

Figure 7.30

STATEMENTS	REASONS
1. $\overleftrightarrow{AP}$ is a tangent and $\overleftrightarrow{PC}$ is a secant forming $\angle P$	1. Given
2. $\overline{AC}$ is a chord	2. By construction
3. $\angle CAD = \angle ACB + \angle P$	2. Ext. $\angle$ of $\triangle$ = sum remote int. $\angle$'s
4. $\angle P = \angle CAD - \angle ACB$	4. Add.-subtr. law
5. $\angle CAD = \frac{1}{2}\overparen{AC}$	5. The $\angle$ formed by tan. and chord $= \frac{1}{2}$ intercepted arc
6. $\angle ACB = \frac{1}{2}\overparen{AB}$	6. Measure of inscribed $\angle$
7. $\angle P = \frac{1}{2}\overparen{AC} - \frac{1}{2}\overparen{AB}$	7. Substitution law
8. $\angle P = \frac{1}{2}(\overparen{AC} - \overparen{AB})$	8. Distributive law

The proof of the next theorem is similar to that for Theorem 7.17 and is left for you to do as an exercise.

THEOREM 7.18

The angle formed by the intersection of two tangents has a measure one-half the difference of the intercepted arcs.

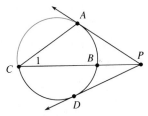

Figure 7.31

E X A M P L E 3   Refer to Figure 7.31. If $\angle 1 = 30°$ and $\angle APB = 34°$, find $\overparen{AC}$.

By Theorem 7.17,

$$\angle APB = \frac{1}{2}(\overparen{AC} - \overparen{AB}).$$

Because $\angle 1 = 30°$, $\overparen{AB} = 60°$. Substituting, we have

$$34° = \frac{1}{2}(\overparen{AC} - 60°)$$

$$68° = \overparen{AC} - 60° \qquad \text{Multiply both sides by 2}$$

$$128° = \overparen{AC}. \qquad \text{Add } 60° \text{ to both sides}$$

Thus, $\overparen{AC} = 128°$. ◪

PRACTICE EXERCISE 2

Refer to Figure 7.31. If $\overparen{ACD} = 236°$, find $\angle APD$.

THEOREM 7.19

Two tangent segments to a circle from the same point have equal lengths.

Given: $\overleftrightarrow{PA}$ and $\overleftrightarrow{PB}$ are tangents to a circle with center O (See Figure 7.32.)

Prove: $PA = PB$

Auxiliary lines: Construct radii $\overline{OA}$ and $\overline{OB}$ and segment $\overline{OP}$

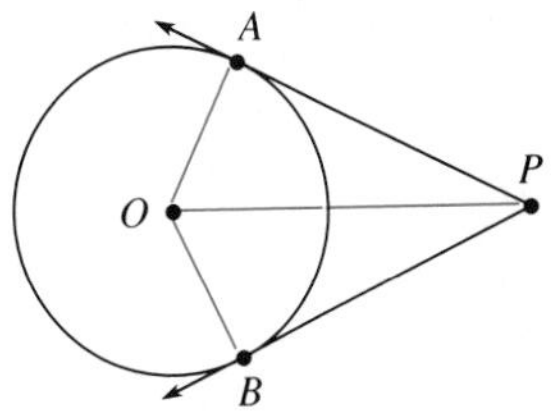

Proof:

Figure 7.32

STATEMENTS	REASONS
1. $\overleftrightarrow{PA}$ and $\overleftrightarrow{PB}$ are tangents to a circle with center O	1. Given
2. $\overline{OA}$ and $\overline{OB}$ are radii	2. By construction
3. $\overline{OA} \perp \overline{AP}$ and $\overline{OB} \perp \overline{PB}$	3. Radii are $\perp$ to tangents
4. $\triangle AOP$ and $\triangle BOP$ are right triangles	4. Def. rt. $\triangle$
5. $OA = OB$	5. Radii are $=$
6. $OP = OP$	6. Reflexive law
7. $\triangle AOP \cong \triangle BOP$	7. HL = HL
8. $PA = PB$	8. cpoctae

> ### THEOREM 7.20
>
> If a secant and a tangent are drawn to a circle from an external point, the length of the tangent segment is the mean proportional between the length of the secant segment and its external segment.

Given: $\overleftrightarrow{AP}$ is a tangent and $\overleftrightarrow{PC}$ is a secant to circle with center O (See Figure 7.33.)

Prove: $\dfrac{PC}{PA} = \dfrac{PA}{PB}$

Auxiliary lines: Construct segments $\overline{AB}$ and $\overline{AC}$

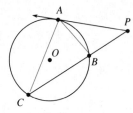

Figure 7.33

Proof:

STATEMENTS	REASONS
1. $\overleftrightarrow{AP}$ is a tangent and $\overleftrightarrow{PC}$ is a secant	1. Given
2. $\angle C = \dfrac{1}{2}\overarc{AB}$	2. Measure of inscribed $\angle$
3. $\angle PAB = \dfrac{1}{2}\overarc{AB}$	3. The $\angle$ formed by tan. and chord $= \dfrac{1}{2}$ intercepted arc
4. $\angle C = \angle PAB$	4. Sym. and Trans. laws
5. $\angle P = \angle P$	5. Reflexive law
6. $\triangle APB \sim \triangle APC$	6. AA ~ AA
7. $\dfrac{PC}{PA} = \dfrac{PA}{PB}$	7. Corr. sides of ~ $\triangle$'s are proportional

EXAMPLE 4 Refer to Figure 7.33. Assume that $PA = 10$ cm and $PB = 6$ cm. Find BC.

Let $x = BC$. Then $PC = BC + PB = x + 6$.
By Theorem 7.20,

$$\frac{PC}{PA} = \frac{PA}{PB}.$$

The gears in a watch illustrate many properties of tangent circles.

Substituting, we obtain

$$\frac{x + 6}{10} = \frac{10}{6}.$$

$6(x + 6) = (10)(10)$ Means-extremes property

$6x + 36 = 100$ Distributive law

$6x = 64$ Subtract 36 from both sides

$x = 10.666 \ldots$ Divide by 6

When a decimal has a repeating block of digits such as the digit 6 in 10.666 . . . , we usually write the answer as $10.\overline{6}$, placing a bar over the repeating digit or digits. Thus, $BC = 10.\overline{6}$ cm. ◨

We have seen that a line and a circle can intersect in one or two points or none at all. A similar situation exists for two circles. If two circles intersect in exactly one point, the circles are said to be **tangent** to each other. Two possibilities exist. In Figure 7.34(a), the circles are **tangent internally** with point of tangency P, and in Figure 7.34(b), the circles are **tangent externally** with point of tangency Q.

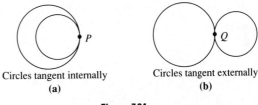

Circles tangent internally   Circles tangent externally
(a)   (b)

Figure 7.34

If two circles do not intersect, the circles can have a **common tangent.** If the circles are located on the same side of a common tangent, as in Figure 7.35(a), the tangent is called a **common external tangent.** If the circles are located on opposite sides of a common tangent, as in Figure 7.35(b), the tangent is called a **common internal tangent.**

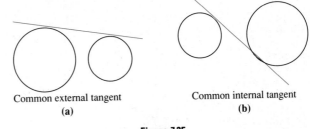

Common external tangent   Common internal tangent
(a)   (b)

Figure 7.35

DEFINITION: LINE OF CENTERS

The line passing through the centers of two circles is called their **line of centers.**

The proof of the following theorem follows from Postulates 7.6 and 2.3 and is left for you to do as an exercise.

> **THEOREM 7.21**
>
> If two circles are tangent internally or externally, the point of tangency is on their line of centers.

The next theorem also involves the line of centers of two circles.

> **THEOREM 7.22**
>
> If two circles intersect in two points, then their line of centers is the perpendicular bisector of their common chord.

Given: Two circles with centers O and P that intersect at points A and B. (See Figure 7.36.)

Prove: $\overline{OP} \perp \overline{AB}$ and $AC = BC$

Auxiliary lines: Construct radii $\overline{OA}$, $\overline{OB}$, $\overline{PA}$, and $\overline{PB}$

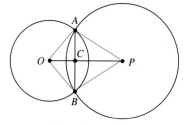

Figure 7.36

Proof:

STATEMENTS	REASONS
1. Two circles centered at O and P intersect at points A and B.	1. Given
2. $OA = OB$ and $PA = PB$	2. Radii are $=$
3. $OP = OP$	3. Reflexive law
4. $\triangle AOP \cong \triangle BOP$	4. SSS = SSS
5. $\angle AOP = \angle BOP$	5. cpoctae
6. $OC = OC$	6. Reflexive law
7. $\triangle AOC \cong \triangle BOC$	7. SAS = SAS
8. $AC = BC$	8. cpoctae
9. $\angle ACO = \angle BCO$	9. cpoctae
10. $\angle ACO$ and $\angle BCO$ are adjacent angles	10. Def. adj. $\angle$'s
11. $\overline{OP} \perp \overline{AB}$	11. Def. of $\perp$ lines

We'll conclude this section with two constructions involving common external and internal tangents to a circle.

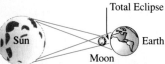

Region of Total Eclipse

Earth
Sun
Moon

A solar eclipse occurs when the moon passes between the earth and the sun and blocks the sun's rays. Some areas of the earth experience no eclipse at all while others experience a partial eclipse, and a small area will be in total eclipse. The lines drawn tangent to the sun and moon in the figure show the areas affected by the eclipse. A total eclipse occurs in the area between the two external tangents.

> **CONSTRUCTION 7.3**
>
> Construct a common external tangent to two given circles that are not congruent.

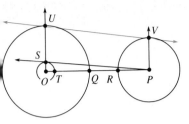

Figure 7.37

Given: Two noncongruent circles with centers O and P (See Figure 7.37.)

To Construct: $\overleftrightarrow{UV}$, an external tangent to both circles

Construction:

1. Construct $\overline{OP}$ obtaining points Q and R, the intersections of the circles and the segment.
2. Construct $\overline{QT}$ on $\overline{QO}$ such that $QT = RP$.
3. Construct a circle centered at O with radius OT.
4. Use Construction 7.2 to construct a tangent to the circle in Step 3 from point P, $\overrightarrow{PS}$.
5. Construct $\overrightarrow{OS}$ intersecting the original circle at point U.
6. Construct $\overrightarrow{PV}$ through P parallel to $\overrightarrow{OU}$ using Construction 4.1.
7. Construct $\overleftrightarrow{UV}$. Because $\square USPV$ is a rectangle, $\overleftrightarrow{UV}$ is the desired tangent.

> ### CONSTRUCTION 7.4
> Construct a common internal tangent to two given circles.

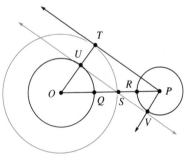

Figure 7.38

Given: Two circles with centers O and P (See Figure 7.38.)

To Construct: $\overleftrightarrow{UV}$, an internal tangent to both circles

Construction:

1. Construct $\overline{OP}$ obtaining points Q and R, the intersections of the circles and the segment.
2. Construct $\overline{QS}$ on $\overline{QP}$ such that $QS = RP$.
3. Construct a circle centered at O with radius OS.
4. Use Construction 7.2 to construct a tangent to this circle from P, $\overleftrightarrow{PT}$.
5. Construct $\overline{OT}$ with point U the intersection of $\overline{OT}$ with the original circle.
6. Use Construction 4.1 to construct line $\overleftrightarrow{PV}$ through P parallel to $\overline{OT}$ intersecting the original circle at V.
7. Construct $\overleftrightarrow{UV}$. Because $\square UTPV$ is a rectangle, $\overleftrightarrow{UV}$ is the desired tangent.

ANSWERS TO PRACTICE EXERCISES: **1.** 877 mi   **2.** 56°

7.3 EXERCISES

Exercises 1–22 refer to the figure below.

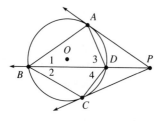

1. If $\overset{\frown}{AD} = 70°$, find $\angle PAD$.
2. If $\overset{\frown}{CD} = 60°$, find $\angle PCD$.
3. If $\angle 1 = 36°$, find $\angle PAD$.
4. If $\angle 2 = 29°$, find $\angle PCD$.
5. If $\angle PAD = 40°$, find $\angle 1$.
6. If $\angle PCD = 35°$, find $\angle 2$.
7. If $\overset{\frown}{AB} = 140°$ and $\overset{\frown}{AD} = 70°$, find $\angle APD$.
8. If $\overset{\frown}{BC} = 85°$ and $\overset{\frown}{CD} = 61°$, find $\angle CPD$.
9. If $\angle 1 = 36°$ and $\angle 3 = 70°$, find $\angle APD$.
10. If $\angle 2 = 30°$ and $\angle 4 = 50°$, find $\angle CPD$.
11. If $\angle APD = 40°$ and $\overset{\frown}{AB} = 138°$, find $\angle 1$.
12. If $\angle CPD = 30°$ and $\overset{\frown}{BC} = 80°$, find $\angle 2$.
13. If $\overset{\frown}{ADC} = 130°$, find $\angle APC$.
14. If $\overset{\frown}{ABC} = 240°$, find $\angle APC$.
15. If $\angle 1 = 35°$ and $\angle 2 = 30°$, find $\angle APC$.
16. If $\angle 3 = 70°$ and $\angle 4 = 50°$, find $\angle APC$.
17. If $AP = 17$ cm, find CP.
18. If $PC = 24$ yd, find PA.
19. If $AP = 12$ cm and $BP = 18$ cm, find DP.
20. If $PC = 8$ ft and $PB = 12$ ft, find PD.
21. If $AP = 15$ yd and $DP = 10$ yd, find BD.
22. If $PC = 24$ cm and $PD = 18$ cm, find BD.
23. If two circles do not intersect and neither is inside the other, how many common internal tangents do the circles have?
24. If two circles do not intersect and neither is inside the other, how many common external tangents do the circles have?
25. *Given:* $\overleftrightarrow{PA}$ and $\overleftrightarrow{PB}$ are tangents to the circle

    Prove: $\angle P + \overset{\frown}{ACB} = 180°$
26. *Given:* $\overleftrightarrow{PA}$ is tangent to the circle with center O

    Prove: $\triangle APB \sim \triangle APC$

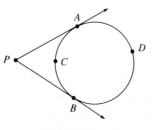

Exercise 25

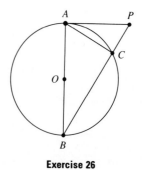

Exercise 26

27. *Given:* $\overleftrightarrow{AB}$ and $\overleftrightarrow{CD}$ are common external tangents to the circles that are not congruent.

    Prove: $AB = CD$

28. *Given:* $\overleftrightarrow{AB}$ and $\overleftrightarrow{CD}$ are common internal tangents to the circles

    Prove: $AB = CD$

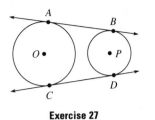

Exercise 27

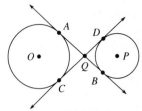

Exercise 28

29. Prove Theorem 7.18.

30. Prove Theorem 7.21.

31. Construct a tangent to a circle at a point on the circle.

32. Construct a tangent to a circle from a point outside the circle.

33. Construct a common external tangent to two circles that do not intersect and have unequal radii.

34. Construct a common external tangent to two circles that do not intersect and have equal radii.

35. Construct a common internal tangent to two circles that do not intersect and have unequal radii.

36. Construct a common internal tangent to two circles that do not intersect and have equal radii.

37. From a balloon 1 mi high, how far away, to the nearest tenth of a mile, is the horizon, the farthest point that can be seen on the surface of the earth?

38. An airplane is flying at an altitude of 6 mi. To the nearest tenth of a mile, how far is the airplane from the horizon?

A large screen television is placed on a shelf on a wall in a sports lounge. Where is the best place to sit to view the screen? If you sit too close or too far, the viewing angle will be decreased as shown in the figure below left. The answer to our question involves finding the largest viewing angle. Suppose we draw a sketch describing the problem as shown below right.

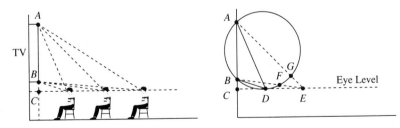

We can show that the best view can be found where the circle through A B is tangent to the eye-level line at point D. Because $\angle ADB = \frac{1}{2}\overparen{AB}$ and at any other point E, $\angle AEB = \frac{1}{2}(\overparen{AB} - \overparen{FG})$, $\angle ADB$ will be greater than $\angle AEB$.

By Theorem 7.20, the length of tangent segment $\overline{CD}$ is the mean proportional between the length of the secant segment $\overline{AC}$ and its external segment $\overline{BC}$. Thus,

$$\frac{AC}{CD} = \frac{CD}{BC} \text{ or } (CD)^2 = (AC)(BC).$$

It then follows that $CD = \sqrt{(AC)(BC)}$. Use this information in Exercises 39–40.

39. The base of a TV screen is 5 ft above eye level and the screen is 4 ft high. How far away should a viewer sit to have the maximum viewing angle?

40. A picture is hanging on a wall in a gallery. If the picture is 3 ft high and the base of the picture is 5 ft above eye level, how far should a viewer stand to have the maximum viewing angle?

7.4 CIRCLES AND REGULAR POLYGONS

In this section we will study relationships between circles and polygons.

> **DEFINITION: INSCRIBED AND CIRCUMSCRIBED CIRCLES AND POLYGONS**
>
> If a polygon has its vertices on a circle, the polygon is **inscribed in the circle** and the circle is **circumscribed around the polygon.** If each side of a polygon is tangent to a circle, the polygon is **circumscribed around the circle** and the circle is **inscribed in the polygon.**

In Figure 7.39(a), quadrilateral $ABCD$ is inscribed in the circle centered at O and the circle centered at O is circumscribed around the quadrilateral. In Figure 7.39(b), $\triangle QRS$ is circumscribed around the circle centered at P and the circle centered at P is inscribed in the triangle.

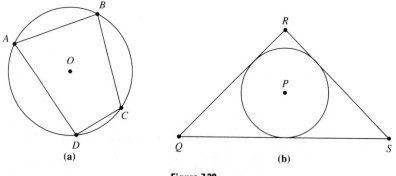

Figure 7.39

Karl Fredrick Gauss (1777–1855)

Karl Fredrick Gauss was one of the greatest mathematicians of all time. Even during his lifetime, he was called "the prince of mathematicians." He made many major contributions to arithmetic, number theory, algebra, astronomy, biology, physics, and, of course, geometry. At the age of 19, Gauss proved by considering points on a circle that a regular polygon with 17 sides can be constructed using a straightedge and a compass. He considered this to be one of his greatest achievements.

THEOREM 7.23

If a quadrilateral is inscribed in a circle, the opposite angles are supplementary.

Given: $ABCD$ is a quadrilateral inscribed in a circle (See Figure 7.40.)

Prove: $\angle A$ and $\angle C$ are supplementary and $\angle B$ and $\angle D$ are supplementary

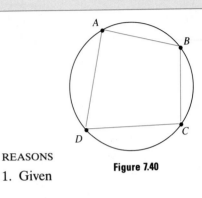

Figure 7.40

Proof:

STATEMENTS	REASONS
1. Quadrilateral $ABCD$ is inscribed in a circle	1. Given
2. $\angle A = \frac{1}{2}\overset{\frown}{BCD}$ and $\angle C = \frac{1}{2}\overset{\frown}{DAB}$	2. Measure of inscribed $\angle$
3. $\overset{\frown}{BCD} + \overset{\frown}{DAB} = 360°$	3. Arc add. post.
4. $\angle A + \angle C = \frac{1}{2}\overset{\frown}{BCD} + \frac{1}{2}\overset{\frown}{DAB}$	4. Add.-subtr. law
5. $\angle A + \angle C = \frac{1}{2}(\overset{\frown}{BCD} + \overset{\frown}{DAB})$	5. Distributive law
6. $\angle A + \angle C = \frac{1}{2}(360°)$	6. Substitution law
7. $\angle A + \angle C = 180°$	7. Simplify
8. $\angle A$ and $\angle C$ are supplementary	8. Def. of supp. $\angle$'s

We could show that $\angle B$ and $\angle D$ are supplementary in a similar manner.

The proof of the next theorem follows from Theorem 7.23 and the definition of a rectangle. The proof is requested in the exercises.

THEOREM 7.24

If a parallelogram is inscribed in a circle, then it is a rectangle.

Circles with regular polygons have many special properties. The following theorems introduce some of them.

THEOREM 7.25

If a circle is divided into n equal arcs, $n > 2$, then the chords formed by these arcs form a regular n-gon.

The proof of Theorem 7.25 follows by noting that equal arcs have equal chords, and that each angle of the *n*-gon formed is one-half the sum of four equal arcs.

THEOREM 7.26

If a circle is divided into *n* equal arcs, *n* > 2, and tangents are constructed to the circle at the endpoints of each arc, then the figure formed by these tangents is a regular *n*-gon.

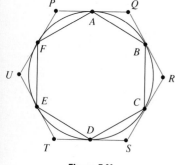

Figure 7.41

The following paragraph-style proof to this theorem shows 6 equal arcs, but note that the same proof can be applied to any number of arcs *n*, *n* > 2. Refer to Figure 7.41.

PROOF: Because $\overset{\frown}{AB} = \overset{\frown}{BC} = \overset{\frown}{CD} = \overset{\frown}{DE} = \overset{\frown}{EF} = \overset{\frown}{FA}$, the corresponding chords are also equal making $AB = BC = CD = DE = EF = FA$. Because the tangent segments to a circle from a point outside the circle are equal, $\triangle AQB$, $\triangle BRC$, $\triangle CSD$, $\triangle DTE$, $\triangle EUF$, and $\triangle FPA$ are all isosceles triangles. Because angles formed by a chord and a tangent each measure one-half the intercepted arc, and because the arcs are all equal, the base angles of the six triangles are all equal. Thus, all six triangles are congruent by ASA = ASA. Hence, $\angle P = \angle Q = \angle R = \angle S = \angle T = \angle U$ because they are corresponding parts of congruent triangles, making all angles of the hexagon equal. Also, because $\overline{PQ}, \overline{QR}, \overline{RS}, \overline{ST}, \overline{TU}$, and $\overline{UP}$ are all formed by adding two equal segments, $PQ = QR = RS = ST = TU = UP$. Thus, because all angles and all sides are equal, $PQRSTU$ is a regular hexagon. ■

If we are given a circle, and we have a way to determine *n* equal arcs on the circle, we can use Theorems 7.25 and 7.26 to inscribe and circumscribe a regular *n*-gon in and around the circle.

E X A M P L E 1   Inscribe a regular octagon in a circle.

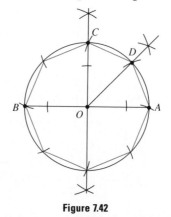

Figure 7.42

Consider the circle in Figure 7.42. Choose a point A on the circle and construct radius $\overline{OA}$, extended to form diameter $\overline{BA}$. Use Construction 2.4 to form the perpendicular bisector of $\overline{BA}$ containing $\overline{OC}$, and use Construction 2.6 to bisect $\angle AOC$ obtaining $\overline{OD}$. Using length AD, mark off eight equal arcs around the circle starting at point A. The chords formed by these arcs give the desired regular octagon. ◪

PRACTICE EXERCISE 1

Circumscribe a regular octagon around the circle given in Example 1.

We now consider the problem of circumscribing a circle around a given regular polygon. You will be asked in the exercises to prove that the construction that follows does indeed result in the required circle.

CONSTRUCTION 7.5

Construct a circle that is circumscribed around a given regular polygon.

The construction uses a regular pentagon, but note that you can use any regular polygon.

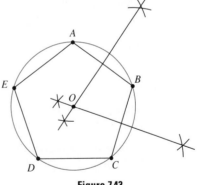

Figure 7.43

Given: Regular pentagon $ABCDE$ (See Figure 7.43.)

To Construct: A circle that is circumscribed around $ABCDE$

Construction:

1. Use Construction 2.4 to form the perpendicular bisectors of two adjacent sides, for instance $\overline{AB}$ and $\overline{BC}$. The bisectors intersect in point O.

2. Use O as the center and AO as the radius and construct a circle. This is the desired circle circumscribed around $ABCDE$.

Next we'll inscribe a circle in a given regular polygon. You will be asked in the exercises to prove that the construction results in the required circle.

CONSTRUCTION 7.6

Construct a circle that is inscribed in a given regular polygon.

The construction uses a regular hexagon, but note that you can use any regular polygon.

Given: Regular hexagon *ABCDEF* (See Figure 7.44.)

To Construct: A circle that is inscribed in *ABCDEF*

Construction:

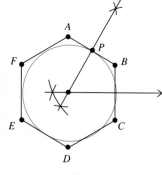

Figure 7.44

1. Use Construction 2.4 to form the perpendicular bisector of two adjacent sides, for instance $\overline{AB}$ and $\overline{BC}$. The bisectors intersect in point *O*.
2. Let *P* be the midpoint of $\overline{AB}$. Use *O* as the center and *OP* as the radius and construct a circle. This is the desired circle inscribed in *ABCDEF*.

DEFINITION: CENTER OF A REGULAR POLYGON

The **center of a regular polygon** is the center of the circle circumscribed around the polygon.

In view of Constructions 7.5 and 7.6, the center of a regular polygon is also the center of the circle that is inscribed in the polygon because the center points coincide.

DEFINITION: RADIUS OF A REGULAR POLYGON

A **radius of a regular polygon** is the segment joining the center of the polygon to one of its vertices.

THEOREM 7.27

All radii of a regular polygon are equal in length.

The proof of Theorem 7.27 is requested in the exercises.

DEFINITION: CENTRAL ANGLE OF A REGULAR POLYGON

A **central angle of a regular polygon** is an angle formed by two radii to two adjacent vertices.

THEOREM 7.28

All central angles of a regular polygon have the same measure.

The proof of Theorem 7.28 is requested in the exercises.

Because there are n equal central angles in a regular n-gon and these angles sum to 360°, the proof of the formula given in the next theorem should be obvious.

THEOREM 7.29

The measure a of each central angle in a regular n-gon is determined with the formula

$$a = \frac{360°}{n}.$$

EXAMPLE 2 Find the measure of each central angle in a regular octagon.

Because an octagon has 8 sides, by Theorem 7.29 we substitute 8 for n in the formula

$$a = \frac{360°}{n}.$$

$$a = \frac{360°}{8} = 45°$$

Thus, in Figure 7.45, central angle $\angle AOB$ has measure 45°. ◪

PRACTICE EXERCISE 2

Find the number of sides in a regular polygon if each central angle measures 24°.

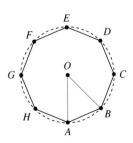

Figure 7.45

ANSWERS TO PRACTICE EXERCISES: **1.** Begin by dividing the circle into 8 equal arcs using the method shown in Example 1. Use Construction 7.1 to construct the tangent to the circle at each of these eight points. The points of intersection of these tangents form the vertices of the circumscribed octagon. **2.** 15 sides

7.4 EXERCISES

Exercises 1–8 refer to the figure below in which quadrilateral *ABCD* is inscribed in the circle centered at *O*.

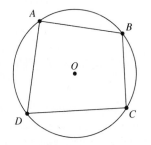

1. If $\angle A = 86°$, what is the measure of $\angle C$?   **2.** If $\angle B = 97°$, what is the measure of $\angle D$?

3. Assume that $\angle A + \angle B + \angle C = 276°$. Find $\angle B$.   **4.** Assume that $\angle B + \angle C + \angle D = 269°$. Find $\angle C$.

5. If $\overline{AB} \parallel \overline{DC}$ and $AB = DC$, find $\angle A$.   **6.** If $AB = DC$ and $AD = BC$, find $\angle B$.

7. If $\angle A = \angle C$, find $\angle C$.   **8.** If $AB = BC = CD = DA$, find $\angle D$.

9. Draw a circle and divide it into four equal arcs by constructing the perpendicular bisector of a diameter. Inscribe a square in the circle.

10. Circumscribe a square around a given circle. [Hint: Refer to Exercise 9.]

11. Draw a circle and divide it into six equal arcs. Inscribe a regular hexagon in the circle.

12. Circumscribe a regular hexagon around a given circle.

13. Draw a circle and divide it into three equal arcs. Inscribe an equilateral triangle in the circle. [Hint: You may want to divide the circle into six equal arcs first. The length of a side of a regular inscribed hexagon is equal to the radius.]

14. Circumscribe an equilateral triangle around a given circle. [Hint: Refer to Exercise 13.]

15. Construct a square and inscribe a circle in it.

16. Construct a square and circumscribe a circle around it.

17. Construct an equilateral triangle and circumscribe a circle around it.

18. Construct an equilateral triangle and inscribe a circle in it.

In Exercises 19–24, find the measure of each central angle of the given regular polygon.

19. equilateral triangle   **20.** square   **21.** regular pentagon

22. regular hexagon   **23.** regular nonagon   **24.** regular decagon

In Exercises 25–28, find the number of sides in a regular polygon if each central angle has the given measure.

25. 45°   **26.** 30°   **27.** 20°   **28.** 15°

29. Prove Theorem 7.24.

30. Prove that Construction 7.5 provides the required circumscribed circle.

31. Prove that Construction 7.6 provides the required inscribed circle.

32. Prove Theorem 7.27.

33. Prove Theorem 7.28.

34. Find the perimeter of a regular hexagon that is circumscribed around a circle with radius 4 cm.

35. Find the length of a side of an equilateral triangle that is inscribed in a circle with radius 4 cm.

36. To inscribe a regular pentagon $ABCDE$ in a circle centered at O (see the figure below), select a point A on the circle, construct radius $\overline{OA}$, and construct radius $\overline{OP} \perp \overline{OA}$. Construct the midpoint Q of $\overline{OP}$, and draw segment $\overline{QA}$. Bisect $\angle OQA$ to obtain R on $\overline{OA}$. Construct $\overline{BR} \perp \overline{OA}$. Then $\overline{AB}$ is one side of the desired pentagon. Use AB to mark off five equal arcs on the circle to complete the pentagon. Draw a circle with radius about 2 inches in length and inscribe a regular pentagon in the circle. Then circumscribe a regular pentagon around the circle by constructing lines perpendicular to radii drawn to each vertex.

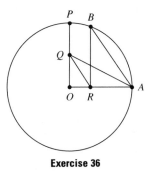

Exercise 36

7.5 Sectors, Arc Length, and Area

If we cut a piece from a pizza as shown in Figure 7.46, the slice is an example of a *sector* of a circle.

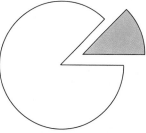

Figure 7.46   Sector of a Circle

> ### DEFINITION: SECTOR
>
> A **sector** of a circle is a region bounded by two radii of the circle and the arc of the circle determined by the radii.

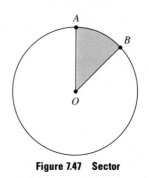

Figure 7.47　Sector

The sector of the circle shown in color in Figure 7.47, denoted by sector AOB, is formed by radii $\overline{OA}$ and $\overline{OB}$ and arc $\overparen{AB}$. Assume that the measure of $\overparen{AB}$ is 45°, that is $\angle AOB = 45°$. Then because 45° is one-eighth of 360°, the area of the sector is one-eighth the area of the circle. In view of this, it is reasonable to accept the next postulate.

> ### POSTULATE 7.7　AREA OF A SECTOR
>
> The area of a sector of a circle with radius r whose arc has measure $m°$ is determined with the formula
>
> $$A = \frac{m}{360}\pi r^2.$$

EXAMPLE 1　What is the area of a slice of pizza cut from a pizza with diameter 20 inches if the arc of the slice measures 30°?

Substitute 30 for m and 10 for r in the formula

$$A = \frac{m}{360}\pi r^2.$$
$$= \frac{30}{360}\pi(10)^2$$
$$= \frac{1}{12}\pi(100)$$
$$= \frac{25}{3}\pi$$

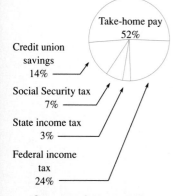

One way to form a mental image of statistical data is to display the data using a *circle graph*. The size of a sector of the circle can visually convey the information presented. The circle graph above shows the distribution of Mr. Whitney's monthly income. You have probably seen graphs like this in newspapers and magazines.

The actual area is $\frac{25}{3}\pi$ in² which can be approximated by 26.2 in², using 3.14 for π and rounding to the nearest tenth of a square inch.   ◨

We defined the measure of an arc to be the measure of its central angle. We now consider the problem of finding the *length* of an arc. In Figure 7.47, if $AB = 45°$, because 45° is one-eighth of 360°, it is reasonable to assume that the length of the arc would be one-eighth the circumference of the circle.

POSTULATE 7.8   ARC LENGTH

The **length of an arc** measuring $m°$ in a circle with radius r is determined with the formula

$$L = \frac{m}{360}2\pi r = \frac{m}{180}\pi r.$$

EXAMPLE 2   What is the length of an arc measuring 30° in a circle with radius 15 cm?

Substitute 30 for m and 15 for r in the formula

$$L = \frac{m}{180}\pi r.$$
$$= \frac{30}{180}\pi(15)$$
$$= \frac{5}{2}\pi$$

The actual length is $\frac{5}{2}\pi$ cm which can be approximated by 7.9 cm, using 3.14 for π and rounding to the nearest tenth of a centimeter.   ◨

Another important construction in a circle is a *segment*.

DEFINITION: SEGMENT OF A CIRCLE

A **segment** of a circle is a region bounded by a chord of the circle and the arc formed by the chord.

The color region in Figure 7.48 is a segment of the circle. To find the area of this segment, find the area of sector AOB and subtract the area of $\triangle AOB$.

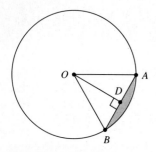

Figure 7.48 Segment of a Circle

E X A M P L E 3   Find the area of the segment of the circle in Figure 7.48 if $\angle AOB = 60°$ and $AO = 20$ cm.

Construct altitude $\overline{OD}$ of $\triangle AOB$. Because $\angle AOB = 60°$, $\angle AOD = 30°$. With $OA = 20$ cm, $DA = 10$ cm and $OD = 10\sqrt{3}$ cm using properties of a 30°-60°-right triangle. Then $AB = 2DA = 20$ cm, so

$$\text{Area } \triangle AOB = \frac{1}{2}(AB)(OD).$$
$$= \frac{1}{2}(20)(10\sqrt{3})$$
$$= 100\sqrt{3} \text{ cm}^2$$

The area of sector AOB is $\frac{60}{360}\pi(20)^2 = \frac{200}{3}\pi$ cm².

Thus, the area of the segment is

$$\left(\frac{200}{3}\pi - 100\sqrt{3}\right) \text{ cm}^2$$

which can be approximated by 36.1 cm².   ◩

Previously we considered areas of triangles and certain quadrilaterals such as parallelograms, rectangles, and trapezoids. We'll now show how to find the area of any regular polygon using the notion of an *apothem* of a regular polygon.

DEFINITION: APOTHEM OF A REGULAR POLYGON

An **apothem of a regular polygon** is a line segment from the center of the polygon perpendicular to one of its sides.

In Figure 7.49, $\overline{OP}$ is an apothem of hexagon $ABCDEF$ inscribed in the circle with center O.

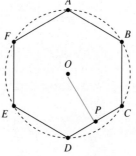

Figure 7.49 Apothem

THEOREM 7.30

Every apothem of a regular polygon has the same length.

The proof of Theorem 7.30 follows from Theorem 7.10 and is left for you to do as an exercise. Because every apothem of a regular polygon has the same length, as stated in Theorem 7.30, we often say that any apothem of a regular polygon is *the* apothem of the regular polygon.

THEOREM 7.31

The apothem of a regular polygon bisects its respective side.

The proof of Theorem 7.31 follows from Theorem 7.8 and is left for you to do as an exercise.

Consider the regular hexagon $ABCDEF$ in Figure 7.50 with apothem $\overline{OP}$. The area of $ABCDEF$ is the sum of the areas of congruent triangles $\triangle ABO$, $\triangle BCO$, $\triangle CDO$, $\triangle DEO$, $\triangle EFO$, and $\triangle FAO$. The area of $\triangle ABO$ is

$$\frac{1}{2}(AB)(OP).$$

Similarly, we can find the area of each of the five remaining triangles. If we add these areas together and simplify, we obtain

$$\frac{1}{2}(OP)(AB + BC + CD + DE + EF + FA).$$

Let p be the perimeter of $ABCDEF$, then

$$p = AB + BC + CD + DE + EF + FA.$$

Let a be the length of the apothem. Then the sum of the areas of the triangles, which is equal to the area of the regular hexagon, simplifies to

$$A = \frac{1}{2}ap.$$

A similar argument can be given for any regular polygon, which would provide the proof of the next theorem.

THEOREM 7.32   AREA OF A REGULAR POLYGON

The area of a regular polygon with apothem of length a and perimeter p is determined with the formula

$$A = \frac{1}{2}ap.$$

 Find the area of a regular hexagon with sides measuring 12 cm.

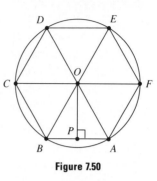

Figure 7.50

Refer to Figure 7.50. The measure of central angle $\angle AOB$ is $\frac{360°}{6} = 60°$ by Theorem 7.29. Then $\angle AOP = 30°$ so $\triangle AOP$ is a 30°-60°-right triangle. Because $AB = 12$ cm and $AP = 6$ cm, hypotenuse $AO = 12$ cm. Thus, the side opposite the 60° angle, the apothem, has length $\frac{\sqrt{3}}{2}(12) = 6\sqrt{3}$ cm. The perimeter of the hexagon is $6(12) = 72$ cm. Substitute $6\sqrt{3}$ for a and 72 for p in the formula

$$A = \frac{1}{2}ap.$$

$$A = \frac{1}{2}(6\sqrt{3})(72)$$
$$= 216\sqrt{3}$$

Thus, the exact area is $216\sqrt{3}$ cm², which can be approximated by 374.1 cm², correct to the nearest tenth of a square centimeter. ▨

PRACTICE EXERCISE 1

Use the formula $A = \frac{1}{2}ap$ to find the area of a square. Compare the result with the formula for the area of a square given in Chapter 4.

In Postulate 7.2, we said that the area of a circle is determined with the formula $A = \pi r^2$. The following example shows how the formula for the area of a regular polygon can lead us to the formula for the area of a circle. Suppose we are given a circle. If we inscribe a square in the circle, the area of the square would give an approximation for the area of the circle. It would not be a very good approximation because the area of the four segments of the circle would

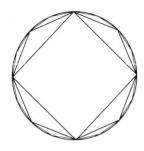

Figure 7.51

not be included. If we bisect the four equal arcs and construct the inscribed octagon as shown in Figure 7.51, the area of the octagon would clearly be a better approximation for the area of the circle.

If we repeat this process forming a regular 16-gon, the area of the 16-gon would be an even better approximation for the area of the circle. Continuing in this manner, we would obtain a better and better approximation of the area of the circle. The area of each regular polygon is

$$A = \frac{1}{2}ap,$$

and as the process continues, a is approaching the radius of the circle r, and the perimeter p is approaching the circumference of the circle, $2\pi r$. Thus, the area is approaching

$$A = \frac{1}{2}r(2\pi r) = \pi r^2.$$

NOTE: Because it is not always easy to find the length of the apothem of a regular polygon with many sides, we often approximate the area of such a polygon with the area of its circumscribed circle. If the apothem is known but the perimeter is not, approximate the perimeter using $r = a$ for the inscribed circle.

ANSWER TO PRACTICE EXERCISE: **1.** Both formulas give $A = s^2$.

7.5 Exercises

In Exercises 1–4, find the area of each grey sector and the length of its arc.

1.

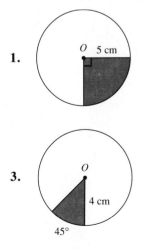

2.

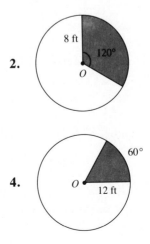

3.

4.

In Exercises 5–10, find the area of each grey region of the circle. Use 3.14 for π and give the answer correct to the nearest tenth.

5.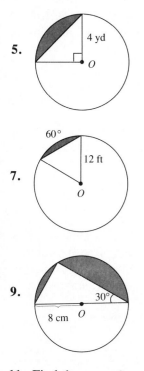

6.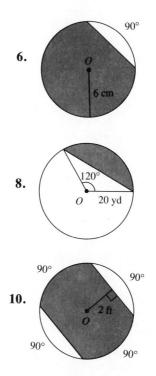

7.

8.

9.

10.

11. Find the area of a sector of a circle if the diameter of the circle is 8.2 cm and the arc of the sector is 30°. Give answer correct to the nearest tenth of a square centimeter.

12. Find the area of a sector of a circle if the diameter of the circle is 21.5 ft and the arc of the sector is 60°. Give answer correct to the nearest tenth of a square foot.

13. The area of a sector of a circle is 24π yd². If the arc of the sector is 60°, find the diameter of the circle.

14. The area of a sector of a circle is 50π cm². If the arc of the sector is 45°, find the diameter of the circle.

15. A regular polygon has perimeter 80 yd and apothem 10 yd. Find its area.

16. A regular polygon has perimeter 144 cm and apothem 18 cm. Find its area.

17. Find the area of a regular hexagon with sides 16 ft.

18. Find the area of a regular hexagon with sides 12 ft.

19. The area of a regular hexagon is $1350\sqrt{3}$ cm². Find the length of a side.

20. The area of a regular hexagon is $864\sqrt{3}$ yd². Find the length of a side.

21. Estimate the area of a regular 16-gon with apothem 12 ft. [Hint: Find the area of the inscribed circle using $A = \pi r^2$.]

22. Estimate the area of a regular 20-gon with apothem 8.5 cm.

In Exercises 23–24, find the area of the grey region. Give answer correct to the nearest tenth.

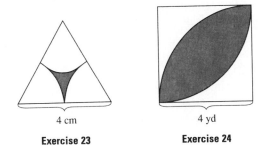

4 cm

Exercise 23

4 yd

Exercise 24

23. The arcs have their centers at the vertices of the equilateral triangle.

24. The arcs have their centers at opposite vertices of the square.

25. Prove Theorem 7.30.

26. Prove Theorem 7.31.

27. In the figure below, the two circles have a common center with radii 60 yd and 70 yd. The grey region corresponds to a jogging track. What is the area of the track, to the nearest tenth of a square yard?

Exercise 27

28. In the figure below, the externally tangent circles with centers O and P have radii 6 ft and 2 ft, respectively. What is the area of the grey region, correct to the nearest tenth of a square foot?

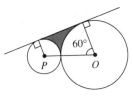

Exercise 28

KEY TERMS AND SYMBOLS

7.1 circle, p. 231
center, p. 231
radius, p. 231
diameter, p. 231
circumference, p. 231
pi (π), p. 231
semicircle, p. 234
major arc, p. 234
minor arc, p. 234
central angle, p. 235
intercepts, p. 235
measure of an arc, p. 235
inscribed angle, p. 235
7.2 chord, p. 241

bisector of an arc, p. 243
secant, p. 247
7.3 tangent, p. 252
point of tangency, p. 252
tangent internally, p. 259
tangent externally, p. 259
common tangent, p. 259
common external tangent,
p. 259
common internal tangent,
p. 259
7.4 inscribed polygon, p. 264
circumscribed polygon,
p. 264

inscribed circle, p. 264
circumscribed circle, p. 264
center of a regular
polygon, p. 268
radius of a regular
polygon, p. 268
central angle of a regular
polygon, p. 269
7.5 sector, p. 272
length of an arc, p. 273
segment (of a circle),
p. 273
apothem, p. 274

PROOF TECHNIQUES

To Prove:

Inscribed Angles Equal

1. Show that they intercept the same arc. (Corollary 7.3, p. 237)
2. Show that each is inscribed in a semicircle and is thus a right angle. (Corollary 7.4, p. 237)

Chords are Equal

1. Show that they are formed by equal arcs. (Theorem 7.7, p. 243)
2. Show that they are equidistant from the center of the circle. (Theorem 7.10, p. 244)

Line is a Tangent

1. Show it is perpendicular to a radius at the point of tangency. (Postulate 7.5, p. 253)

Figure Is a Regular *n*-gon

1. Show that it is inscribed in a circle using chords determined by dividing the circle into *n* equal arcs. (Theorem 7.25, p. 265)
2. Show that it is circumscribed around a circle using tangent lines determined by dividing the circle into *n* equal arcs. (Theorem 7.26, p. 266)

Section 7.1

1. Find the diameter of the circle with radius 2.6 cm.

2. Find the radius of the circle with diameter $\frac{4}{5}$ yd.

3. Find the approximate circumference and area of the circle with diameter 9.2 ft. Use 3.14 for π.

4. A machine part is in the shape of an equilateral triangle 10 inches on a side. A hole with diameter 3 inches is drilled in the center of the part. To the nearest tenth, what is the area of remaining metal?

5. Refer to the figure to the right to answer each question.
  (a) What is $\angle AOB$ called with respect to the circle?
  (b) What is $\angle ACB$ called with respect to the circle?
  (c) What is the measure of $\overset{\frown}{AB}$?
  (d) What is the measure of $\overset{\frown}{ACB}$?
  (e) What is the measure of $\angle ACB$?

6. Prove that vertical angles with vertices at the center of a circle intercept equal arcs on the circle.

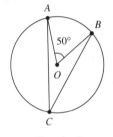

Exercise 5

Section 7.2

Exercises 7–10 refer to the figure below.

7. If $\overset{\frown}{BC} = 56°$ and $\overset{\frown}{AD} = 132°$, find $\angle BPC$.   **8.** If $\overset{\frown}{CD} = \overset{\frown}{DA}$ and $\overset{\frown}{CD} = 135°$, find $\overset{\frown}{DA}$.

9. If $\overline{BD} \perp \overline{CA}$ and $\overline{BD}$ passes through the center of the circle, find CP if $AP = 17$ cm.

10. If $AP = 5$ ft, $CP = 6$ ft, and $BP = 3$ ft, find PD.

11. If two chords in a circle are equal, what can be said about their distance from the center of the circle?

Exercises 12–14 refer to the figure below.

12. If $\overset{\frown}{BD} = 68°$ and $\overset{\frown}{AC} = 28°$, find $\angle P$.   **13.** If $\overset{\frown}{AC} = 30°$ and $\angle P = 24°$, find $\overset{\frown}{BD}$.

14. If $PB = 12$ ft, $PA = 3$ ft, and $PC = 4$ ft, find PD.

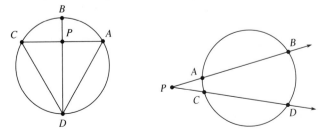

Exercises 7–10   **Exercises 12–14**

15. *Given: ABCD is a rectangle*
Prove: $\overset{\frown}{AB} = \overset{\frown}{CD}$

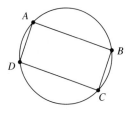

Exercise 15

16. *Given:* $\overline{AD} \perp \overline{BC}$ *and* $\overline{AD}$ *contains the center O*
Prove: $\overset{\frown}{AB} = \overset{\frown}{AC}$

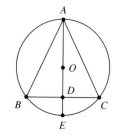

Exercise 16

Section 7.3

Exercises 17–22 refer to the figure below.

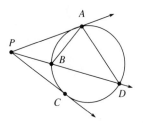

17. If $\overset{\frown}{AB} = 66°$, find $\angle PAB$.
19. If $\overset{\frown}{AD} = 130°$ and $\overset{\frown}{AB} = 64°$, find $\angle APD$.
21. If $AP = 38$ cm, find CP.
23. *Given:* $\overset{\frown}{BD} = 2\overset{\frown}{AC}$
    Prove: $PC = BC$

18. If $\angle ADB = 35°$, find $\angle PAB$.
20. If $\overset{\frown}{ABC} = 130°$, find $\angle APC$.
22. If $AP = 20$ ft and $DP = 30$ ft, find BP.
24. *Given:* $\overline{AD}$ and $\overline{DC}$ are tangent to the circle and $ABCD$ is a parallelogram
    Prove: ABCD is a rhombus

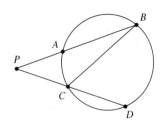

Exercise 23

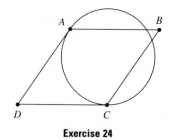

Exercise 24

25. *Given:* $\overline{AB}$ is a common internal
tangent to the circles centered at
O and P, and $\overleftrightarrow{OP}$ is the line of
centers

Prove: $\angle O = \angle P$

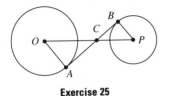

Exercise 25

26. Prove that if two circles are tangent externally, tangents to the circles
from a point on their common internal tangent are equal.

27. Construct a tangent to a circle at a point on the circle.

28. If an airplane is flying at an altitude of 5 mi, how far to the nearest tenth
of a mile is the airplane from the horizon? Use 4000 mi for the radius of
the earth.

Section 7.4

Exercises 29–30 refer to the figure below.

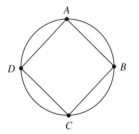

29. If $\angle D = 88°$, find $\angle B$.

30. If $\overline{AD} \parallel \overline{BC}$ and $AD = BC$, find $\angle D$.

31. Inscribe a square in a circle. Bisect a side of the square and use the result
to inscribe a regular octagon in the circle.

32. Construct a regular hexagon and inscribe a circle in it.

33. What is the measure of each central angle in a regular 18-gon?

34. How many sides does a regular polygon have if each central angle has
measure 10°?

35. Find the length of a side of a regular hexagon that is inscribed in a circle
with radius 14 cm.

36. Prove that a radius of a regular polygon bisects an interior angle of the
polygon.

Section 7.5

37. Find the area of a sector of a circle, correct to the nearest tenth of a square inch if the radius of the circle is 11.4 inches and the arc of the sector is 18°. Use 3.14 for π.

38. Find the area of a segment of a circle formed by two radii measuring 10 cm that form a central angle of 60°. Give the area correct to the nearest tenth of a square centimeter. Use 3.14 for π.

39. Find the area of a regular hexagon with sides 30 yd. Give the area correct to the nearest tenth of a square yard.

40. Estimate the area of a regular 30-gon with apothem 15.5 ft. Use 3.14 for π.

41. Find the area of the grey region in the figure below. The arcs are semicircles with centers at the midpoints of the sides of the square. Leave your answer in terms of π.

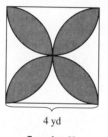

4 yd

Exercise 41

42. To the nearest tenth, what is the length of an arc in a circle of radius 5.2 ft formed by a central angle measuring 40°?

PRACTICE TEST

1. Find the circumference and area of a circle with diameter 12.6 cm. Use 3.14 for π and give answers correct to the nearest tenth.

Problems 2–5 refer to the figure below in which O is the center of the circle, $\angle BOD = 130°$, $\angle ADC = 30°$, $BE = 6$ cm, $EC = 2$ cm, $AE = 4$ cm.

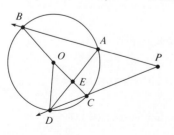

2. Find $\angle BAD$.   **3.** Find $\angle P$.   **4.** Find DE.   **5.** Find $\angle AEC$.

284

Problems 6–7 refer to the figure below in which O is the center of the circle.

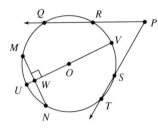

6. If $\overline{MN} \perp \overline{UV}$ and $MW = 4$ cm, find NW.

7. If $PQ = 12$ cm, $PR = 6$ cm, and $PT = 9$ cm find PS.

8. If $\overline{PA}$ and $\overline{PB}$ are two tangents to a circle at points A and B from common point P outside the circle, and $PA = 15$ ft, find PB.

9. Construct a tangent to a circle from a point outside the circle.

10. Quadrilateral $ABCD$ is inscribed in a circle with $\angle A$ opposite $\angle C$. If $\angle B = 100°$, find $\angle D$.

11. Construct an equilateral triangle and inscribe a circle in it.

12. What is the measure of each central angle in a regular octagon?

13. Find the area of a regular polygon with perimeter 56 yd and apothem 7 yd.

14. Find the area of a segment of a circle formed by the chord joining the endpoints of two radii each measuring 8 cm and forming a central angle of 60°.

15. Prove that the tangents to a circle at the ends of a diameter are parallel.

8

*I*NEQUALITIES

Up to now our study of geometry has involved proving line segments, angles, and arcs, equal. In this chapter we study relations between unequal line segments, unequal angles, and unequal arcs. We begin by reviewing properties of inequalities from beginning algebra. Then we consider properties of inequalities related to triangles and circles.

The following applied problem depends on an understanding of topics introduced in this chapter, and its solution appears in Section 8.1, Example 3.

A water-diversion project in Arizona uses a canal to bring water from the Colorado River to central Arizona. One portion of the canal, located south of two towns, is in a straight line running west to east. An engineer wants to build a pumping station on the canal at a point from which water can be supplied to the two towns. To minimize the cost of building the two pipe lines, he must find the point on the canal at which the sum of the distances from that point to the two towns is the least possible. Explain how the engineer can locate this point.

8.1 Inequalities Involving Triangles

You probably studied properties of inequalities in beginning algebra. Recall that the symbol for "is less than" is $<$ and the symbol for "is greater than" is $>$. For example,

$$5 < 8, \quad 2 < 15, \quad 4 > 1, \quad \text{and} \quad 7 > 0$$

are all true statements involving inequalities.

For real numbers a and b, we could give a precise definition of the inequality $a < b$ and go on to prove several properties of inequalities. Instead, we'll assume a familiarity with inequalities and present these properties as postulates.

> **POSTULATE 8.1   TRICHOTOMY LAW**
>
> If a and b are real numbers, exactly one of the following is true:
>
> $$a < b, a = b, \text{ or } a > b.$$

The trichotomy law states that given any two real numbers, either they are equal or one is less than the other.

> **POSTULATE 8.2   TRANSITIVE LAW**
>
> If a, b, and c are real numbers with $a < b$ and $b < c$, then $a < c$.

For example, if we know that $x < 5$ and $5 < w + 1$, then $x < w + 1$.

> **POSTULATE 8.3   ADDITION PROPERTIES OF INEQUALITIES**
>
> If a, b, c, and d are real numbers with $a < b$ and $c < d$, then
>
> $$a + c < b + c \quad \text{and} \quad a + c < b + d.$$

The first addition property states that if the same quantity is added to both sides of an inequality, the sums are unequal in the same order. The second addition property states that if unequal quantities are added to unequal quantities in the same order, the sums are also unequal in the same order.

> **POSTULATE 8.4   SUBTRACTION PROPERTIES OF INEQUALITIES**
>
> If a, b, c, and d are real numbers with $a < b$ and $c = d$, then
>
> $$a - c < b - c, a - c < b - d, \quad \text{and} \quad c - a > d - b.$$

The first two subtraction properties are similar to the addition properties. The third property, however, states that if unequal quantities are subtracted from equal quantities, the differences are unequal but in the *reverse* order. For example, $5 < 12$ and $20 = 20$, so

$$20 - 5 > 20 - 12$$
$$15 > 8.$$

POSTULATE 8.5   MULTIPLICATION PROPERTIES OF INEQUALITIES

If a, b, and c are real numbers with $a < b$, then

$$ac < bc \text{ if } c > 0 \quad \text{and} \quad ac > bc \text{ if } c < 0.$$

The first multiplication property states that when both sides of an inequality are multiplied by a *positive* number, then the products are unequal in the *same* order. For example, $4 < 7$ and $3 \cdot 4 < 3 \cdot 7$ because $12 < 21$. The second property states that if both sides of an inequality are multiplied by a *negative* number, then the products are unequal in the *reverse* order. For example, $4 < 7$ and $(-3)(4) > (-3)(7)$ because $-12 > -21$.

POSTULATE 8.6   DIVISION PROPERTIES OF INEQUALITIES

If a, b, and c are real numbers with $a < b$, then

$$\frac{a}{c} < \frac{b}{c} \text{ if } c > 0 \quad \text{and} \quad \frac{a}{c} > \frac{b}{c} \text{ if } c < 0.$$

The division property is similar to the multiplication property. Notice that when both sides of an inequality are divided by the same *negative* number, the inequality symbol is *reversed*.

POSTULATE 8.7   THE WHOLE IS GREATER THAN ITS PARTS

If a, b, and c are real numbers with $c = a + b$ and $b > 0$, then $c > a$.

In algebra class, Burford was told that a and b are two counting numbers with $a < b$. He then gave the following "proof" that $a > b$. What is wrong with Burford's "proof"?

$$a < b$$
$$a - b < 0$$
$$(a - b)^2 > 0$$
$$a^2 - 2ab + b^2 > 0$$
$$a^2 - 2ab > -b^2$$
$$a^2 + a^2 - 2ab > a^2 - b^2$$
$$2a^2 - 2ab > a^2 - b^2$$
$$2a(a - b) > (a + b)(a - b)$$
$$2a > a + b$$
$$2a - a > a - a + b$$
$$a > b$$

NOTE: In many algebra texts, Postulate 8.7 is presented as a definition of "less than" or "greater than" and the remaining postulates are proved as theorems. Note that although the postulates were stated using either $<$ or $>$, they are true for both inequality symbols.   □

The next example shows how the properties of inequalities are used to solve simple inequalities.

EXAMPLE 1   Solve the inequality.

$$2x - 3 < 3x + 5$$
$$2x - 3 + 3 < 3x + 5 + 3 \qquad \text{Add 3 to both sides (Post. 8.3)}$$
$$2x < 3x + 8 \qquad \text{Simplify}$$
$$2x - 3x < 3x - 3x + 8 \qquad \text{Subtract } 3x \text{ from both sides (Post. 8.4)}$$
$$-x < 8$$
$$(-1)(-x) > (-1)(8) \qquad \text{Multiply both sides by } -1 \text{ and reverse the inequality symbol (Post. 8.5)}$$
$$x > -8$$

Thus, the given inequality is true when x is any number greater than -8. We usually give the solution simply as $x > -8$. ◪

PRACTICE EXERCISE 1

Solve $5(2x - 1) > 12x + 7$.

The next postulate seems simple, but it will be needed to prove the following theorems involving inequalities and triangles.

POSTULATE 8.8

Each angle of a triangle is greater than 0°.

THEOREM 8.1

An exterior angle of any triangle is greater than each remote interior angle.

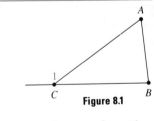

Figure 8.1

Given: $\triangle ABC$ (See Figure 8.1.)

Prove: $\angle 1 > \angle B$

Proof:

STATEMENTS	REASONS
1. $\triangle ABC$	1. Given
2. $\angle 1 = \angle B + \angle A$	2. Ext. $\angle$ = sum of remote int. $\angle$'s
3. $\angle A > 0°$	3. An $\angle$ of a $\triangle > 0°$
4. $\angle 1 > \angle B$	4. Whole is > each part

We could show $\angle 1 > \angle A$ in a similar manner.

We know that the angles opposite equal sides in a triangle are equal. The next theorem considers angles opposite unequal sides.

> **THEOREM 8.2**
>
> If two sides of a triangle are unequal, then the angles opposite those sides are unequal in the same order.

Given: $\triangle ABC$ with $AB > AC$
  (See Figure 8.2.)

Prove: $\angle ACB > \angle B$

Auxiliary line: Construct $\overline{AD}$ on $\overline{AB}$
  such that $AD = AC$ and draw $\overline{DC}$

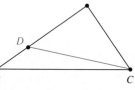

Figure 8.2

Proof:

STATEMENTS	REASONS
1. $AB > AC$	1. Given
2. $AD = AC$	2. By construction
3. $\angle ADC = \angle ACD$	3. Angles opp. $=$ sides are $=$
4. $\angle ACB = \angle ACD + \angle DCB$	4. Angle add. post.
5. $\angle DCB > 0°$	5. An $\angle$ of a $\triangle > 0°$
6. $\angle ACB > \angle ACD$	6. Whole is $>$ each part
7. $\angle ACB > \angle ADC$	7. Substitution law
8. $\angle ADC > \angle B$	8. Ext. $\angle$ of $\triangle$ is $>$ a remote int. $\angle$
9. $\angle ACB > \angle B$	9. Transitive law

The converse of Theorem 8.2 is also true.

> **THEOREM 8.3**
>
> If two angles of a triangle are unequal, then the sides opposite those angles are unequal in the same order.

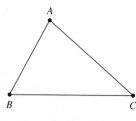

Figure 8.3

We give a paragraph-style indirect proof of this theorem making use of the trichotomy law and referring to $\triangle ABC$ in Figure 8.3.

PROOF: Suppose we are given that $\angle B > \angle C$. We must prove that $AC > AB$. By the trichotomy law, one of the following is true.

$$AC < AB \quad \text{or} \quad AC = AB \quad \text{or} \quad AC > AB.$$

If we show that the first two possibilities lead to contradictions, we know that $AC > AB$.

Assume that $AC < AB$. Then by Theorem 8.2, $\angle B < \angle C$, a contradiction, because we are given that $\angle B > \angle C$.

Assume that $AC = AB$. Then $\angle B = \angle C$ because angles opposite equal sides in a triangle are equal. This too is a contradiction because we are given that $\angle B > \angle C$.

Thus, $AC > AB$ because it is the only possibility that remains using the trichotomy law. ∎

E X A M P L E 2 In $\triangle ABC$, $AB = 12$ cm, $BC = 10$ cm, and $AC = 14$ cm. Which angle of the triangle is the smallest? the largest?

By Theorem 8.2, because $AC > AB > BC$, and $\angle B$ is opposite $\overline{AC}$, $\angle C$ is opposite $\overline{AB}$, and $\angle A$ is opposite $\overline{BC}$, $\angle B > \angle C > \angle A$. Thus, $\angle A$ is the smallest angle and $\angle B$ is the largest angle. ◻

PRACTICE EXERCISE 2

In $\triangle PQR$, $\angle P = 75°$ and $\angle Q = 65°$. Which side is the shortest? the longest?

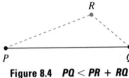

Figure 8.4 $PQ < PR + RQ$

In finding the shortest distance between two points such as P and Q in Figure 8.4, most of us would agree that it is a shorter distance to go from P to Q along the segment $\overline{PQ}$ than to go through a point off of $\overline{PQ}$. In other words, if R is a point *not* on PQ, $PQ < PR + RQ$ for any location of R. This example is a direct result of the next theorem.

THEOREM 8.4 THE TRIANGLE INEQUALITY THEOREM

The sum of the lengths of any two sides of a triangle is greater than the length of the third side.

Given: $\triangle ABC$ (See Figure 8.5.)

Prove: $AB + BC > AC$

Auxiliary lines: Construct $\overline{BD}$ by extending $\overline{AB}$ and locating D on $\overleftrightarrow{AB}$ such that $BD = BC$; then construct $\overline{CD}$

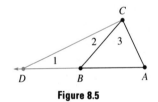

Figure 8.5

Proof:

STATEMENTS	REASONS
1. $\triangle ABC$ is a triangle	1. Given
2. Point D is on $\overleftrightarrow{AB}$ with $BD = BC$	2. By construction
3. $\triangle DBC$ is a triangle	3. Def. of $\triangle$
4. $\angle 1 = \angle 2$	4. $\angle$'s opp. = sides are =
5. $\angle ACD = \angle 2 + \angle 3$	5. Angle add. post.
6. $\angle 3 > 0°$	6. An $\angle$ of a $\triangle > 0°$
7. $\angle ACD > \angle 2$	7. Whole is > each part
8. $\angle ACD > \angle 1$	8. Substitution law
9. In $\triangle ACD$, $AD > AC$	9. Sides opp. $\neq$ $\angle$'s are $\neq$ in same order

10. $AD = AB + BD$ 10. Segment add. post
11. $AB + BD > AC$ 11. Substitution law
12. $AB + BC > AC$ 12. Substitution law

We could also show that $AC + CB > AB$ and $BA + AC > BC$ in a similar manner.

We can now solve the applied problem given in the chapter introduction.

E X A M P L E 3   A water-diversion project in Arizona uses a canal to bring water from the Colorado River to central Arizona. One portion of the canal, located to the south of two towns, is in a straight line running west to east. An engineer wants to build a pumping station on the canal at a point from which water can be supplied to the two towns. To minimize the cost of building the two pipe lines, he must find the point on the canal at which the sum of the distances from that point to the two towns is the least possible. Explain how the engineer can locate this point.

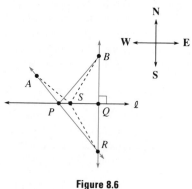

Figure 8.6

Figure 8.6 shows the information given in the problem with ℓ as the canal and A and B as the two towns. To locate the point P for the pumping station, the engineer constructs the line through B perpendicular to ℓ, locating point Q. He then determines point R on $\overleftrightarrow{BQ}$ such that $QR = BQ$. He draws line $\overleftrightarrow{AR}$ determining point P on ℓ, and constructs $\overline{PB}$. He then concludes that $\overline{AP}$ and $\overline{BP}$ determine the two pipe lines.

To verify this, he chooses any other point on ℓ, for instance S, and shows that $AS + BS > AP + BP$. By the triangle inequality theorem, in $\triangle ASR$

$$AS + SR > AR = AP + PR.$$

But because ℓ is the perpendicular bisector of $\overline{BR}$, and S and P are on ℓ, $SR = BS$ and $BP = PR$. Substituting,

$$AS + BS > AP + BP.$$

Thus, P is the desired location for the pumping station.   ◪

The next theorem considers inequalities relative to two triangles.

> **THEOREM 8.5 SAS INEQUALITY THEOREM**
>
> If two sides of one triangle are equal to two sides of another triangle, and the included angle of the first is greater than the included angle of the second, then the third side of the first triangle is greater than the third side of the second triangle.

This theorem is proved in three cases depending on whether the endpoint of a constructed segment is located on, inside, or outside the given triangle. You will be asked to verify each case in the exercises.

> **PRACTICE EXERCISE 3**
>
> In $\triangle ABC$, $AB = 8$ ft, $BC = 10$ ft, and $\angle B = 36°$. In $\triangle DEF$, $DE = 8$ ft, $EF = 10$ ft, and $\angle E = 42°$. What is the relationship between AC and DF?

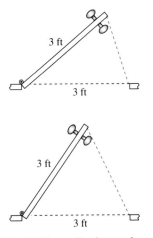

3 ft

3 ft

3 ft

3 ft

The SAS inequality theorem is sometimes called the Hinge Theorem. Consider the two top views of a door shown above. The door and the frame are each 3 ft wide. As the door opens, the angle at the hinge (the vertex of the angle) increases, and the opening between the edge of the door and the frame also increases.

> **THEOREM 8.6 SSS INEQUALITY THEOREM**
>
> If two sides of one triangle are equal to two sides of another triangle, and the third side of the first is greater than the third side of the second, then the included angle of the first triangle is greater than the included angle of the second triangle.

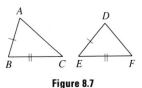

Figure 8.7

PROOF: The proof of Theorem 8.6 is indirect and uses the trichotomy law. Refer to Figure 8.7.

Assume that $\triangle ABC$ and $\triangle DEF$ are given with $AB = DE$, $BC = EF$, and $AC > DF$ as shown in Figure 8.7. We must show that $\angle B > \angle E$. By the trichotomy law,

$$\angle B < \angle E, \quad \angle B = \angle E, \quad \text{or} \quad \angle B > \angle E.$$

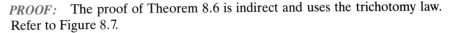

If we can show that the first two possibilities lead to contradictions, then we know that $\angle B > \angle E$, the desired conclusion.

Assume that $\angle B < \angle E$. By the SAS Inequality Theorem we have $AC < DF$, a contradiction.

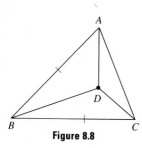

Figure 8.8

Assume that $\angle B = \angle E$, then $\triangle ABC \cong \triangle DEF$ by SAS $\cong$ SAS making $AC = DF$ by cpoctae, again a contradiction.

Because two of the possibilities cannot occur, we know that the third possibility, $\angle B > \angle E$, must be true by the trichotomy law. ■

E X A M P L E   4   In Figure 8.8, $AB = BC$ and $AD > DC$. What is the relationship between $\angle ABD$ and $\angle DBC$?

We can use the SSS Inequality Theorem on $\triangle ADB$ and $\triangle DBC$. Because $AB = BC$ and $BD = BD$, $AD > DC$ implies that $\angle ABD > \angle DBC$. ▨

ANSWERS TO PRACTICE EXERCISES: **1.** $x < -6$ **2.** The shortest side is $\overline{PQ}$, and the longest side is $\overline{QR}$. **3.** $DF > AC$

8.1 EXERCISES

In Exercises 1–10, name the postulate that explains why each statement is true.

1. If $w < x$ and $x < 2$ then $w < 2$.

2. If $u < v$ then $u + 5 < v + 5$.

3. If $x < 12$ then $\dfrac{x}{2} < 6$.

4. If a is a real number, then $a < 10$, $a = 10$, or $a > 10$.

5. If $y = 5$, then $y - 2 > 5 - 4$.

6. If $w > 8$, then $5w > 40$.

7. If $x = z + 3$, then $x > z$.

8. If $\overline{AB}$ and $\overline{CD}$ are two segments, then $AB = CD$, $AB < CD$, or $AB > CD$.

9. If $\angle A > \angle B$ and $\angle B > 90°$, then $\angle A > 90°$.

10. If $\overline{AB}$, $\overline{BC}$, and $\overline{AC}$ are segments with $BC > 0$ and $AC = AB + BC$, then $AC > AB$.

In Exercises 11–18, solve each inequality and give a reason for each step in the solution.

11. $x + 3 < 7$

12. $-2y > 10$

13. $1 - 3x < 8$

14. $2y + 3 > 5y - 3$

15. $\dfrac{2x - 1}{3} < 5$

16. $\dfrac{y + 3}{-2} < 17$

17. $3(2x + 8) < 4(x - 3)$

18. $5(y + 3) + 1 > y - 4$

Exercises 19–30 refer to the figure below. Answer *true* or *false*. If the answer is false, explain why.

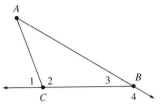

19. $\angle A = \angle 1$.   **20.** $\angle A < \angle 1$.   **21.** $\angle 4 > \angle 2$.   **22.** $\angle 4 < \angle A$.

23. If $AC = 7$ cm, $AB = 12$ cm, and $\angle 2 = 110°$, then $\angle 3 < 110°$.

24. If $AC = 6$ ft, $BC = 7$ ft, and $\angle A = 33°$, then $\angle 3 < 33°$.

25. If $\angle A = 40°$, $\angle 3 = 38°$, and $AC = 16$ yd, then $BC < 16$ yd.

26. If $\angle 3 = 31°$, $\angle 2 = 98°$, and $AB = 22$ cm, then $AC < 22$ cm.

27. If $AB = 25$ ft, $AC = 15$ ft, and $BC = 20$ ft, then $\angle C > \angle A > \angle 3$.

28. If $\angle A = 34°$ and $\angle 3 = 36°$, then $AB < AC < BC$.

29. If $AB = 10$ cm and $AC = 7$ cm, then $BC < 17$ cm.

30. If $AC = 7.2$ yd and $CB = 6.3$ yd, then $AB > 13.5$ yd.

Exercises 31–34 refer to the figure below.

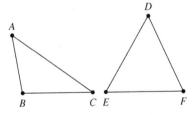

31. If $AB = EF$, $BC = DF$, and $\angle B > \angle F$, what is the relationship between AC and DE?

32. If $AC = DE$, $BC = DF$, and $\angle C > \angle D$, what is the relationship between AB and EF?

33. If $AC = 12$ cm, $BC = 8$ cm, $AB = 7$ cm, $DE = 12$ cm, $DF = 8$ cm, and $EF = 6$ cm, what is the relationship between $\angle C$ and $\angle D$?

34. If $AC = 10$ ft, $BC = 7$ ft, $AB = 6$ ft, $DE = 10$ ft, $DF = 8$ ft, and $EF = 6$ ft, what is the relationship between $\angle A$ and $\angle E$?

35. Explain why it is impossible to construct a triangle with sides 3 inches, 4 inches, and 8 inches.

36. The foreman of a ranch told his son Cal to measure the sides of a triangular pasture. Cal returned and told his father the sides are 10 mi, 12 mi, and 25 mi. Why was Cal sent to do the job again?

Exercises 37–39 present the proof of Theorem 8.5 by considering three cases.

Given: $\triangle ABC$, $\triangle DEF$, $AB = DE$,
  $BC = EF$, and $\angle B > \angle E$

Auxiliary line: Construct $\overrightarrow{BP}$ such
  that $\angle PBC = \angle E$ and locate
  point Q on $\overrightarrow{BP}$ such that
  $BQ = ED$

37. Assume that Q is on $\overline{AC}$. Refer to the figure below.

  Prove: $AC > DF$

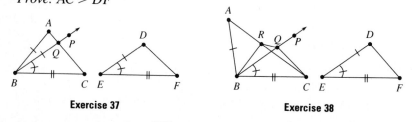

Exercise 37   **Exercise 38**

38. Assume that Q is not on $\overline{AC}$ but outside $\triangle ABC$. Refer to the figure above. Construct the bisector of $\angle ABQ$ and call the point of intersection with $\overline{AC}$ point R. Construct $\overline{RQ}$ and $\overline{QC}$.

  Prove: $AC > DF$

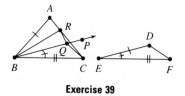

Exercise 39

39. Assume that Q is not on $\overline{AC}$ but inside $\triangle ABC$. Refer to the figure above. Construct the bisector of $\angle ABQ$ and call the point of intersection with $\overline{AC}$ point R. Construct $\overline{RQ}$ and $\overline{QC}$.

  Prove: $AC > DF$

40. *Given:* $\overrightarrow{CD}$ bisects $\angle ACB$

  Prove: $\angle 1 > \angle 2$

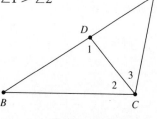

Exercise 40

41. *Given:* $\angle A > \angle B$ and $\angle D > \angle E$

  Prove: $BE > AD$

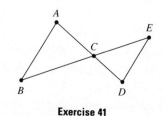

Exercise 41

42. Prove that a diagonal of a rectangle is longer than any side.

43. Prove that the difference between the lengths of two sides of a triangle is less than the third side.

44. Prove that the perimeter of a quadrilateral is greater than the sum of its diagonals.

45. Prove that the sum of the lengths of the line segments drawn from any point inside a triangle to the vertices is greater than one-half the perimeter of the triangle.

46. Two cabins are located to the west of a stream that flows north to south as shown in the figure below. The owners of the cabins want to build a pumping station on the streambank so that the sum of the distances from the station to the cabins is the least possible. Explain how the owners should locate the point at which to build the pumping station.

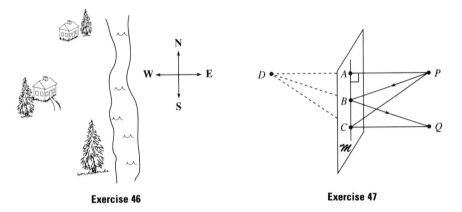

Exercise 46   Exercise 47

47. A light source at point P is reflected off the mirror M from P to Q. Prove that the path of the ray of light from P to B to Q is shorter than the path from P to C to Q where C is any other point on line $\overline{AB}$ in the mirror. [*Hint: PA = AD* because *D* is the perpendicular reflection of *P* in the mirror. Show $DB + BQ = DQ < DC + CQ.$]

48. Two sides of a triangular pasture are 3 mi and 5 mi. What is the possible range of values for the length of the third side? [*Hint:* Let x be the length of the third side, and solve the system of three inequalities obtained using the triangle-inequality theorem.]

49. The air distances from Phoenix to Denver, San Francisco, and Dallas are 589 mi, 651 mi, and 868 mi, respectively. Use the triangle-inequality theorem to find minimum and maximum air distances between
(a) Denver and San Francisco, (b) San Francisco and Dallas, and
(c) Denver and Dallas.

8.2 *Inequalities Involving Circles*

In Section 8.1 we considered inequalities related to triangles. In this section we will study inequalities that involve circles.

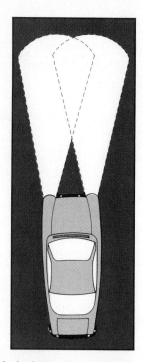

In the design of the headlights of an automobile, engineers use properties of arcs intercepted by central angles to determine the size of the illuminated area.

THEOREM 8.7

In the same circle or in congruent circles, the greater of two central angles intercepts the greater of two arcs.

Given: Circle with center O, central angles $\angle 1$ and $\angle 2$, with $\angle 1 > \angle 2$

Prove: $\overset{\frown}{AB} > \overset{\frown}{CD}$

Proof:

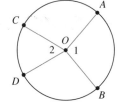

Figure 8.9

STATEMENTS	REASONS
1. $\angle 1$ and $\angle 2$ are central angles of the circle with center O and $\angle 1 > \angle 2$	1. Given
2. $\overset{\frown}{AB} = \angle 1$ and $\overset{\frown}{CD} = \angle 2$	2. Def. of measure of an arc
3. $\overset{\frown}{AB} > \overset{\frown}{CD}$	3. Substitution law

The proof of Theorem 8.7 for the case involving congruent circles is left for you to do as an exercise. The converse of this theorem is also true.

THEOREM 8.8

In the same circle or in congruent circles, the greater of two arcs is intercepted by the greater of two central angles.

The proof of Theorem 8.8 is left for you to do as an exercise.

THEOREM 8.9

In the same circle or in congruent circles, the greater of two chords forms the greater arc.

Given: Circle centered at O with chords $\overline{AB}$ and $\overline{CD}$ such that $AB > CD$ (See Figure 8.10.)

Prove: $\overset{\frown}{AB} > \overset{\frown}{CD}$

Auxiliary lines: Construct radii $\overline{OA}$, $\overline{OB}$, $\overline{OC}$, and $\overline{OD}$

Proof:

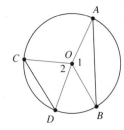

Figure 8.10

STATEMENTS	REASONS
1. $\overline{AB}$ and $\overline{CD}$ are chords of the circle with center O such that $AB > CD$	1. Given

2. $OA = OB = OC = OD$ 2. Radii are all $=$
3. $\angle 1 > \angle 2$ 3. SSS ineq. thm.
4. $\overset{\frown}{AB} > \overset{\frown}{CD}$ 4. The $>$ of 2 central $\angle$'s intercepts
                                                        the $>$ arc

The proof of Theorem 8.9 for the case involving congruent circles is left for you to do as an exercise. The converse of this theorem is also true.

THEOREM 8.10

In the same circle or in congruent circles, the greater of two arcs has the greater chord.

PRACTICE EXERCISE 1

Complete the proof of Theorem 8.10.

Given: Arcs $\overset{\frown}{AB}$ and $\overset{\frown}{CD}$ in the circle with center O such that $\overset{\frown}{AB} > \overset{\frown}{CD}$

Prove: $AB > CD$

Auxiliary lines: Construct radii $\overline{OA}$, $\overline{OB}$, $\overline{OC}$, and $\overline{OD}$ and chords $\overline{AB}$ and $\overline{CD}$

Proof:

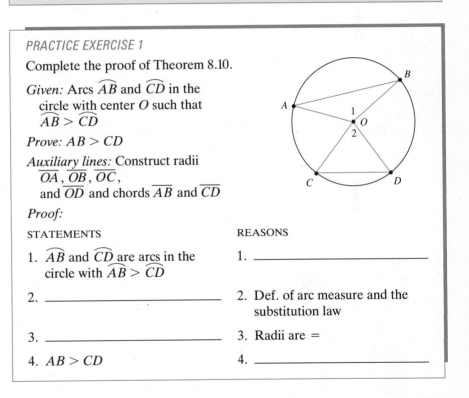

STATEMENTS	REASONS
1. $\overset{\frown}{AB}$ and $\overset{\frown}{CD}$ are arcs in the circle with $\overset{\frown}{AB} > \overset{\frown}{CD}$	1. _____
2. _____	2. Def. of arc measure and the substitution law
3. _____	3. Radii are $=$
4. $AB > CD$	4. _____

The proof of the remaining part of Theorem 8.10 for congruent circles is left for you to do as an exercise.

THEOREM 8.11

In the same circle or in congruent circles, the greater of two unequal chords is nearer the center of the circle.

For Theorem 8.11, we will give the proof for congruent circles and leave the proof for the same circle for you to do as an exercise.

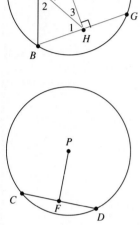

Given: Congruent circles with centers O and P, chords $\overline{AB}$ and $\overline{CD}$ with $AB > CD$, $\overline{OE} \perp \overline{AB}$ and $\overline{PF} \perp \overline{CD}$ (See Figure 8.11.)

Prove: $OE < PF$

Auxiliary lines: Construct chord $\overline{BG}$ with $BG = CD$, construct $\overline{OH}$ such that $\overline{OH} \perp \overline{BG}$, and construct $\overline{EH}$

Proof:

Figure 8.11

STATEMENTS	REASONS
1. $\overline{AB}$ and $\overline{CD}$ are chords in congruent circles with $AB > CD$, $\overline{OE} \perp \overline{AB}$, and $\overline{PF} \perp \overline{CD}$	1. Given
2. $\overline{BG}$ is a chord with $BG = CD$, $\overline{OH} \perp \overline{BG}$, and $\overline{EH}$ is a segment	2. By construction
3. $AB > BG$	3. Substitution law
4. $\frac{1}{2}AB > \frac{1}{2}BG$	4. Mult. prop. of $>$
5. $EB = \frac{1}{2}AB$ and $BH = \frac{1}{2}BG$	5. Line $\perp$ to chord through the center bisects the chord
6. $EB > BH$	6. Substitution law
7. $\angle 1 > \angle 2$	7. Using Statement 6 and $\angle$'s opp $\neq$ sides are $\neq$ in same order
8. $\angle OHB = \angle OEB$	8. Rt. $\angle$'s are $=$
9. $\angle 1 + \angle 3 = \angle 2 + \angle 4$	9. $\angle$ add. post. and substitution law
10. $\angle 3 < \angle 4$	10. Using Statements 7 and 9 and the subtr. prop. of $>$
11. $OE < OH$	11. Sides opp $\neq$ $\angle$'s are $\neq$ in same order
12. $OH = PF$	12. Equal chords are $=$ distance from center
13. $OE < PF$	13. Substitution law

The converse of Theorem 8.11 is also true and can be proved by reversing the order of the statements in the proof above. The proof is left for you to do as an exercise.

> **THEOREM 8.12**
>
> In the same circle or in congruent circles, if two chords have unequal distances from the center of the circle, the chord nearer the center is greater.

NOTE: Theorem 8.11 can also be stated using "the *smaller* of two unequal chords is *farther from* the center"; and Theorem 8.12 using "the chord *farther from* the center is *smaller.*" □

The next corollary follows directly from Theorem 8.12 using the fact that a diameter passes through the center of a circle making its distance from the center zero.

> **COROLLARY 8.13**
>
> Every diameter of a circle is greater than any other chord that is not a diameter.

ANSWER TO PRACTICE EXERCISE: **1.** 1. Given 2. $\angle 1 > \angle 2$ 3. $OA = OB = OC = OD$ 4. SAS ineq. thm.

8.2 EXERCISES

In Exercises 1–8, answer *true* or *false*. If the answer is false, explain why.

1. If $\angle A$ and $\angle B$ are two central angles in a circle with $\angle A > \angle B$, then the arc intercepted by $\angle A$ is greater than the arc intercepted by $\angle B$.

2. If $\overline{XY}$ and $\overline{WZ}$ are two chords in a circle with $WZ > XY$, then $\overparen{XY} > \overparen{WZ}$.

3. If $\overline{AB}$ and $\overline{CD}$ are two chords in a circle with $AB > CD$, then $\overline{AB}$ is nearer the center than $\overline{CD}$.

4. If $\overline{PQ}$ and $\overline{RS}$ are two chords in a circle with $\overline{PQ}$ nearer the center, then $PQ > RS$.

5. If $\overparen{AB}$ and $\overparen{CD}$ are two arcs in a circle with $\overparen{AB} > \overparen{CD}$, then chord $\overline{CD}$ is nearer the center of the circle.

6. The length of any chord in a circle is less than or equal to the length of a diameter.

7. If a central angle in a circle measures 45°, the chord determined by the angle is shorter than a radius. [*Hint:* What angle determines a chord equal to a radius?]

8. If a central angle in a circle measures 70°, the chord determined by the angle is shorter than a radius.

Exercises 9–16 refer to the figure at the right.

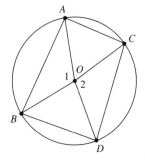

9. If $\angle 1 = 130°$ and $\angle 2 = 110°$, what is the relationship between $\overset{\frown}{AB}$ and $\overset{\frown}{CD}$?

10. If $\overset{\frown}{AC} = 60°$ and $\overset{\frown}{BD} = 70°$, what is the relationship between $\angle BOD$ and $\angle COA$?

11. If $AB = 13.5$ cm and $CD = 12.5$ cm, what is the relationship between $\overset{\frown}{AB}$ and $\overset{\frown}{CD}$?

12. If $\overset{\frown}{BD} = 6.5$ ft and $\overset{\frown}{AC} = 6.0$ ft, what is the relationship between $\overline{AC}$ and $\overline{BD}$?

13. If $\overset{\frown}{AB} = 125°$ and $\overset{\frown}{CD} = 120°$, what is the relationship between $\overline{AB}$ and $\overline{CD}$?

Exercises 9–16

14. If $AC = 5.5$ cm and $BD = 6.2$ cm, which of $\overline{AC}$ and $\overline{BD}$ is nearer center O?

15. If $AB = 10.2$ yd and $CD = 9.8$ yd, which of $\overline{AB}$ and $\overline{CD}$ is farther from center O?

16. If $\overline{BC}$ is a diameter of the circle and $\angle 1 < 180°$, which of $\overline{AB}$ and $\overline{BC}$ is nearer the center O?

17. Prove Theorem 8.7 for congruent circles.

18. Prove Theorem 8.8 for the same circle.

19. Prove Theorem 8.8 for congruent circles.

20. Prove Theorem 8.9 for congruent circles.

21. Prove Theorem 8.10 for congruent circles.

22. Prove Theorem 8.11 for the same circle.

23. Prove Theorem 8.12.

24. Prove that in a circle a chord determined by an arc of 180° is twice as long as a chord determined by an arc of 60°.

25. If a square and an equilateral triangle are inscribed in a circle, prove that the apothem of the square is greater than the apothem of the triangle.

26. If $\triangle ABC$ is an isosceles triangle with base $\overline{AB}$ inscribed in a circle and $\angle A > 60°$, prove that $\overline{BC}$ is nearer the center than $\overline{AB}$.

27. If $\triangle ABC$ is inscribed in a circle and $\angle A > \angle B$, prove that $\overset{\frown}{BC} > \overset{\frown}{AC}$.

28. *Given:* $\overline{AB}$ and $\overline{CD}$ are chords in a circle centered at O, $AB > CD$, and $\overline{AB}$ and $\overline{CD}$ intersect at P. $\overline{EF}$ is a diameter containing P

Prove: $\angle 1 < \angle 2$

Auxiliary lines: Construct $\overline{RS}$ through P with $\angle RPO = \angle 2$.

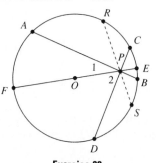

Exercise 28

PROOF TECHNIQUES

To Prove:

Two Angles Unequal

1. Show one is an exterior angle of a triangle and the other is a remote interior angle. (Theorem 8.1, p. 289)
2. Show they are angles opposite unequal sides in a triangle. (Theorem 8.2, p. 290)
3. Show they are angles in two triangles included between equal corresponding sides but opposite unequal sides. (Theorem 8.6, p. 293)
4. Show they are central angles in the same circle or congruent circles that intercept unequal arcs. (Theorem 8.8, p. 298)

Two Segments Unequal

1. Show they are sides of a triangle opposite unequal angles of the triangle. (Theorem 8.3, p. 290)
2. Show they are third sides of two triangles in which the remaining two sides of one are equal to the remaining two sides of the other, and the included angle of one is unequal to the included angle of the other. (Theorem 8.5, p. 293)
3. Show they are chords of the same or congruent circles that form unequal arcs. (Theorem 8.10, p. 299)
4. Show they are chords in the same circle or in congruent circles whose distances from the center of the circle(s) are unequal. (Theorem 8.12, p. 301)

Two Arcs Unequal

1. Show they are arcs in the same circle or in congruent circles intercepted by unequal central angles. (Theorem 8.7, p. 298)
2. Show they are arcs in the same circle or in congruent circles formed by unequal chords. (Theorem 8.9, p. 298)

REVIEW EXERCISES

Section 8.1

In Exercises 1–5, name the postulate that explains why each statement is true.

1. If $\angle A < \angle B$ and $\angle B < \angle C$ then $\angle A < \angle C$.
2. If $x < 8$, then $x - 3 < 8 - 3$.
3. If $-3x < 9$, then $x > -3$.

4. If $\overline{AB}$ and $\overline{PQ}$ are segments, then $AB = PQ$, $AB < PQ$, or $AB > PQ$.

5. If $\angle A = \angle B + \angle C$ and $\angle C = 20°$, then $\angle A > \angle B$.

In Exercises 6 and 7, solve each inequality and give a reason for each step in the solution.

6. $\dfrac{x + 3}{-2} > 7$

7. $2(y - 1) < 3y + 5$

Exercises 8–12 refer to the figure below. Answer *true* or *false*. If the answer is false, explain why.

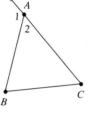

8. $\angle B > \angle 1$

9. $\angle 1 > \angle C$

10. If $AC = 10.2$ cm, $AB = 9.6$ cm, and $\angle C = 42°$, then $\angle B > 42°$.

11. If $\angle C = 53°$ and $\angle 2 = 64°$, then $BC < AC < AB$.

12. If $AC = 11.5$ yd and $BC = 10.8$ yd, then $AB < 22.3$ yd.

13. In $\triangle ABC$ and $\triangle DEF$, $AB = EF$, $BC = DF$, and $\angle B < \angle F$. What is the relationship between AC and DE?

14. Is it possible to have a triangle with sides measuring 10 ft, 12 ft, and 23 ft?

15. In $\triangle ABC$ median $\overline{CP}$ makes $\angle APC > \angle BPC$. Prove that $AC > BC$.

Section 8.2

Exercises 16–20 refer to the figure below. Answer *true* or *false*. If the answer is false, explain why.

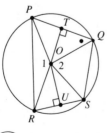

16. If $\angle 1 > \angle 2$, then $\overparen{PR} > \overparen{QS}$.

17. If $\angle 1 > \angle 2$, then $PR < QS$.

18. If $PQ > RS$, then $TO > UO$.

19. If $\overparen{PQ} > \overparen{RS}$, then $PQ < RS$.

20. If $TO > UO$, then $\overparen{PQ} < \overparen{RS}$.

21. In a circle, if $\overarc{AB} = 2\overarc{CD}$ is $AB = 2CD$?

22. If $\overline{AB}$ is a diameter of a circle, C is a point on the circle intersecting ray $\overrightarrow{AC}$ of $\angle BAC$, and $\angle BAC = 35°$, prove that $BC < AC$.

PRACTICE TEST

1. Name the postulate that explains why the following statement is true: If $30° < \angle B < \angle A$ then $\angle B - 10° < \angle A - 10°$.

2. Solve $4(x - 3) > 5(x - 2)$.

3. If $\angle E$ is an exterior angle of $\triangle ABC$ adjacent to $\angle A$, what is the relationship between $\angle E$ and $\angle B$?

4. In $\triangle ABC$ and $\triangle DEF$, $AC = DF$, $BC = EF$, and $AB < DE$. What is the relationship between $\angle C$ and $\angle F$?

5. In $\triangle ABC$, if $AB = 10$ ft and $BC = 8$ ft, then AC must be less than ___?___ ft and more than ___?___ ft.

Problems 6–8 refer to the figure below.

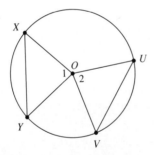

6. If $\overarc{XY} > \overarc{UV}$, what is the relationship between $\angle 1$ and $\angle 2$?

7. If $\overarc{XY} < \overarc{UV}$, which of $\overline{XY}$ or $\overline{UV}$ is nearer center O?

8. If $\angle 1 > \angle 2$, what is the relationship between $\overline{XY}$ and $\overline{UV}$?

9. Prove that the diagonals of a rhombus that is not a square are unequal.

10. If $\triangle ABC$ is an isosceles triangle with base $\overline{AB}$ inscribed in a circle and $\angle A < 60°$, prove that $\overline{AB}$ is nearer the center than $\overline{BC}$.

9

SOLID GEOMETRY

To this point, we have discussed only figures that lie completely in a single plane. In this chapter we consider planes, lines, and figures that can be formed in space.

One of our main considerations will be determining the surface area and volume of solid figures. Many applications, such as the one that follows, are based on surface area and volume. The solution for this application appears in Section 9.4, Example 2.

A geodesic dome is a structure composed of many triangular pieces that form a spherical shape. As a result, the surface area and volume of such a dome can be approximated using properties of a sphere. Find the number of square meters of synthetic material required to cover a geodesic dome that has a hemisphere 42 m in diameter. What is the volume enclosed by the structure?

9.1   PLANES AND THE POLYHEDRON

In a plane, either two distinct lines are parallel or they intersect. This is not true for lines in space, but a similar property does exist for a line and a plane. Figure 9.1 illustrates.

DEFINITION: LINE PARALLEL TO A PLANE

A line is **parallel** to a plane if it does not intersect the plane.

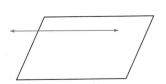

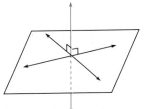

Figure 9.1   Line Parallel to Plane   **Figure 9.2   Line Perpendicular to Plane**

If a line does intersect a plane, it can be *perpendicular* to the plane. See Figure 9.2.

DEFINITION: LINE PERPENDICULAR TO A PLANE

A line is **perpendicular** to a plane if each line in the plane that passes through the point of intersection is perpendicular to the line.

We can also define parallel planes. Refer to Figure 9.3.

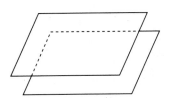

Figure 9.3   Parallel Planes

DEFINITION: PARALLEL PLANES

Two planes are **parallel** if they do not intersect.

Before we define perpendicular planes, consider the following postulate about intersecting planes.

POSTULATE 9.1

The intersection of two distinct planes is a line.

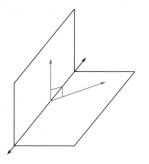

Figure 9.4   Perpendicular Planes

Refer to Figure 9.4 for the following definition.

DEFINITION: PERPENDICULAR PLANES

Two planes are **perpendicular** if either plane contains a line that is perpendicular to the other plane.

There are, of course, planes that intersect but that are not perpendicular.

DEFINITION: OBLIQUE PLANES

If two planes or a line and a plane intersect but are not perpendicular, they are called **oblique.**

With this background, we can now consider solid figures formed by the intersection of planes.

DEFINITION: POLYHEDRON

A solid figure formed by the intersection of planes is called a **polyhedron.**

Some of the simplest solid figures are polyhedrons. For example, the box shown in Figure 9.5 fits the definition of a polyhedron.

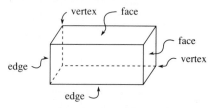

Figure 9.5   Polyhedron

The plane surfaces of a polyhedron are called **faces,** the lines of intersection of the planes are called **edges,** and the intersections of edges are called **vertices** (plural of **vertex**).

The great mathematician, Leonard Euler (1707–1783) discovered a relationship between the number of faces f, the number of vertices v, and the number of edges e for any polyhedron. The equation is

$$f + v = e + 2.$$

That is, the sum of the faces and vertices is always two more than the number of edges. For example in Figure 9.5, $f = 6$, $v = 8$, and $e = 12$. Thus

$$f + v = 14 \quad \text{and} \quad e + 2 = 14.$$

EXAMPLE 1 Find the number of vertices of a polyhedron that has 12 faces and 30 edges.

Substitute $f = 12$ and $e = 30$ in the formula and solve for v.

$$f + v = e + 2$$
$$12 + v = 30 + 2$$
$$v = 32 - 12$$
$$v = 20$$

Thus the polyhedron has 20 vertices. ◪

DEFINITION: REGULAR POLYHEDRON

A **regular polyhedron** is a solid figure in which all faces are congruent regular polygons.

By considering the nature of the intersecting faces in polyhedrons, it can be shown that there are only five possible regular polyhedrons. They are shown and labeled in Figure 9.6.

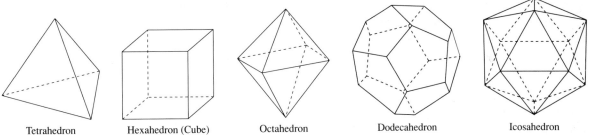

Tetrahedron   Hexahedron (Cube)   Octahedron   Dodecahedron   Icosahedron

Figure 9.6 Regular Polyhedrons

The following table gives a summary of the regular polyhedrons including the name, polygon face, number of faces f, number of vertices v, and number of edges e. You can check that each figure satisfies the equation $f + v = e + 2$.

Name	Polygon Face	f	v	e
Tetrahedron	Triangle	4	4	6
Hexahedron (Cube)	Square	6	8	12
Octahedron	Triangle	8	6	12
Dodecahedron	Pentagon	12	20	30
Icosahedron	Triangle	20	12	30

9.1 EXERCISES

In the figure below, line ℓ is the intersection of planes P, Q and R. Lines k and m are in P, lines n and s are in R, $m \perp \ell$, $m \perp n$, $\ell \parallel k$, and $\ell \parallel s$. Use this information to answer true or false in Exercises 1–12.

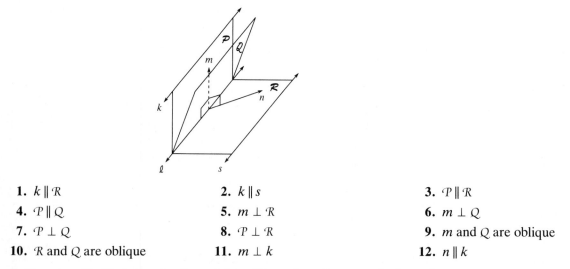

1. $k \parallel R$ **2.** $k \parallel s$ **3.** $P \parallel R$

4. $P \parallel Q$ **5.** $m \perp R$ **6.** $m \perp Q$

7. $P \perp Q$ **8.** $P \perp R$ **9.** m and Q are oblique

10. R and Q are oblique   **11.** $m \perp k$ **12.** $n \parallel k$

In Exercises 13–18, determine the number of faces f, vertices v, and edges e in each polyhedron. Check to see that $f + v = e + 2$.

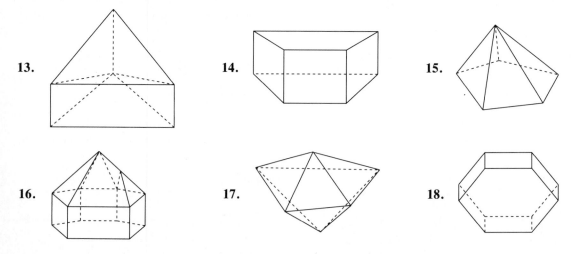

13.   **14.**   **15.**

16.   **17.**   **18.**

19. A polyhedron has 12 vertices and 30 edges. How many faces does it have?

20. A polyhedron has 14 faces and 9 vertices. How many edges does it have?

Solve each applied problem in Exercises 21–22.

21. A regular dodecahedron is to be constructed using metal tubing for the edges. If each edge is to be 3 m long, how much tubing will be required for the project?

22. A storage tank in the shape of a cube is to be insulated using material that costs $2.50 per square foot. If the edge of the cube is 12 ft, how much will the insulation cost?

A **diagonal** of a polyhedron is a line segment joining two vertices that are not on the same face. Use this definition in Exercises 23–24.

23. How many diagonals does a cube have?

24. Use the Pythagorean Theorem to derive a formula for the length of the diagonal of a cube with edge of length e.

25. Can a vertex of a regular polyhedron be formed using six equilateral triangles?

26. Can a vertex of a regular polyhedron be formed using three hexagons?

9.2 PRISMS AND PYRAMIDS

In this section we will define two types of polyhedrons and determine formulas for calculating their surface area and volume. We first consider the *prism*. Refer to Figure 9.7.

> **DEFINITION: PRISM**
>
> The solid figure formed by joining two congruent polygonal regions in parallel planes is called a **prism.** The polygonal regions are called **bases** and the other surfaces are **lateral faces.**

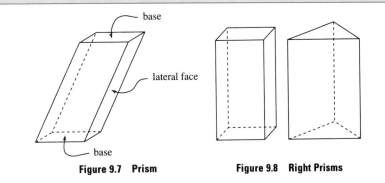

Figure 9.7   Prism   **Figure 9.8   Right Prisms**

Notice that the lateral faces are parallelograms. If the lateral faces are rectangles as in Figure 9.8, the prism is a **right prism;** otherwise it is called an **oblique prism.**

To determine the surface area of a prism, we note that the **total surface area** *SA* is the sum of twice the **base area** *B* and the **lateral area** *LA*.

$$SA = 2B + LA$$

To develop a formula for the lateral area, we must restrict ourselves to right prisms. Consider the right prism shown in Figure 9.9 where *h* is the **height,** the length of an **altitude,** and *a, b, c, d,* and *e* are lengths of the sides of a base. Because the sides of a right prism are rectangles, the lateral area is given by

$$\begin{aligned} LA &= ah + bh + ch + dh + eh \\ &= (a + b + c + d + e)h \\ &= ph \end{aligned}$$

where *p* is the perimeter of a base. This is an informal proof of the following theorem.

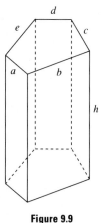

Figure 9.9

> ### THEOREM 9.1   LATERAL AREA OF A RIGHT PRISM
>
> The lateral area of a right prism is determined with the formula
> $$LA = ph,$$
> where *p* is the perimeter of a base and *h* is the height of the prism.

E X A M P L E   1   How many square inches of glass are needed to make the fish tank in Figure 9.10 if the top is left open?

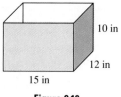

10 in
12 in
15 in

Figure 9.10

We must find the lateral area and the area of one of the bases. The base area is

$$(15 \text{ in})(12 \text{ in}) = 180 \text{ in}^2.$$

To find the lateral area, first determine the perimeter.

$$\begin{aligned} p &= 15 \text{ in} + 12 \text{ in} + 15 \text{ in} + 12 \text{ in} \\ &= 54 \text{ in} \\ LA &= ph \\ &= (54 \text{ in})(10 \text{ in}) \\ &= 540 \text{ in}^2 \end{aligned}$$

Thus, the total surface area of glass required is

$$\begin{aligned} SA &= 180 \text{ in}^2 + 540 \text{ in}^2 \\ &= 720 \text{ in}^2. \end{aligned}$$

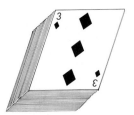

We've now considered how to determine the volume of a right prism but not that of an oblique prism. The picture above shows two decks of cards, one forming a right prism and the other an oblique prism. These suggest that the two have the same volume and that the same formula works for both right and oblique prisms.

PRACTICE EXERCISE 1

The outside walls of a workshop are to be covered with siding that costs $1.50 per square foot. How much will it cost if the building is 16 ft wide, 18 ft long, and 8 ft high?

In our study of plane geometry we used units such as 1 cm and 1 inch to measure length and 1 square centimeter (cm²) and 1 square inch (in²) to measure area. For measuring the volume of a solid, we'll need cubic units such as 1 cubic centimeter (cm³) and 1 cubic inch (in³). Figure 9.11 illustrates these units using one of the most basic prisms, the cube.

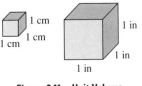

Figure 9.11   Unit Volume

To see how we find the volume of other prisms in cubic units, consider Figure 9.12. The prism in Figure 9.12(a) has volume 20 cm³ because there are 10 cm³ on the bottom layer and 10 cm³ on the top layer. In Figure 9.12(b) there is one more layer (10 cm³) than in Figure 9.12(a), and the volume is 30 cm³.

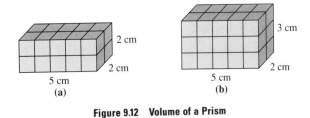

Figure 9.12   Volume of a Prism

The volumes in Figure 9.12(a) and 9.12(b) can be found by multiplying the area of the base of each prism by its height.

$$20 \text{ cm}^3 = \underbrace{5 \text{ cm} \cdot 2 \text{ cm}}_{\text{area of base}} \cdot \underbrace{2 \text{ cm}}_{\text{height}}$$

$$30 \text{ cm}^3 = \underbrace{5 \text{ cm} \cdot 2 \text{ cm}}_{\text{area of base}} \cdot \underbrace{3 \text{ cm}}_{\text{height}}$$

Theorem 9.2 follows from this information.

> **THEOREM 9.2   VOLUME OF A RIGHT PRISM**
>
> The volume of a right prism is determined with the formula
> $$V = Bh,$$
> where B is the area of a base and h is the height.

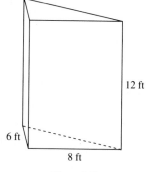

12 ft

6 ft

8 ft

Figure 9.13

EXAMPLE 2   A chemical is to be stored in the tank (a right prism) shown in Figure 9.13. The base is a right triangle with legs 6 ft and 8 ft, and the tank is 12 ft deep. How many cubic feet of chemical will the tank hold?

Because the base is a right triangle,

$$B = \frac{1}{2}(6 \text{ ft})(8 \text{ ft}) = 24 \text{ ft}^2.$$

Thus, the volume of the right prism is

$$V = Bh$$
$$= (24 \text{ ft}^2)(12 \text{ ft}) = 288 \text{ ft}^3.$$

The tank will hold 288 ft³ of chemical.   ◼

> **PRACTICE EXERCISE 2**
>
> A highway construction truck has a bed 5.5 yd long, 2.5 yd wide, and 2.0 yd deep. How many cubic yards of gravel will it hold?

We now consider another polyhedron, the *pyramid*.

> **DEFINITION: PYRAMID**
>
> The solid figure formed by connecting a polygon with a point not in the plane of the polygon is called a **pyramid.** The polygonal region is called the **base** and the point is the **vertex.** The line segment from the vertex perpendicular to the plane of the base is the **altitude.**

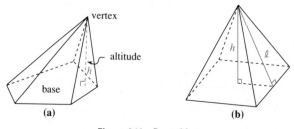

(a)   (b)

Figure 9.14   Pyramids

Figure 9.14(a) shows an **oblique pyramid** and Figure 9.14(b) shows a *regular pyramid* like those built by the Egyptians. A **regular pyramid** has a regular polygon for a base, congruent isosceles triangles for the lateral surfaces, and an altitude passing through the center of the base.

In Figure 9.14(b), the distance ℓ is called the **slant height** of the lateral surfaces of a regular pyramid. The measure ℓ is also the height of the triangular face of the pyramid. Thus, we can determine the area of each triangle using one-half the base multiplied by the slant height. But if we add all the bases of the triangles, we obtain the perimeter of the base of the pyramid, which leads to the following theorem.

THEOREM 9.3   LATERAL AREA OF A REGULAR PYRAMID

The lateral area of a regular pyramid is determined with the formula

$$LA = \frac{1}{2}p\ell,$$

where p is the perimeter of the base and ℓ is the slant height.

Of course, the total surface area of a pyramid is the lateral area plus the area of the base.

$$SA = \frac{1}{2}p\ell + B$$

E X A M P L E   3   Find the total surface area of a regular pyramid if each side of its square base is 42 m and the altitude of the pyramid has length 28 m.

The area of the square base is $(42)^2$ square meters, but before we can find the lateral surface area, we must use the height and base dimensions to find the slant height. In Figure 9.15, the right triangle with ℓ as its hypotenuse will have one leg 28 m (the height) while the other leg is one-half the length of the side of the base, 21 m. By the Pythagorean Theorem,

$$\ell^2 = (28)^2 + (21)^2$$
$$\ell = \sqrt{(28)^2 + (21)^2}$$
$$= \sqrt{784 + 441}$$
$$= \sqrt{1225} = 35 \text{ m}.$$

The perimeter of the base is

$$p = 4(42) = 168 \text{ m}.$$

Finally, the total surface area is

$$SA = \frac{1}{2}p\ell + B$$
$$= \frac{1}{2}(168)(35) + (42)^2$$
$$= 2940 + 1764 = 4704 \text{ m}^2. \quad \blacksquare$$

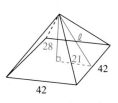

Figure 9.15

The Transamerica Tower in San Francisco, California, was designed using the form of a pyramid.

PRACTICE EXERCISE 3

Find the total surface area of a pyramid that has four equilateral triangles as faces and an edge $3\sqrt{3}$ m long.

The formula for the volume of a pyramid can be derived by comparing the prism and the pyramid. Because of the complexity level, we will state the following theorem without proof.

THEOREM 9.4　VOLUME OF A PYRAMID

The volume of a pyramid is determined with the formula

$$V = \frac{1}{3}Bh,$$

where B is the area of the base and h is the height.

The volume formula is true for all pyramids but we will continue to stress regular pyramids.

EXAMPLE 4　The rock used to build the pyramid in Example 3 weighs 1200 kg per cubic meter. What is the total weight of the pyramid?

We must first calculate the volume of the pyramid and then multiply by 1200. Recall from Example 3 that the square base had a side 42 m and height 28 m. Thus,

$$V = \frac{1}{3}Bh$$

$$= \frac{1}{3}(42)^2(28)$$

$$= 16{,}464 \text{ m}^3.$$

To find the total weight, multiply by 1200.

$$\text{Total Weight} = (16{,}464)(1200)$$
$$= 19{,}756{,}800 \text{ kilograms}$$

Thus, the pyramid weighs approximately 20,000,000 kg.　▨

ANSWERS TO PRACTICE EXERCISES:　**1.** $816　**2.** 27.5 yd³
3. $(27\sqrt{3} + 27)$ m²

9.2 EXERCISES

In Exercises 1–4, determine whether the statement is true or false.

1. For a prism with rectangular base of dimensions ℓ and w and with height h, the total surface area is $SA = 2\ell w + 2\ell h + 2wh$.

2. For a prism with rectangular base of dimensions ℓ and w and with height h, the volume is $V = \ell wh$.

3. The lateral surfaces of a right prism are rectangles.

4. If a prism and a pyramid have the same base and altitude, the volume of the pyramid is one-half the volume of the prism.

Determine the total surface area of each right prism described in Exercises 5–12.

5. cube with edge 5 ft

6. cube with edge 1.5 m

7. rectangular base 2 ft by 6 ft; height 5 ft

8. rectangular base 6.4 cm by 15.5 cm; height 20.2 cm

9. equilateral triangle base with side 2.8 yd; height 1.5 yd

10. right isosceles triangular base with legs 6 cm; height 11 cm

11. regular hexagon base with side 26 inches; height 14 inches

12. right triangle base with one leg 9 mm and hypotenuse 14 mm; height 22 mm

Determine the volume of each right prism described in Exercises 13–20.

13. cube with edge 8 m

14. cube with edge 10.2 cm

15. rectangular base 2.3 cm by 1.2 cm; height 4.5 cm

16. rectangular base 3.4 inches by 9.5 inches; height 8.75 inches

17. right isosceles triangular base with legs 10.5 ft; height 20.4 ft

18. equilateral triangle base with side 8 cm; height 24 cm

19. right triangle base with hypotenuse 32 m and leg 20 m; height 9 m

20. regular hexagon base with side 3.6 yd; height 4.8 yd

Find the lateral surface area of each regular pyramid described in Exercises 21–24.

21. equilateral triangle base with side 22 m; slant height 28 m

22. square base with side 6.8 ft; slant height 4.5 ft

23. square base with side 124 cm; height 86 cm

24. regular hexagon base with side 17.8 inches; height 32.2 inches

Find the volume of each regular pyramid described in Exercises 25–28.

25. square base with side 22 yd; height 15 yd

26. equilateral triangle base with side 8.2 m; height 4.6 m

27. regular hexagon base with side 16 cm; slant height 24 cm

28. square base with side 30.6 ft; slant height 46.8 ft

Solve the applied problems in Exercises 29–34.

29. A highway construction truck has a bed 6.5 yd long, 3.0 yd wide, and 2.4 yd deep. How many cubic yards of gravel will it hold?

30. A tank is in the shape of a cube. How many grams of water will it hold if the edge of the cube is 22 cm and water weighs 1 g per cubic centimeter?

31. How many square centimeters of glass will it take to form the bottom and sides of the tank in Exercise 30?

32. Elizabeth Wright owns a rectangular building 30 ft long, 12 ft wide, and 8 ft high. How much will it cost to put siding on the building if the siding costs $0.75 per square foot?

33. A sculpture in a park has the shape of a pyramid with a square base 100 ft on a side. The height of the structure is 25 ft. If one gallon of paint will cover 220 ft², how many gallons will be required to paint the lateral exterior surface of the structure?

34. A perfume bottle is a pyramid with square base. If the base has side 4 cm and slant height 3 cm, how many cubic centimeters of perfume will the bottle hold?

35. The rectangular base of a right prism has dimensions ℓ and w and height h. Derive a formula for the length d of a diagonal. (A diagonal joins vertices that are not on the same face.)

36. Use the formula derived in Exercise 35 to find the length of the diagonal of a rectangular box that is 17.0 inches by 11.0 inches by 14.0 inches.

9.3 CYLINDERS AND CONES

A cylinder has the same general shape as a prism but has circles for its bases instead of polygons.

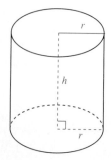

Figure 9.16   Right Cylinder

> **DEFINITION: CYLINDER**
>
> The solid figure formed by joining two congruent circles in parallel planes is called a **cylinder.**

The cylinder shown in Figure 9.16 is a **right cylinder** because the line joining the centers of the circles is an altitude. We will restrict our work to right cylinders and from now on will use the term "cylinder" to mean "right cylinder."

To find the surface area of a cylinder, we'll first consider a can without top or bottom as shown in Figure 9.17(a). If it is cut along the seam and pressed flat, as in Figure 9.17(b), a rectangle is formed. The length of the rectangle is the circumference of the base circle and the width, *h,* is the length of the altitude.

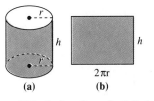

Figure 9.17   Surface Area of a Cylinder

Thus, the surface area of the side of the can is $2\pi rh$. Because the area of the top of the can is πr^2, the area of both bases is $2\pi r^2$, which leads to the following theorem.

THEOREM 9.5   SURFACE AREA OF A CYLINDER

The total **surface area** of a cylinder is determined with the formula
$$SA = 2\pi rh + 2\pi r^2,$$
where *r* is the radius of a base and *h* is the **height,** the length of the **altitude.**

Felix Klein (1849–1925)

Felix Klein, a German mathematician, was instrumental in formulating a theory that unified the concepts of Euclidean and non-Euclidean geometries. Three famous problems in geometry served as the basis for a series of lectures Klein gave to acquaint students with new developments in geometry. The three problems: the duplication of a cube (constructing a cube with volume twice that of a given cube); the quadrature of a circle (constructing a square with the same area as a given circle); and the trisection of an angle. All three problems proved impossible. Klein is perhaps most famous for his work in topology, sometimes called rubber-sheet geometry.

Notice that the lateral area is the same as for a prism,
$$LA = (\text{perimeter of base})(\text{height}),$$
because for a circular base the perimeter is $2\pi r$.

E X A M P L E   1   Find to the nearest square foot the total surface area of a cylinder with radius 7 ft and height 9 ft. Use 3.14 for π. (The calculator value of π will give a slightly different answer.)

$$\begin{aligned} SA &= 2\pi rh + 2\pi r^2 \\ &\approx 2(3.14)(7)(9) + 2(3.14)(7)^2 \\ &\approx 703 \text{ ft}^2 \quad \blacksquare \end{aligned}$$

PRACTICE EXERCISE 1

Find the total surface area of a cylinder with height 22.3 cm and radius 12.8 cm. Use 3.14 for π and give your answer to the nearest tenth of a square centimeter.

E X A M P L E   2   The outside and top of a cylindrical water tank are to be painted. The tank has radius 3 m and height 8 m. Approximately how many liters of paint will be required if a liter covers 8 m²?

All warm-blooded animals lose heat during sleep in the same amount per square unit of surface area of skin. As a result, the necessary amount of food that must be consumed is proportional to the animal's total surface area and not its weight (or volume). This is why a small animal must consume large amounts of food, sometimes more than half its weight, on a daily basis.

To find the lateral surface area, find $2\pi rh$.

$$2\pi rh \approx 2(3.14)(3)(8)$$
$$\approx 151 \text{ m}^2$$

To find the area of the top, find πr^2.

$$\pi r^2 \approx (3.14)(3)^2$$
$$\approx 28 \text{ m}^2$$

Thus, the approximate area to be painted is

$$151 \text{ m}^2 + 28 \text{ m}^2 = 179 \text{ m}^2.$$

To find the number of liters of paint, divide by 8 m² per liter.

$$\frac{179}{8} \approx 22.4 \text{ liters}$$

The job will require between 22 and 23 liters. ▨

By using prisms to approximate the cylinder, we can show that the volume of a cylinder is the area of the base multiplied by the height.

THEOREM 9.6   VOLUME OF A CYLINDER

The **volume** of a cylinder is determined with the formula

$$V = \pi r^2 h,$$

where r is the radius of a base and h is the height.

E X A M P L E   3   A can of blueberry pie filling has diameter 3.0 inches and height 4.5 inches. How many cans of filling are needed to fill a 9-inch diameter pie pan 1 inch deep?

First find the volume of pie filling in one can.

$V = \pi r^2 h$
$\quad \approx (3.14)(1.5)^2(4.5)$ $r = 1.5$ inches because the diameter is 3 inches
$\quad \approx 31.8 \text{ in}^3$ Volume of can

Now find the volume of the 9-inch diameter (4.5-inch radius) pie pan.

$V = \pi r^2 h$
$\quad \approx (3.14)(4.5)^2(1)$ Filling is 1-inch deep
$\quad \approx 63.6 \text{ in}^3$ Volume of pie pan

Now divide by 31.8 in³.

$$\frac{63.6}{31.8} = 2 \text{ cans}$$

It will take 2 cans of filling for the pie. ▨

PRACTICE EXERCISE 2
Water is stored in a cylindrical tank with base diameter 12 m and height 10 m. How many water trucks having cylindrical tanks with diameter 1.6 m and length 5 m can be filled from the storage tank?

A solid figure that is closely related to the pyramid is the *cone.*

DEFINITION: CONE
The solid figure formed by connecting a circle with a point not in the plane of the circle is called a **cone.**

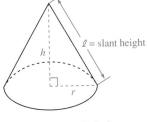

Figure 9.18   Right Cone

Our work with cones will emphasize **right cones** like the one shown in Figure 9.18. Notice that the altitude passes through the center of the base circle. Also, the **slant height** is the distance ℓ indicated in Figure 9.18.

By inscribing in a cone a regular pyramid with an ever-increasing number of sides for the base, it can be shown that for a cone

$$LA = \frac{1}{2}(\text{perimeter})(\text{slant height})$$

$$= \frac{1}{2}(2\pi r)\ell = \pi r \ell.$$

Also,

$$V = \frac{1}{3}(\text{area of base})(\text{height})$$

$$= \frac{1}{3}\pi r^2 h.$$

THEOREM 9.7   LATERAL AREA AND VOLUME OF A RIGHT CONE
The lateral area of a right cone is determined with the formula
$$LA = \pi r \ell,$$
where r is the radius of the base, h is the height, and $\ell = \sqrt{r^2 + h^2}$ is the slant height. Also, the volume is

$$V = \frac{1}{3}\pi r^2 h.$$

E X A M P L E 4   A pyramid with square base and height 12 cm is inscribed in a cone. If the square base has side 10 cm, find the total surface area
and the volume of the cone. Refer to Figure 9.19.

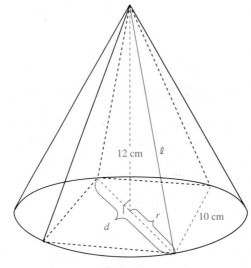

Figure 9.19

The height of the cone is the same as the height of the pyramid, 12 cm. The
radius of the base of the cone will be one-half the length of a diagonal d of the
square. By the Pythagorean Theorem,

$$d^2 = 10^2 + 10^2.$$
$$d = \sqrt{10^2 + 10^2}$$
$$= \sqrt{2 \cdot 10^2}$$
$$= 10\sqrt{2}$$
$$\frac{d}{2} = r = 5\sqrt{2} \text{ cm}$$

With h and r known, we can find the volume of the cone.

$$V = \frac{1}{3}\pi r^2 h$$

$$\approx \frac{1}{3}(3.14)(5\sqrt{2})^2(12)$$

$$= \frac{1}{3}(3.14)(25)(2)(12)$$

$$= 628 \text{ cm}^3$$

We can also determine the area of the base of the cone.

$$B = \pi r^2$$
$$\approx (3.14)(5\sqrt{2})^2$$
$$= (3.14)(25)(2)$$
$$= 157 \text{ cm}^2$$

Before we can find the lateral area, we must determine the slant height. Use the Pythagorean Theorem.

$$\ell^2 = r^2 + 12^2$$
$$\ell = \sqrt{r^2 + 12^2}$$
$$= \sqrt{(5\sqrt{2})^2 + (12)^2}$$
$$= \sqrt{50 + 144}$$
$$= \sqrt{194}$$
$$\approx 14 \text{ cm}$$

Now find the lateral area.

$$LA = \pi r \ell$$
$$\approx 3.14(5\sqrt{2})(14)$$
$$\approx 311 \text{ cm}^2$$

The total surface area is

$$SA = B + LA$$
$$\approx 157 + 311$$
$$= 468 \text{ cm}^2. \quad \blacksquare$$

PRACTICE EXERCISE 3

A conical structure has height 16.2 ft and a base radius of 7.4 ft. How many square feet of plastic will be required to cover its lateral area?

ANSWERS TO PRACTICE EXERCISES: **1.** 2821.5 cm² **2.** 112.5 **3.** 413.8 ft²

9.3 Exercises

In Exercises 1–2, determine whether the statement is true or false.

1. If a cylinder and a cone have the same radius and height, the volume of the cylinder is three times the volume of the cone.
2. If the height of a cylinder is doubled and the radius kept constant, the total surface area of the cylinder is doubled.

Find the total surface area of each cylinder in Exercises 3–8. Use 3.14 for π.

3. $r = 6$ cm, $h = 5$ cm
4. $r = 9$ cm, $h = 15$ cm
5. $r = 12.2$ in, $h = 30.0$ in
6. $r = 3.4$ m, $h = 10.5$ m
7. $d = 11.4$ m, $h = 4.4$ m
8. $d = 20.4$ in, $h = 8.5$ in

Find the volume of each cylinder in Exercises 9–14. Use 3.14 for π.

9. $r = 7$ cm, $h = 8$ cm   **10.** $r = 14$ cm, $h = 19$ cm   **11.** $r = 14.5$ in, $h = 35.5$ in

12. $r = 4.9$ m, $h = 17.8$ m   **13.** $d = 12.6$ m, $h = 16.2$ m   **14.** $d = 28.6$ in, $h = 10.7$ in

Find the total surface area of each cone in Exercises 15–20. Use 3.14 for π.

15. $r = 3$ yd, $\ell = 5$ yd   **16.** $r = 7$ cm, $\ell = 17$ cm   **17.** $r = 28.4$ in, $\ell = 42.6$ in

18. $r = 17.8$ m, $\ell = 58.5$ m   **19.** $r = 16$ ft, $h = 23$ ft   **20.** $r = 8.8$ cm, $h = 17.2$ cm

Find the volume of each cone in Exercises 21–26. Use 3.14 for π.

21. $r = 3$ in, $h = 4$ in   **22.** $r = 19$ m, $h = 29$ m   **23.** $r = 6.9$ cm, $h = 14.2$ cm

24. $r = 18.5$ yd, $h = 27.2$ yd   **25.** $r = 20$ m, $\ell = 40$ m   **26.** $r = 8.9$ in, $\ell = 21.6$ in

27. If a cylinder has volume 26.8 cm³, what is the volume of a cylinder with the same radius but twice as high? with the same height but radius twice as long?

28. If a cylinder has volume 26.8 cm³, what is the volume of a cylinder with the same height but one-half the radius? with the same radius but one-half the height?

29. Find the total surface area and volume of a cylinder that has a prism with square base inscribed in it. The side of the base of the prism is 24 inches and the height is 16 inches. Use 3.14 for π.

30. Find the total surface area and volume of a cone that has a pyramid with square base inscribed in it. The side of the base of the pyramid is 6.8 m and the height is 9.8 m. Use 3.14 for π.

Solve each applied problem in Exercises 31–36. Use 3.14 for π.

31. A can of cherry pie filling has diameter 8 cm and height 12 cm. How many cans are needed to fill a 20-cm diameter pie pan 3 cm deep?

32. A lab stores mercury in a cylinder of radius 16 cm and height 20 cm. How many cylindrical tubes with radius 0.5 cm and height 100 cm can be filled from a full supply?

33. A cylindrical storage tank has radius 3.8 ft and height 9.8 ft. How many gallons of paint are needed to paint the tank (including top and bottom) if one gallon covers 150 ft²?

34. A cylindrical tank is made from sheet metal. If the tank is to have radius 2.2 ft and height 12.0 ft, what is the total cost if the price of the metal is $16.50 per square foot?

35. An inverted cone used as a funnel for grain has radius 4.6 ft and height 22.2 ft. How many gallons of paint are required to paint the outside lateral surface if one gallon of paint covers 180 ft²?

36. A cone is filled with a liquid chemical to a depth of 3.2 m above the base. If the radius of the base is 3.7 m and the height is 7.4 m, how many cubic meters of liquid are in the tank?

9.4 SPHERES AND COMPOSITE FIGURES

In this section, we will consider a new figure—the *sphere*—and work with applications that involve composition of the figures we have studied.

> **DEFINITION: SPHERE**
> A **sphere** is the set of all points in space a given distance, called the **radius,** from a given point, the **center.**

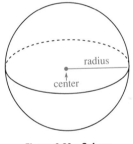

Figure 9.20   Sphere

Figure 9.20 shows a sphere. If we consider all the points in the interior of the sphere, we have a solid figure that is also called a sphere. This is the figure to which we refer when we discuss the volume of a sphere. The following postulate presents formulas for determining the surface area and volume of a sphere.

> **POSTULATE 9.2   SURFACE AREA AND VOLUME OF A SPHERE**
> For a sphere with radius r, the surface area is
> $$SA = 4\pi r^2$$
> and the volume is
> $$V = \frac{4}{3}\pi r^3.$$

The planet Saturn is spherical in shape. Its most distinguishing features, however, are its rings. Scientists have discovered seven major rings but many smaller ones probably also exist. It is believed that the rings are composed of ice and dust particles.

In Example 1, we'll work with the **diameter** of a sphere, which is twice its radius.

E X A M P L E 1 Find the surface area and volume of a sphere with diameter 18 cm. Refer to Figure 9.21.

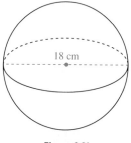

Figure 9.21

Because the diameter is 18 cm, the radius is $r = \dfrac{18}{2}$ cm = 9 cm. We will need this number to determine both the surface area and volume.

$$SA = 4\pi r^2$$
$$\approx 4(3.14)(9)^2$$
$$\approx 1017 \text{ cm}^2$$

The surface area is approximately 1017 cm².

$$V = \frac{4}{3}\pi r^3$$
$$\approx \frac{4}{3}(3.14)(9)^3$$
$$\approx 3052 \text{ cm}^3$$

Thus, the volume is 3052 cm³. ◪

We now consider the application given in the chapter introduction. The example introduces a **hemisphere,** which is one-half a sphere.

E X A M P L E 2 A geodesic dome is a structure composed of many triangular pieces that form a spherical shape. As a result, the surface area and volume of such a dome can be approximated using properties of a sphere. Find the number of square meters of synthetic material required to cover a geodesic dome that has a hemisphere 42 m in diameter. What is the volume enclosed by the structure?

Because the dome is a hemisphere, the surface area is one-half the surface of a sphere.

$$SA = \frac{1}{2}(4\pi r^2)$$
$$\approx 2(3.14)(21)^2 \qquad \text{The radius is one-half the diameter}$$
$$\approx 2769 \text{ m}^2$$

Thus, it will take 2769 m² of material to cover the dome.

The volume of the dome is one-half the volume of a sphere.

$$V = \frac{1}{2}\left(\frac{4}{3}\pi r^3\right)$$

$$\approx \frac{2}{3}(3.14)(21)^3$$

$$\approx 19{,}386 \text{ m}^3 \quad \blacksquare$$

PRACTICE EXERCISE 1

A spherical tank with radius 3.2 m is to be insulated. If the material costs $2.50 per square meter, how much will it cost to cover the surface of the sphere?

We now consider applications that involve composite solid figures.

The volume of gas in a hot-air balloon can be calculated using properties of a sphere. The volume of gas in the balloon varies directly with the temperature of the gas and inversely with the pressure on the surface of the balloon. Thus, as the temperature of the gas increases and the pressure on the surface decreases when the balloon rises, the volume of the balloon increases.

E X A M P L E   3   A grain storage silo is a cylindrical structure topped with a hemisphere. The radius of both the cylinder and the hemisphere is 12.3 ft, and the total height of the structure is 87.4 ft. How many cubic feet of grain will the silo hold?

We must find the volume of the cylinder and add the volume of the hemisphere. The height of the cylinder must be found by subtracting the radius of the hemisphere (12.3 ft) from the total height of the structure.

$$h = 87.4 - 12.3 = 75.1 \text{ ft}$$

Now we know that for the cylinder $r = 12.3$ ft and $h = 75.1$ ft.

$$V_{cyl} = \pi r^2 h$$

$$\approx (3.14)(12.3)^2(75.1)$$

$$\approx 35{,}676.3 \text{ ft}^3$$

Find the volume of the hemisphere as one-half the volume of a sphere with radius 12.3 ft.

$$V_{hsph} = \left(\frac{1}{2}\right)\left(\frac{4}{3}\right)\pi r^3$$

$$\approx \left(\frac{1}{2}\right)\left(\frac{4}{3}\right)(3.14)(12.3)^3$$

$$\approx 3895.4 \text{ ft}^3$$

The total number of cubic feet of grain the silo will hold is approximately

$$V = 35{,}676.3 + 3895.4 = 39{,}571.7 \text{ ft}^3 \quad \blacksquare$$

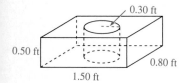

0.30 ft

0.50 ft

0.80 ft

1.50 ft

Figure 9.22

$E\ X\ A\ M\ P\ L\ E$ 4   Find the volume and total surface area of a block of wood with a hole bored in it as shown in Figure 9.22.

To find the volume, we must determine the volume of the prism and subtract the volume of the cylinder.

$$V_{prism} = Bh$$
$$= (1.50)(0.80)(0.50)$$
$$= 0.60 \text{ ft}^3$$

For the cylinder $r = 0.30$ ft and $h = 0.50$ ft, the height of the prism.

$$V_{cyl} = \pi r^2 h$$
$$\approx (3.14)(0.30)^2(0.50)$$
$$\approx 0.14 \text{ ft}^3$$

The approximate volume of the block with the hole is

$$V = 0.60 - 0.14 = 0.46 \text{ ft}^3.$$

Now determine total surface area by first finding the surface area of the prism, subtracting the area of both bases of the cylinder, and adding the lateral area of the cylinder.

$$SA_{prism} = 2(1.50)(0.80) + [2(1.50) + 2(0.80)](0.50)$$
$$= 4.70 \text{ ft}^2$$
$$2B_{cyl} \approx 2(3.14)(0.30)^2$$
$$\approx 0.57 \text{ ft}^2$$
$$LA_{cyl} \approx 2(3.14)(0.30)(0.50)$$
$$\approx 0.94 \text{ ft}^2$$
$$SA \approx 4.70 - 0.57 + 0.94$$
$$= 5.07 \text{ ft}^2 \quad \blacksquare$$

ANSWER TO PRACTICE EXERCISE: **1.** $321.54

9.4 *EXERCISES*

Find the surface area of each sphere in Exercises 1–6. Use 3.14 for π.

1. $r = 12$ m

2. $r = 30$ in

3. $r = 0.75$ ft

4. $r = 0.40$ cm

5. $r = 5.6$ yd

6. $r = 9.8$ yd

Find the volume of each sphere in Exercises 7–12. Use 3.14 for π.

7. $r = 7$ ft

8. $r = 11$ in

9. $r = 0.90$ m

10. $r = 0.64$ cm

11. $r = 13.8$ yd

12. $r = 21.6$ cm

13. The surface area of a sphere is 288 cm². What is the surface area of a sphere with one-half the radius?

14. The volume of a sphere is 288 cm³. What is the volume of a sphere with one-half the radius?

15. A sphere is inscribed in a cube with edge 8 cm. What is the volume of the cube not occupied by the sphere? Use 3.14 for π.

16. A cube with edge 8 cm is inscribed in a sphere. What is the volume of the sphere not occupied by the cube? Use 3.14 for π.

17. A cube and a sphere each have a surface area of 150 in². Find the volume of each and determine which has the larger volume. Use 3.14 for π.

18. A cube and a sphere each have a volume of 216 m³. Find the surface area of each and determine which has the larger surface area. Use 3.14 for π.

19. If the surface area and volume of a sphere have the same numerical value, find the radius of the sphere.

20. A sphere and a cylinder have the same radius and the same surface area. Find the radius in terms of the height h of the cylinder.

Solve each applied problem in Exercises 21–30. Use 3.14 for π.

21. A spherical tank with radius 32 ft is to be filled with gas. How many cubic feet of gas will it hold?

22. If one liter is 1000 cm³, how many liters of a liquid can be stored in a sphere with radius 65 cm?

23. A spherical tank has radius 8.6 m. If a rust-preventing material costs $1.50 per square meter, what is the cost to rustproof the tank?

24. A spherical tank with diameter 36 ft is to be insulated. If insulation costs $0.50 per square foot, how much will the job cost?

25. A chemical storage tank is a cylinder with a hemisphere cap on each end. If the height of the cylindrical portion is 16.2 ft and the radius of the cylinder and hemispheres is 2.8 ft, how many cubic feet of chemical will the tank hold?

26. The chemical tank in Exercise 25 is to be insulated with material costing $1.25 per square foot. What will be the total cost of the insulating material?

27. A silo is a cylinder with a hemisphere of the same radius on top. The total height of the silo is 23.5 m and the radius is 3.8 m. Find the number of cubic meters of grain the silo will hold.

28. An ice-cream cone is filled and topped with a hemisphere of ice cream. The radius of the cone and the hemisphere are both 1.2 inches and the overall height is 6.4 inches. Find the volume of ice cream served.

29. A machine part is a right prism with base 6.4 cm by 5.8 cm and height 2.3 cm. There is a cylindrical hole with radius 1.8 cm drilled vertically through the center of the prism. If the metal weighs 1.5 g per cubic centimeter, what is the weight of the machine part?

30. The finishing operation on the machine part in Exercise 29 costs $0.26 per square centimeter of surface. What is the total cost of the finishing operation?

KEY TERMS AND SYMBOLS

9.1 parallel line and plane, p. 307
perpendicular line and plane, p. 307
parallel planes, p. 307
perpendicular planes, p. 308
oblique planes, p. 308
polyhedron, p. 308
faces, p. 308
edges, p. 308
vertex (vertices), p. 308
regular polyhedron, p. 309

9.2 prism, p. 311
base of prism, p. 311
lateral face of prism, p. 311
right prism, p. 311

oblique prism, p. 311
total surface area (SA), p. 312
base area (B), p. 312
lateral area (LA), p. 312
height of a prism, p. 312
altitude of a prism, p. 312
pyramid, p. 314
base of a pyramid, p. 314
vertex, p. 314
altitude of a pyramid, p. 314
oblique pyramid, p. 315
regular pyramid, p. 315
slant height of a pyramid, p. 315

9.3 cylinder, p. 318

right cylinder, p. 318
surface area of a cylinder, p. 319
height of a cylinder, p. 319
altitude of a cylinder, p. 319
cone, p. 321
right cone, p. 321
slant height of a cone, p. 321

9.4 sphere, p. 325
radius, p. 325
center, p. 325
diameter, p. 325
hemisphere, p. 326

REVIEW EXERCISES

Section 9.1

In the figure below, line s is the intersection of planes P and Q, and line v is the intersection of Q and R. Line t is in P, u is in Q, w is in R, $t \perp u$, $u \perp s$, $u \perp v$, and $u \perp w$. Use this information to answer true or false in Exercises 1–6.

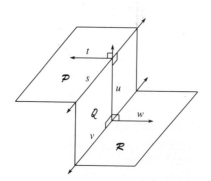

1. $s \parallel R$

2. $w \parallel P$

3. $P \parallel R$

4. $Q \perp P$

5. Q and R are oblique.

6. u and P are oblique.

Determine the number of faces f, vertices v, and edges e in each polyhedron in Exercises 7–8. Check to see that $f + v = e + 2$.

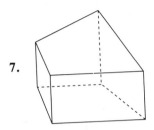

7.

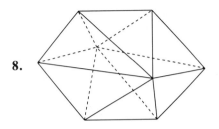

8.

9. A polyhedron has 14 vertices and 36 edges. How many faces does it have?

10. The edges of a regular octahedron are to be constructed with tubing that costs $1.65 per foot. If each edge is 2.5 ft in length, how much will all the tubing required for the project cost?

Section 9.2

Find the total surface area of each right prism described in Exercises 11–12.

11. equilateral triangle base with side 12.6 cm; height 21.3 cm

12. rectangular base with sides 9.6 ft and 5.8 ft; height 2.6 ft

Determine the volume of each right prism described in Exercises 13–14.

13. cube with edge 18.6 in

14. regular hexagon base with side 5.2 m; height 7.4 m

Find the lateral surface area of each regular pyramid described in Exercises 15–16.

15. equilateral triangle base with side 251 cm; slant height 314 cm

16. regular hexagon base with side 9.8 ft; height 7.2 ft

Find the volume of each regular pyramid described in Exercises 17–18.

17. equilateral triangle base with side 16.8 in; height 25.5 in

18. square base with side 42.6 m; slant height 61.4 m

Solve each applied problem in Exercises 19–20.

19. A metal rectangular box with no top is to be made with a base 10 cm by 25 cm and height 8 cm. How many square centimeters of metal are required?

20. A decorative bottle has the shape of a regular pyramid with triangular base. If the base has side 2.2 inches and the height is 4.1 inches, how many cubic inches of liquid will the bottle hold?

Section 9.3

Find the total surface area of each cylinder in Exercises 21–22. Use 3.14 for π.

21. $r = 2.3$ ft, $h = 6.5$ ft

22. $r = 21.2$ cm, $h = 19.5$ cm

Find the volume of each cylinder in Exercises 23–24. Use 3.14 for π.

23. $r = 165$ in, $h = 214$ in

24. $r = 42.6$ m, $h = 15.7$ m

Find the total surface area of each cone in Exercises 25–26. Use 3.14 for π.

25. $r = 6.2$ cm, $\ell = 9.5$ cm

26. $r = 18.9$ ft, $h = 26.8$ ft

Find the volume of each cone in Exercises 27–28. Use 3.14 for π.

27. $r = 12.8$ m, $h = 19.2$ m

28. $r = 25.8$ in, $\ell = 39.2$ in

29. Find the total surface area and volume of a cone that is inscribed in a pyramid with square base. The side of the base of the pyramid is 10.8 yd and the height is 17.2 yd. Use 3.14 for π.

30. Find the volume of a storage tank that consists of a cylinder topped by a cone. The radius of both cone and cylinder is 3.6 m, the height of the cylinder is 10.8 m, and the height of the cone is 1.3 m. Use 3.14 for π.

Section 9.4

Find the surface area of each sphere in Exercises 31–32. Use 3.14 for π.

31. $r = 3.9$ ft

32. $r = 26.8$ m

Find the volume of each sphere in Exercises 33–34. Use 3.14 for π.

33. $r = 0.92$ cm

34. $r = 12.9$ in

35. The volume of a sphere is 26.1 cm³. What is the volume of a sphere with twice the radius?

36. A sphere has a surface area of 100 m². Find the volume to the nearest unit. Use 3.14 for π.

Solve each applied problem in Exercises 37–40. Use 3.14 for π.

37. A spherical tank with radius 15.8 cm is to be filled with a liquid weighing 0.9 grams per cubic centimeter. How many grams of the liquid will the tank hold?

38. A geodesic dome is a hemisphere with radius 23.5 ft. How many square feet of synthetic material will be required to cover the structure?

39. A water tank is a cylinder with a hemisphere cap on each end. The overall height of the tank is 9.9 m. If the radius of the cylinder and hemispheres is 1.8 m, how many kg of water will the tank hold if water weighs 1000 kg per cubic meter?

40. A machine part is in the shape of a cone with a hemisphere covering its base. The radius of the cone and sphere is 2.3 cm and the slant height of the cone is 12.2 cm. How much will it cost to finish the part if the finishing operation costs $2.65 per square centimeter?

PRACTICE TEST

In the figure below, line u is the intersection of $\mathcal{P}$ and $\mathcal{R}$, line v is the intersection of $\mathcal{Q}$ and $\mathcal{R}$, m and n are in $\mathcal{Q}$, k is in $\mathcal{R}$, ℓ is in $\mathcal{P}$, $k \perp u$, $k \perp \ell$, $k \perp v$, $k \perp n$, and $n \perp v$. Use this information to answer true or false in Exercises 1–4.

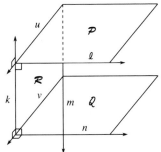

1. $\mathcal{P} \parallel \mathcal{Q}$

2. $\mathcal{R} \perp \mathcal{Q}$

3. $m \parallel \mathcal{P}$

4. m and $\mathcal{R}$ are oblique.

5. A polyhedron has 12 faces and 10 vertices. How many edges does it have?

6. Find the total surface area of a right prism with height 8.2 cm and an equilateral triangle base with side 12.6 cm.

7. Find the volume of a cone with radius 28.6 inches and slant height 42.5 inches.

8. Find the lateral area of a regular pyramid with height 18.3 meters and a square base with side 32.2 meters.

9. Find the total surface area of a cylinder with radius 1.2 ft and height 6.7 ft.

10. Find the volume of a geodesic dome in the shape of a hemisphere with radius 12.8 yd.

11. The surface area of a sphere is 468 in². What is the surface area of a sphere with one-third the radius?

12. A silo is a cylinder with a hemisphere on top. The radius of the cylinder and hemisphere is 9.8 m and the overall height of the silo is 57.2 m. How many cubic meters of grain will the silo hold?

10

GEOMETRIC LOCI AND CONCURRENCY THEOREMS

In this chapter we study ways to describe the location of various geometric figures as they relate to other given figures or conditions. The ability to make these descriptions has many applications including the one presented below, which is solved in Section 10.2, Example 1.

A gardener wishes to install in a flower garden a single sprinkler with a circular water pattern. The garden has the shape of a triangle with sides 20 ft, 16 ft, and 18 ft. At what point should the sprinkler be placed to best reach all three vertices of the garden?

10.1 LOCUS AND BASIC THEOREMS

In Chapter 7, we defined a circle as the set of all points in a plane a given distance r, the radius, from a point O, the center of the circle. Another way to describe a circle is to say "a circle is the *locus of points* in a plane a given distance r from a fixed point O."

DEFINITION: LOCUS OF POINTS

A **locus** consists of all those points and only those points that satisfy one or more conditions.

Unless we say otherwise, we will agree to restrict loci (plural of locus) to a plane. In each example, we will make a sketch and find a few points on the locus before determining all the points on the locus.

E X A M P L E 1   What is the locus of all points (in a plane) equidistant from two parallel lines?

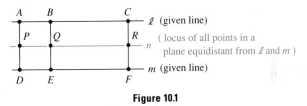

Figure 10.1

Referring to Figure 10.1, we can reason that points such as P, Q, and R that are each equidistant from the two parallel lines ℓ and m will lie on a line between the given lines ℓ and m. Also, because $PA = PD$, $QB = QE$, and $RC = RF$, the desired locus is the line n parallel to both ℓ and m and midway between ℓ and m.   ◨

Notice that Figure 10.1 can also be used in a discussion of the locus of all points equidistant from a given line. For this locus, let n be the given line. Then, ℓ and m would be the required locus because the points equidistant from the given line n form two lines parallel to n at the same distance on either side of n.

We can also give an example of a locus of points involving circles, but first consider the following definition.

DEFINITION: DISTANCE FROM A POINT TO A CIRCLE

Consider a circle with center O, a point P not on the circle, and point Q that is the intersection of $\overrightarrow{OP}$ and the circle. The **distance from P to the circle** is PQ.

E X A M P L E 2   Determine the locus of all points equidistant from two **concentric circles** with the same center O, and radii r and R.

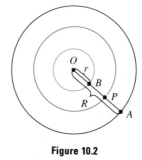

Figure 10.2

In Figure 10.2, the circles with radii $OA = R$ and $OB = r$ are the given circles. If $AP = BP$ then P is a point on the desired locus. All such points form a circle with center at O and radius OP. OP can be determined in terms of r and R.

$$OP = OB + BP$$
$$= r + \frac{1}{2}(R - r)$$
$$= r + \frac{1}{2}R - \frac{1}{2}r$$
$$= \frac{1}{2}(r + R)$$

Thus, the locus of all points equidistant from two concentric circles centered at O with radii r and R is a circle with center O and radius $\frac{1}{2}(r + R)$.   ☑

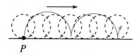

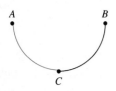

P

An example of an interesting locus is a *cycloid*, the path of a fixed point on a circle as the circle rolls along a straight line. As the circle rolls from left to right in the figure above, point P traces out a cycloid.

The figure above shows one arch of a cycloid (inverted). If an object is released from point *A*, it will travel to point *C* along the curve in less time than along any other path from *A* to *C*. This is why a cycloid is referred to as the "curve of fastest descent."

> **PRACTICE EXERCISE 1**
>
> Determine the locus of all points a given distance d from a circle with center O and radius r where $r > d$.

Notice that in the definition, a locus consists of *all those points and only those points* satisfying given conditions. This means that to prove that a given figure is a locus we must show that any point satisfying the conditions is on the figure *and* that any point on the figure satisfies the conditions. Thus, each of the following proofs is in two parts.

> **THEOREM 10.1**
>
> The locus of all points equidistant from two given points A and B is the perpendicular bisector of $\overline{AB}$.

Part I:

Given: C is any point equidistant from A and B; D is the midpoint of $\overline{AB}$ (See Figure 10.3.)

Prove: $\overline{CD}$ is the perpendicular bisector of $\overline{AB}$

Construction: Draw $\overline{AC}$ and $\overline{BC}$

Proof:

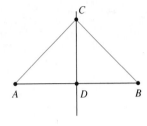

Figure 10.3

STATEMENTS	REASONS
1. C is equidistant from A and B	1. Given
2. $AC = BC$	2. By construction and the definition of "equidistant from 2 pts"
3. $CD = CD$	3. Reflexive law
4. D is the midpoint of $\overline{AB}$	4. Given
5. $AD = BD$	5. Def. of midpt. of seg.
6. ACD and BCD are triangles	6. Def. of $\triangle$
7. $\triangle ACD \cong \triangle BCD$	7. SSS = SSS
8. $\angle ADC = \angle BDC$	8. cpoctae
9. $\angle ADC$ and $\angle BDC$ are adjacent angles	9. Def. of adj. $\angle$'s
10. $\overline{CD} \perp \overline{AB}$	10. From Statements 8 and 9 and def. of $\perp$ lines
11. $\overline{CD}$ is the $\perp$ bisector of $\overline{AB}$	11. From Statements 4 and 10 and def. of $\perp$ bisector

Thus, we have shown that any point equidistant from A and B is on the perpendicular bisector of $\overline{AB}$. Now we must show that any point on the perpendicular bisector of $\overline{AB}$ is equidistant from A and B.

Part II:

Given: C is any point on the perpendicular bisector $\overline{CD}$ of $\overline{AB}$

Prove: C is equidistant from A and B

Proof:

STATEMENTS	REASONS
1. $\overline{CD}$ is $\perp$ bisector of $\overline{AB}$	1. Given
2. $AD = BD$	2. Def. of bisector
3. $\angle ADC$ and $\angle BDC$ are adjacent angles	3. Def. of adj. $\angle$'s
4. $\angle ADC = \angle BDC$	4. Def. of $\perp$ lines
5. $CD = CD$	5. Reflexive law
6. $\triangle ACD \cong \triangle BCD$	6. SAS = SAS
7. $AC = BC$	7. cpoctae
8. C is equidistant from A and B	8. Def. of "equidistant from 2 pts."

> ### THEOREM 10.2
>
> The locus of all points equidistant from the sides of an angle is the angle bisector.

Part I:

Given: D is any point equidistant from $\overrightarrow{BA}$ and $\overrightarrow{BC}$, the sides of ∠ABC (See Figure 10.4.)

Prove: $\overrightarrow{BD}$ is the bisector of ∠ABC

Construction: Construct $\overline{AD} \perp \overline{AB}$ and $\overline{CD} \perp \overline{BC}$

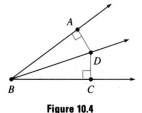

Figure 10.4

Proof:

STATEMENTS	REASONS
1. $\overline{AD} \perp \overline{AB}$ and $\overline{CD} \perp \overline{BC}$	1. By construction
2. $AD = CD$	2. Def. of "equidistant from 2 lines"
3. ∠BAD and ∠BCD are right angles	3. ⊥ lines form rt. ∠'s
4. △ADB and △CDB are right triangles	4. Def. of rt. △
5. $BD = BD$	5. Reflexive law
6. △ADB ≅ △CDB	6. HL = HL
7. ∠ABD = ∠CBD	7. cpoctae
8. $\overrightarrow{BD}$ is the bisector of ∠ABC	8. Def. of ∠ bisector

Thus, we have shown that any point equidistant from the sides of ∠ABC is on the bisector of ∠ABC. Now we must show that any point on the bisector of ∠ABC is equidistant from the sides of ∠ABC.

> ### PRACTICE EXERCISE 2
>
> Complete the proof of Theorem 10.2.
>
> Part II:
>
> *Given:* D is any point on $\overrightarrow{BD}$, the bisector of ∠ABC
>
> *Prove:* D is equidistant from $\overrightarrow{BA}$ and $\overrightarrow{BC}$
>
> *Construction:* Construct $\overline{AD} \perp \overline{AB}$ and $\overline{CD} \perp \overline{BC}$
>
> *Proof:*
>
> | STATEMENTS | REASONS |
> |---|---|
> | 1. _____ | 1. Given |

A telephone jack is located in the center of a wall that is 20 ft long. What is the locus of the positions to place a telephone if the connecting cable is 7 ft long?

2. $\angle ABD = \angle CBD$
3. $\overline{AD} \perp \overline{AB}$ and $\overline{CD} \perp \overline{BC}$
4. $\angle DAB$ and $\angle DCB$ are right angles
5. $\triangle ABD$ and $\triangle CDB$ are right triangles
6. _____
7. $\triangle ABD \cong \triangle CDB$
8. _____
9. _____

2. _____
3. By construction
4. _____
5. _____
6. Reflexive law
7. AAS = AAS
8. cpoctae
9. Def. of "equidistant from 2 lines" with Statements 3 and 8

The definition of a locus includes the possibility of more than one condition being given to determine the locus. The next example illustrates.

EXAMPLE 3   Describe the locus of points that is on both a circle of radius r and a circle with a larger radius R.

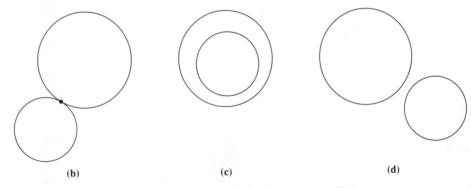

(a)   (b)   (c)   (d)

Figure 10.5   Locus of Points Common to Two Circles

Figure 10.5 illustrates the possibilities. Figure 10.5(a) shows exactly two points in the locus; Figure 10.5(b) shows one point in the locus; and Figure 10.5(c) and Figure 10.5(d) show no point in the locus. Thus, depending on the relationship of the circles, the locus will contain exactly two points, exactly one point or no points.   ∎

ANSWERS TO PRACTICE EXERCISES: **1.** The points on the circles centered at O, with radii $r - d$ and $r + d$.   **2.** 1. $\overrightarrow{BD}$ is the bisector of $\angle ABC$ 2. Def. of $\angle$ bisector   4. $\perp$ lines form rt. $\angle$'s   5. Def. of rt. $\triangle$ 6. $BD = BD$ 8. $AD = CD$ 9. D is equidistant from $\overrightarrow{BA}$ and $\overrightarrow{BC}$

10.1 EXERCISES

Describe and draw a sketch of each locus in Exercises 1–12.

 1. All points 5 units from a given point P.

 2. All points 5 units from a given line m.

 3. All points 3 units from a circle with radius 4 units.

 4. All points 4 units from a circle with radius 3 units.

 5. All points equidistant from parallel lines that are 3 units apart.

 6. All points equidistant from two concentric circles with radii 6 and 8 units.

 7. All points that lie on a given line m and that are also on a given circle.

 8. All points that lie on a given triangle and that are also on a given line m.

 9. In a circle, all points that are midpoints of all chords that are parallel to a given chord.

10. In a circle, all points that are midpoints of all chords of a given length.

11. All points that are centers of circles tangent to both of two parallel lines and that are also on a line intersecting the parallel lines.

12. All points that are centers of circles tangent to both sides of $\angle ABC$ and that are also on the circle that has radius 5 units.

13. Prove that the locus of all points that are centers of circles tangent to both of two parallel lines is the line equidistant from the two given lines.

14. Prove that the locus of all points that are centers of circles tangent to both sides of $\angle ABC$ is the bisector of the angle.

15. Prove that the locus of all points that are the right-angle vertices of a right triangle with fixed hypotenuse is a circle with diameter the hypotenuse. See the figure below.

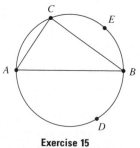

Exercise 15

16. The city council wishes to place a statue in a rectangular park. The statue is to be equidistant from the corners of the park. What is the locus of all points satisfying this condition?

17. A TV station has a broadcast range of 70 mi. What is the locus of all points that can receive the signal from the station?

10.2 TRIANGLE CONCURRENCY THEOREMS

The locus of all points in the intersection of two lines is a single point. Three or more lines can also intersect in one point.

> **DEFINITION: CONCURRENT LINES**
>
> Two or more lines are **concurrent** if they intersect in one and only one point.

We will show that for any triangle, the angle bisectors are concurrent, as are the perpendicular bisectors of the sides, the altitudes, and the medians. First consider the angle bisectors.

> **THEOREM 10.3**
>
> The bisectors of the angles of a triangle meet at a point equidistant from the sides of the triangle.

Given: $\triangle ABC$ with angle bisectors $\overrightarrow{AD}$ and $\overrightarrow{BD}$ concurrent at D (See Figure 10.6.)

Prove: $\overrightarrow{CD}$ bisects $\angle ACB$, and D is equidistant from sides $\overline{AC}$, $\overline{BC}$, and $\overline{AB}$

Construction: Construct $\overline{PD} \perp \overline{AC}$, $\overline{QD} \perp \overline{BC}$, and $\overline{RD} \perp \overline{AB}$

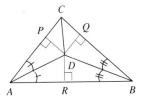

Figure 10.6   Angle Bisectors

Proof:

STATEMENTS	REASONS
1. $\overrightarrow{AD}$ bisects $\angle CAB$ and $\overrightarrow{BD}$ bisects $\angle ABC$	1. Given
2. $\overline{PD} \perp \overline{AC}$, $\overline{QD} \perp \overline{BC}$, and $\overline{RD} \perp \overline{AB}$	2. By construction
3. $PD = RD$ and $RD = QD$	3. $\angle$ bisectors are equidistant from sides
4. $PD = QD$	4. Transitive law
5. $CD = CD$	5. Reflexive law
6. $\angle CPD$ and $\angle CQD$ are right angles	6. $\perp$ lines form rt. $\angle$'s
7. $\triangle CPD$ and $\triangle CQD$ are right triangles	7. Def. of rt. $\triangle$'s
8. $\triangle CPD \cong \triangle CQD$	8. HL = HL

Describing a locus when the points are not in a plane can be difficult. A Möbius strip can illustrate. It is an example of a locus with one side and one edge. One branch of geometry, topology, is concerned with the study of figures and surfaces such as the Möbius strip. Cut a strip of paper and number the corners as shown below.

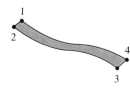

Make a half twist in the paper and tape the two ends together making points 1 and 3 and points 2 and 4 coincide.

Before the strip was taped together to form the Möbius strip, it was impossible to get from a point on the top of the strip to the bottom side without crossing an edge of the paper. Show that any two points can be joined by a line that does not cross an edge on the Möbius strip.

9. $\angle PCD = \angle DCQ$ 9. cpoctae

10. $\overrightarrow{CD}$ bisects $\angle ACB$ 10. Def. of $\angle$ bisector

11. D is equidistant from $\overline{AC}$, $\overline{BC}$,   11. From Statements 3 and 4,
    and $\overline{AB}$ $PD = QD = RD$

The point of intersection of the angle bisectors is called the **incenter** of the triangle. From Theorem 10.3 we see that the incenter is the center of the **inscribed circle.**

CONSTRUCTION 10.1

Inscribe a circle in a given triangle.

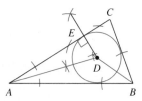

Figure 10.7 Inscribed Circle

Given: $\triangle ABC$ (See Figure 10.7.)

To Construct: The circle inscribed in $\triangle ABC$

Construction: In $\triangle ABC$ construct bisectors of $\angle CAB$ and $\angle ABC$ that intersect at D, the incenter. Construct $\overline{DE}$ perpendicular to $\overline{AC}$ from point D. Construct the circle with center D and radius DE.

THEOREM 10.4

The perpendicular bisectors of the sides of a triangle intersect at a point that is equidistant from the vertices of the triangle.

Given: $\triangle ABC$ with $\overline{PD}$ the perpendicular bisector of $\overline{AC}$, $\overline{RD}$ the perpendicular bisector of $\overline{AB}$, Q the midpoint of $\overline{BC}$ (See Figure 10.8.)

Prove: $\overline{QD}$ is the perpendicular bisector of $\overline{BC}$, and $AD = BD = CD$, that is, D is equidistant from vertices A, B, and C

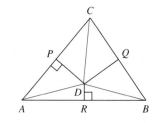

Figure 10.8 Perpendicular Bisectors of Sides

Proof:

STATEMENTS	REASONS
1. $\overline{PD}$ is the perpendicular bisector of $\overline{AC}$, and $\overline{RD}$ is the perpendicular bisector of $\overline{AB}$	1. Given
2. $CD = AD$ and $AD = BD$	2. Pts. on $\perp$ bisector are equidistant from endpts.
3. $CD = BD$	3. Transitive law
4. D is equidistant from A, B, and C	4. From Statements 2 and 3, $AD = BD = CD$
5. D is on the perpendicular bisector of $\overline{BC}$	5. From Statement 4, D is equidistant from B and C

6. Q is the midpoint of $\overline{BC}$
6. Given

7. Q is on the perpendicular bisector of $\overline{BC}$
7. Midpt. of a seg. is on all bisectors

8. $\overline{QD}$ is the perpendicular bisector of $\overline{BC}$
8. Q and D are both on $\perp$ bisector

The point of intersection of the three perpendicular bisectors of the sides of a triangle is called the **circumcenter** of the triangle. This point is the center of the **circumscribed circle** that passes through the three vertices of the triangle. This fact allows us to solve the application in the chapter introduction.

EXAMPLE 1   A gardener wishes to install in a flower garden a single sprinkler with a circular water pattern. The garden has the shape of a triangle with sides 20 ft, 16 ft, and 18 ft. At what point should the sprinkler be placed to best reach all three vertices of the garden?

Theorem 10.4 tells us that the sprinkler should be installed at the intersection of the perpendicular bisectors of the sides of the garden, that is, at the circumcenter of the triangle. As shown in Figure 10.9, we first construct the perpendicular bisectors of two of the sides to find the point P. If the sprinkler is set at P and has a radius of spray equal to PQ, then the whole garden will be watered.   ◼

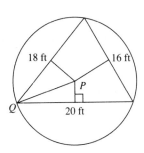

Figure 10.9

THEOREM 10.5

The altitudes of a triangle are concurrent.

Given: $\triangle ABC$ with altitudes $\overline{CR}$, $\overline{AQ}$, and $\overline{BP}$ (See Figure 10.10.)

Prove: $\overline{CR}$, $\overline{AQ}$, and $\overline{BP}$ are concurrent

Construction: $\overline{DE} \parallel \overline{BC}$, $\overline{DF} \parallel \overline{AB}$, $\overline{EF} \parallel \overline{AC}$

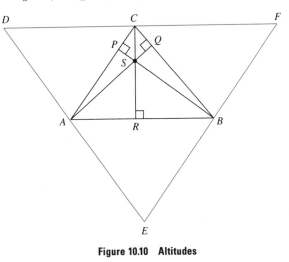

Figure 10.10   Altitudes

PRACTICE EXERCISE 1

Complete the proof of Theorem 10.5.

Proof:

STATEMENTS	REASONS
1. $\triangle ABC$ with altitudes $\overline{CR}$, $\overline{AQ}$, and $\overline{BP}$	1. _____
2. $\overline{AB} \perp \overline{CR}$, $\overline{BC} \perp \overline{AQ}$, $\overline{AC} \perp \overline{BP}$	2. _____
3. $\overline{DE} \parallel \overline{BC}$, $\overline{DF} \parallel \overline{AB}$, $\overline{EF} \parallel \overline{AC}$	3. By construction
4. $\overline{DF} \perp \overline{CR}$, $\overline{DE} \perp \overline{AQ}$, $\overline{EF} \perp \overline{BP}$	4. _____
5. $ABCD$ is a parallelogram	5. _____
6. $AB = CD$	6. _____
7. $ABFC$ is a parallelogram	7. _____
8. $AB = CF$	8. _____
9. $CD = CF$	9. _____
10. $AD = AE$	10. Use $\square ABCD$ and $\square AEBC$ and same steps as above
11. $BE = BF$	11. Use $\square ABFC$ and $\square AEBC$ and same steps as above
12. $\overline{CR}$ is the $\perp$ bisector of $\overline{DF}$	12. _____
13. $\overline{AQ}$ is the $\perp$ bisector of $\overline{DE}$	13. _____
14. $\overline{BP}$ is the $\perp$ bisector of $\overline{EF}$	14. _____
15. $\overline{CR}$, $\overline{AQ}$, and $\overline{BP}$ are concurrent	15. _____

The intersection of the altitudes of a triangle is called the **orthocenter** of the triangle.

THEOREM 10.6

The medians of a triangle meet at a point that is two-thirds the distance from the vertex to the midpoint of the opposite side.

We outline the proof of Theorem 10.6 here and request a complete two-column proof in the exercises. Consider Figure 10.11 in showing the medians are concurrent.

Leonard Euler (1707–1783)

Swiss mathematician Leonard Euler (pronounced "Oy'-ler") proved that the orthocenter, circumcenter, and centroid of a triangle all lie on the same line. The German mathematician Karl Wilhelm Feuerbach (1800–1834) showed that in any triangle the three midpoints of the sides, the three feet of the altitudes, and the three midpoints of the segments joining the circumcenter to each vertex all lie on a circle, called the nine-point circle.

Given: $\overline{AF}$ and $\overline{BD}$ are medians of $\triangle ABC$ intersecting at P, E is the point of intersection of $\overrightarrow{CP}$ and $\overrightarrow{AB}$

Prove: E is the midpoint of $\overline{AB}$

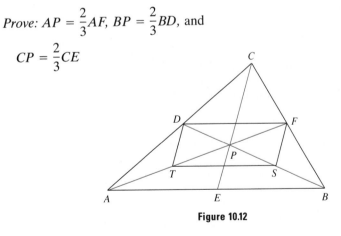

Figure 10.11

Construct Q the midpoint of $\overline{CP}$. Show $PFQD$ is a parallelogram. (Use $AD = DC$ and $PQ = CQ$ to show $\overline{PF} \parallel \overline{DQ}$, and similarly for $\overline{DP} \parallel \overline{QF}$.) Use similar triangles to show $AE = 2DR$ and $BE = 2RF$ making $AE = BE$. Thus, $\overline{CE}$ is the median from C.

We now outline the remaining part of the proof of Theorem 10.6 using Figure 10.12.

Given: $\overline{AF}$, $\overline{BD}$, and $\overline{CE}$ are medians of $\triangle ABC$ concurrent at P

Prove: $AP = \frac{2}{3}AF$, $BP = \frac{2}{3}BD$, and

$$CP = \frac{2}{3}CE$$

Figure 10.12

In Figure 10.12, construct T the midpoint of $\overline{AP}$ and S the midpoint of $\overline{BP}$. Use similar triangles to show $\overline{DF} \parallel \overline{TS}$ and $DF = TS$. Thus, $TSFD$ is a parallelogram. Then use $AT = TP = PF$ to show $AP = \frac{2}{3}AF$. Use similar arguments to show $BP = \frac{2}{3}BD$. A similar figure can be used to show $CP = \frac{2}{3}CE$.

The point at which the medians meet is called the **centroid** of the triangle. If a triangle is cut from a piece of uniform material, it can be balanced on a point placed at the centroid because under these conditions the center of mass is at the centroid.

ANSWER TO PRACTICE EXERCISE: **1.** 1. Given   2. Def. of alt. of △   4. ‖ lines ⊥ to same line   5. Opposite sides ‖ makes quadrilateral a ▱   6. Opposite sides of ▱ are =   7. Same as Reason 5   8. Opposite sides of ▱ are =   9. Sym. and trans. laws   12. Statements 4 and 9   13. Statements 4 and 10   14. Statements 4 and 11   15. ⊥ bisectors of sides of a △ are concurrent

10.2 EXERCISES

In the figure below, *E* is the incenter, *G* is the circumcenter, and *F* is the centroid of △*ABC*. Use this figure for Exercises 1–4.

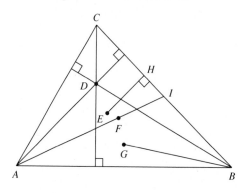

1. What is the radius of the inscribed circle?
2. What is the radius of the circumscribed circle?
3. If *FI* = 3 units, what is the measure of *AF*?
4. What is the orthocenter of △*ABC*?
5. Describe how to construct a circle passing through points *A, B,* and *C* if they are not colinear.
6. If △*ABC* is equilateral, what can be said about the incenter, circumcenter, orthocenter, and centroid?

For Exercises 7–8, consider an equilateral triangle with altitude 12 units.

7. What is the radius of the circumscribed circle?
8. What is the radius of the inscribed circle?

For Exercises 9–10, consider an equilateral triangle with sides 12 units.

9. What is the radius of the inscribed circle?

10. What is the radius of the circumscribed circle?

11. What is the length of the side of an equilateral triangular-shaped flower bed whose sides are tangent to a circular pool with diameter 10 m? Refer to the figure below.

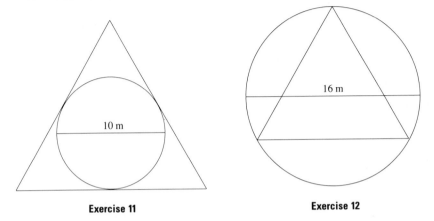

Exercise 11   **Exercise 12**

12. What is the length of the side of the largest equilateral triangular-shaped flower bed that can be placed in a circular lawn with diameter 16 m? Refer to the figure above.

13. Give a two-column proof that the medians of a triangle are concurrent.

14. Give a two-column proof that the medians of a triangle meet at a point that is two-thirds of the distance from the vertex to the midpoint of the opposite side.

15. The owner of three stores wishes to build a distribution center at a point equidistant from the three stores. If the three stores do not lie in the same straight line, explain how the owner should determine the location of the distribution center.

CHAPTER 10 REVIEW

KEY TERMS AND SYMBOLS

10.1 locus, p. 335
distance (point to circle), p. 335
concentric circles, p. 336

10.2 concurrent lines, p. 341
incenter, p. 342
inscribed circle, p. 342
circumcenter, p. 342

circumscribed circle, p. 343
orthocenter, p. 345
centroid, p. 346

REVIEW EXERCISES

Section 10.1

Determine the locus of points described in Exercises 1–4.

1. All points 2 units from a circle with radius 10 units.
2. All points 6 units from a circle with radius 4 units.
3. All points equidistant from the sides of $\angle ABC$ that are also on a circle with center at B.
4. All points that are on a square and also on a line m.
5. Prove that the locus of points that are midpoints of chords of a given length is a circle concentric with the given circle.

Section 10.2

6. Given $\triangle ABC$. Describe how to construct a circle passing through A, B, and C.
7. Given $\triangle ABC$ such that $AB = BC = CA = 30$ cm. Find the radius of the inscribed circle and the radius of the circumscribed circle.
8. $\triangle ABC$ is an isosceles triangle with vertex $\angle C = 90°$. If the base is 16 units, what is the radius of the circumscribed circle?
9. What is the radius of the circumscribed circle for a right triangle with sides 6, 8, and 10 units?
10. Prove that the incenter and the circumcenter of an equilateral triangle coincide.

PRACTICE TEST

Determine the locus of points in Problems 1 and 2.

1. All points equidistant from parallel lines 6 units apart.
2. All points that are on a circle of radius 4 units and also on a circle of radius 6 units.
3. Prove that the locus of points that are midpoints of parallel chords is a diameter of the circle.

4. Given $\triangle ABC$, an equilateral triangle with altitude 21 cm. What is the radius of the inscribed circle? What is the radius of the circumscribed circle?

5. $\triangle ABC$ is an isosceles triangle with vertex $\angle C = 90°$. If the altitude is 3 cm, what is the radius of the circumscribed circle?

6. Prove that the medians from the base angles of an isosceles triangle are equal.

7. What is the locus of points that a dog can reach when it is tied at the center of a 30-foot fence with an 8-foot leash?

11

*I*NTRODUCTION TO *A*NALYTIC *G*EOMETRY

In this chapter we combine algebra with our study of geometry. Algebra and geometry remained separate areas of study until René Descartes, a seventeenth-century French mathematician, joined them into what is now called analytic geometry.

The following example presents an applied problem that involves both algebra and geometry. Its solution appears in Section 11.2, Example 4.

The total sales of a newly formed manufacturing company amounted to $55,000 in 1981. During the year 1988, the total sales were $111,000. Assuming that the total sales y in year x can be approximated by a linear equation, find this equation and use it to approximate the total sales expected in 1995.

11.1 THE CARTESIAN COORDINATE SYSTEM

We introduced and worked with the number line in Chapter 1. When two number lines are placed together, as in Figure 11.1, so that the origins coincide and the lines are perpendicular, the result is a **Cartesian,** or **rectangular, coordinate system** or a **coordinate plane.**

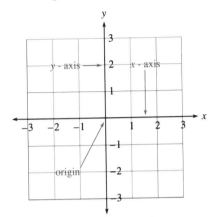

Figure 11.1   Cartesian Coordinate System

The horizontal number line is called the *x*-**axis,** and the vertical number line is called the *y*-**axis.** The point of intersection of the axes is called the **origin.**

Recall that there is one and only one point on a number line associated with each real number. A similar situation exists for points in a plane and **ordered pairs** of numbers.

> POSTULATE 11.1
>
> Associated with each point in the plane there is one and only one ordered pair of numbers.

The positions taken by the members of a marching band during a half-time performance can be shown using a coordinate system superimposed on a football field. The yard lines can serve to give one coordinate.

The ordered pair (2, 3) can be identified with a point in the coordinate plane in the following way. The first number 2, called the *x*-**coordinate** of the point, is associated with a value on the horizontal axis, or *x*-axis. The second number 3, called the *y*-**coordinate** of the point, is associated with a value on the vertical axis, or *y*-axis. The pair (2, 3) is associated with the point at which the vertical line through 2 on the *x*-axis intersects the horizontal line through 3 on the *y*-axis, as shown in Figure 11.2. The ordered pairs (3, 2), (−1, 3), (−3, −2), and (2, −2) are plotted in Figure 11.3.

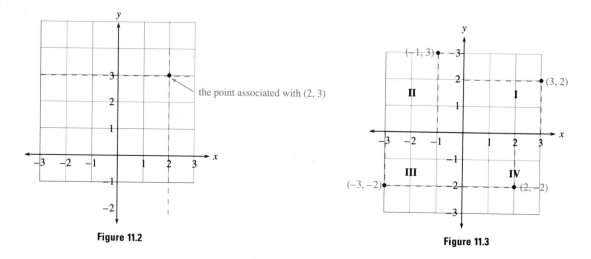

Figure 11.2   Figure 11.3

The axes in a coordinate system separate the plane into four sections called **quadrants.** The first, second, third, and fourth quadrants are identified by the Roman numerals I, II, III, and IV, respectively, in the coordinate plane, as shown in Figure 11.3. The signs of the x-coordinate (listed first) and y-coordinate (listed second) in the various quadrants are as follows.

$$\text{I}: (+, +), \quad \text{II}: (-, +) \quad \text{III}: (-, -), \quad \text{IV}: (+, -)$$

We often use (x, y) to refer to a general ordered pair of numbers. Thus, the point in the plane associated with the pair (x, y) has x as its x-coordinate and y as its y-coordinate. When we identify the point in a plane associated with the given pair, we say that we **plot** the point. When referring to "the point P corresponding to the ordered pair (x, y)," we sometimes simply say "the point (x, y)," or write $P(x, y)$.

In analytic geometry, we plot points that satisfy an equation in two variables in order to construct the geometric figure that corresponds to the equation. We start our discussion with equations of the form

$$ax + by + c = 0$$

which are associated with a line and are called **linear** or **first-degree equations.** An ordered pair (x, y) that satisfies an equation is called a **solution.** If we could plot all the solutions to a linear equation, we would construct the **graph** of the equation in a coordinate system.

Suppose we graph the equation

$$3x - y - 2 = 0.$$

The ordered pair $(2, 4)$ is one solution because if x is replaced by 2 and y is replaced by 4, the resulting equation is true.

René Descartes (1596–1650)

Although the contributions of René Descartes to philosophy and science were well received in his day, he is best remembered for his work in mathematics. The Cartesian coordinate system was the basis of his discovery of analytic geometry. By combining algebra with geometry, he was able to broaden the scope of this ancient discipline.

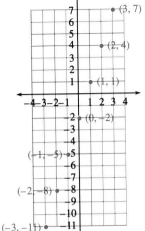

Figure 11.4

$$3x - y - 2 = 0$$
$$3(2) - (4) - 2 = 0 \qquad x = 2 \text{ and } y = 4$$
$$6 - 4 - 2 = 0$$
$$0 = 0$$

You can verify that $(0, -2)$ and $(-1, 5)$ are also solutions whereas $(-2, 4)$ is not.

One way to display a collection of solutions to a linear equation is to make a **table of values.** We solve the equation for y, choose several values for x, substitute these values into the equation, and compute the corresponding values of y. For example, if we solve $3x - y - 2 = 0$ for y, we get

$$y = 3x - 2.$$

We can then construct the table of values.

Substitution	*Result in $y = 3x - 2$*
$x = 0$	$y = 3(0) - 2 = -2$
$x = 1$	$y = 3(1) - 2 = 1$
$x = -1$	$y = 3(-1) - 2 = -5$
$x = 2$	$y = 3(2) - 2 = 4$
$x = -2$	$y = 3(-2) - 2 = -8$
$x = 3$	$y = 3(3) - 2 = 7$
$x = -3$	$y = 3(-3) - 2 = -11$

Table of Values

x	y
0	-2
1	1
-1	-5
2	4
-2	-8
3	7
-3	-11

The table above lists seven solutions to the equation $y = 3x - 2$.

$$(0, -2), \quad (1, 1), \quad (-1, -5), \quad (2, 4), \quad (-2, -8), \quad (3, 7), \quad (-3, -11)$$

Now we can plot the points that correspond to these ordered-pair solutions in a Cartesian coordinate system, as shown in Figure 11.4. It appears that all seven points do lie on a line. Thus, the graph of this equation is the line passing through these seven points in Figure 11.5.

Knowing that the graph of a linear equation is a line, we need only two solutions to graph it. In most cases, the two pairs that are easiest to determine are the **intercepts.** The points at which a line crosses the x-axis and the y-axis are called the **x-intercept** and the **y-intercept,** respectively. Because the y-intercept is a point on the y-axis, it has x-coordinate 0. Similarly, the x-intercept is a point on the x-axis and has y-coordinate 0.

Figure 11.5

EXAMPLE 1   Graph $3x - 5y - 15 = 0$.

Find the x- and y-intercepts by completing the following table.

x	y
0	
	0

When $x = 0$, $-5y = 15$, so that $y = -3$. When $y = 0$, $3x = 15$, so that $x = 5$. The completed table

x	y
0	-3
5	0

shows the y-intercept $(0, -3)$ and the x-intercept $(5, 0)$. If we plot these two points and draw the line through them, we have the graph of $3x - 5y - 15 = 0$ as shown in Figure 11.6.   ◢

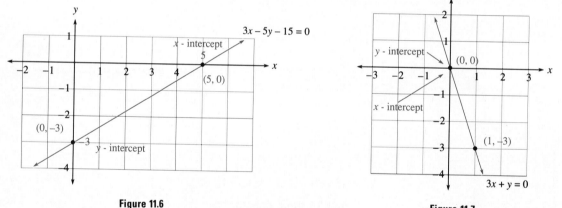

Figure 11.6

Figure 11.7

PRACTICE EXERCISE 1
Graph $2x - 3y = 6$.

There are special cases in which using only the intercepts will not give the two points necessary to graph the equation. For example,

$$3x + y = 0$$

has only one intercept, $(0, 0)$. In an equation like this we must determine another point. The graph of $3x + y = 0$ is shown in Figure 11.7 where the additional point $(1, -3)$ is plotted.

Other special cases include horizontal lines whose equations do not contain an x value, for example,

$$3y + 5 = 0.$$

Also, vertical lines have equations with no y value, such as

$$x - 4 = 0.$$

The graphs of these equations are shown in Figure 11.8 and Figure 11.9.

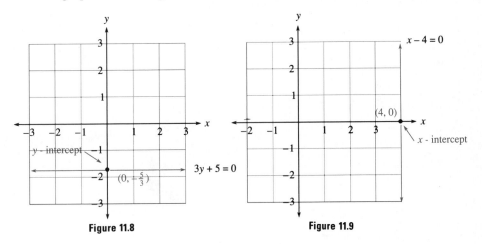

Figure 11.8   Figure 11.9

ANSWER TO PRACTICE EXERCISE: **1.** The graph is a straight line with intercepts $(3, 0)$ and $(0, -2)$.

11.1 E*XERCISES*

1. Plot the points corresponding to the ordered pairs $A(1, 4)$, $B(4, -2)$, $C(-3, 2)$, $D(-3, 0)$, $E(3, 0)$, $F(0, 0)$, $G(-3, -3)$, and $H(0, -2)$.

2. Plot the points corresponding to the ordered pairs $A(2, 5)$, $B(3, -1)$, $C(-4, 1)$, $D(-2, 0)$, $E(4, 0)$, $F(-2, -2)$, $G(0, 0)$, and $H(0, -5)$.

3. Give the coordinates of the points A, B, C, D, E, F, G, and H.

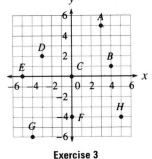

Exercise 3

4. Give the coordinates of the points A, B, C, D, E, F, G, and H.

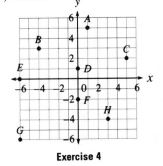

Exercise 4

5. Plot the points associated with the ordered pairs $J\left(\dfrac{1}{2}, 2\right)$,

$K\left(-\dfrac{5}{2}, 1\right)$, $L\left(-2, -\dfrac{7}{4}\right)$, and $M\left(3, -\dfrac{3}{4}\right)$.

6. Plot the points associated with the ordered pairs

$P\left(-\dfrac{1}{2}, 3\right)$, $Q\left(\dfrac{7}{2}, 4\right)$, $R\left(-3, -\dfrac{3}{4}\right)$, and $S\left(2, -\dfrac{7}{3}\right)$.

7. In what quadrants are the points J, K, L, and M of Exercise 5 located?

8. In what quadrants are the points P, Q, R, and S of Exercise 6 located?

In Exercises 9–14, find the missing coordinate so that each ordered pair is a solution to the equation.

9. $x + y + 2 = 0$; (a) $(0, \ )$; (b) $(\ , 0)$; (c) $(1, \ )$; (d) $(\ , -2)$

10. $2x - y + 1 = 0$; (a) $(0, \ )$; (b) $(\ , 0)$; (c) $(-1, \ )$; (d) $(\ , 3)$

11. $x + 3 = 0$; (a) $(0, \ )$; (b) $(\ , 0)$; (c) $(2, \ )$; (d) $(\ , -4)$

12. $2x = 5$; (a) $(0, \ )$; (b) $(\ , 0)$; (c) $(-2, \ )$; (d) $(\ , 1)$

13. $3y + 1 = 0$; (a) $(0, \ )$; (b) $(\ , 0)$; (c) $(3, \ )$; (d) $(\ , -1)$

14. $y = -4$; (a) $(0, \ )$; (b) $(\ , 0)$; (c) $(-5, \ )$; (d) $(\ , 2)$

Determine the intercepts and graph each linear equation in Exercises 15–30.

15. $x + y + 2 = 0$ 16. $x + y - 2 = 0$ 17. $3x + y = 6$ 18. $x + 2y = 4$

19. $x - y = 2$ 20. $y - x = 1$ 21. $x - y = 0$ 22. $y - x = 0$

23. $3x - 7 = 0$ 24. $2x + 3 = 0$ 25. $y = -1$ 26. $y = 2$

27. $3x - 2y = 0$ 28. $2y - 3x = 0$ 29. $3x + 2y = 6$ 30. $3x - 2y = 6$

31. What are the intercepts of the line $x = 0$? What is its graph?

32. What are the intercepts of the line $y = 0$? What is its graph?

33. Graph $y = 2x + 1$, $y = \dfrac{1}{2}x + 1$, $y = 0 \cdot x + 1$, $y = -\dfrac{1}{2}x + 1$, and $y = -2x + 1$ in the same coordinate system. What common characteristic do all of the lines possess?

34. Graph $y = 2x + 3$, $y = 2x + 1$, $y = 2x + 0$, $y = 2x - 1$, and $y = 2x - 3$ in the same coordinate system. What common characteristic do all the lines possess?

35. Mr. Kirk owns a small business that manufactures picture frames. He has determined that the weekly cost in dollars y of making a number of frames x is determined with the equation

$$y = 5x + 100.$$

Find the cost of making 75 picture frames during one week. How many frames were made during a week when the costs were $1100? Graph the equation.

36. A retail store owner estimates that the daily profit in dollars y that she can make on the sale of x dresses is determined with the equation

$$y = 30x - 50.$$

Find the profit she made on a day when 8 dresses were sold. How many dresses were sold on a day when her profit was $1000? Graph the equation.

11.2 SLOPE, DISTANCE, AND MIDPOINT FORMULAS

We can expand our discussion of linear equations by defining the *slope* of a line. Suppose that the points $P(x_1, y_1)$ and $Q(x_2, y_2)$ are two points on a line, shown in Figure 11.10. (The symbol x_1, read "x sub one," is the name given to an x-coordinate.) The **rise** is the vertical change, $y_2 - y_1$, in the line, and the **run** is the horizontal change, $x_2 - x_1$, in the line. We use these terms to define a line's slope.

> **DEFINITION: THE SLOPE OF A LINE**
>
> Let $P(x_1, y_1)$ and $Q(x_2, y_2)$ be two points on a *nonvertical* line. The **slope** of the line is
>
> $$m = \frac{y_2 - y_1}{x_2 - x_1} = \frac{\text{rise}}{\text{run}} = \frac{\text{change in } y\text{-coordinates}}{\text{change in } x\text{-coordinates}}.$$

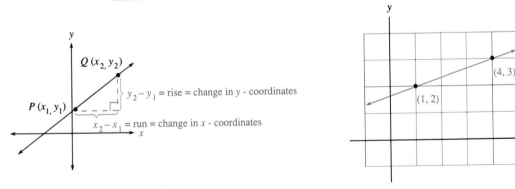

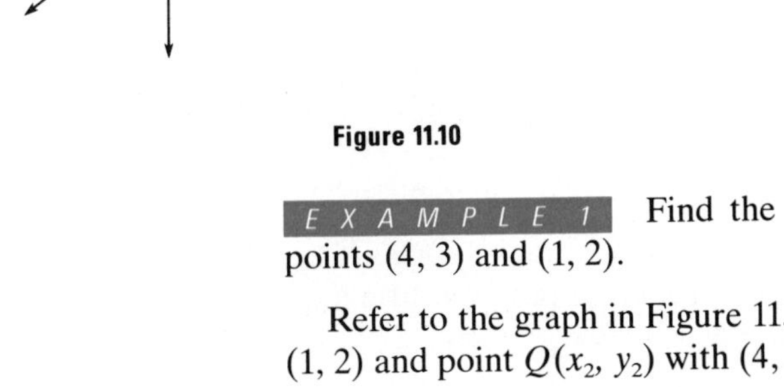

Figure 11.10

Figure 11.11

EXAMPLE 1   Find the slope of the line passing through the two points $(4, 3)$ and $(1, 2)$.

Refer to the graph in Figure 11.11. Suppose we identify point $P(x_1, y_1)$ with $(1, 2)$ and point $Q(x_2, y_2)$ with $(4, 3)$. The slope will then be determined with

$$m = \frac{y_2 - y_1}{x_2 - x_1} = \frac{3 - 2}{4 - 1} = \frac{1}{3}.$$

What happens if we identify $P(x_1, y_1)$ with $(4, 3)$ and $Q(x_2, y_2)$ with $(1, 2)$? In this case we have

$$m = \frac{y_2 - y_1}{x_2 - x_1} = \frac{2 - 3}{1 - 4} = \frac{-1}{-3} = \frac{1}{3}.$$

Thus, we see that the slope is the same regardless of how the two points are identified.   ◪

E X A M P L E   2   Find the slope of the line passing through the two points $(-3, 2)$ and $(1, -1)$.

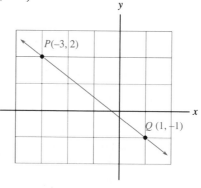

Figure 11.12

Let us identify $P(x_1, y_1)$ with $(-3, 2)$ and $Q(x_2, y_2)$ with $(1, -1)$ as in Figure 11.12. The slope is determined with

$$m = \frac{y_2 - y_1}{x_2 - x_1} = \frac{(-1) - (2)}{(1) - (-3)} \qquad \text{Watch the signs}$$

$$= \frac{-3}{1 + 3} = -\frac{3}{4}. \quad ◪$$

PRACTICE EXERCISE 1

Find the slope of the line passing through the points $(1, -3)$ and $(-2, 4)$.

Finding the slope of a line can be useful in many practical situations. In the design of a drawbridge, the amount of "slope" of the bridge when in an open position determines clearance for ships that pass beneath it. Can you think of other examples in which slope plays a role in construction or architecture?

The line passing through the points $(3, 2)$ and $(-1, 2)$ is horizontal and its slope is

$$m = \frac{y_2 - y_1}{x_2 - x_1} = \frac{2 - 2}{-1 - 3} = \frac{0}{-4} = 0.$$

If we try to calculate the slope of the vertical line passing through $(-1, 4)$ and $(-1, -2)$, we find that it is not defined.

$$m = \frac{y_2 - y_1}{x_2 - x_1} = \frac{-2 - 4}{-1 - (-1)} = \frac{-6}{-1 + 1} = \frac{-6}{0} \qquad \text{Undefined because we cannot divide by zero}$$

The length of a ski run can be computed using the notion of the slope of a line. It is interesting and relevant to note that we often use the term "ski slope" to describe a ski facility.

We can use slope to define two more forms of the equation of a line. In Section 11.1, we defined a linear equation using the **general form**

$$ax + by + c = 0.$$

We now consider the *point-slope form* and the *slope-intercept form*.

DEFINITION: POINT-SLOPE FORM OF THE EQUATION OF A LINE

The equation of a line with slope m and passing through the point (x_1, y_1) is determined using the **point-slope form**

$$y - y_1 = m(x - x_1).$$

E X A M P L E 3   Find the point-slope form of the equation of the line with slope $\frac{1}{2}$ passing through the point $(-3, 1)$. Also, write the general form of the equation of this line.

We have $(x_1, y_1) = (-3, 1)$ and $m = \frac{1}{2}$, so by substituting we obtain the point-slope form.

$$y - y_1 = m(x - x_1)$$
$$y - 1 = \frac{1}{2}(x - (-3)) \qquad \text{Watch the sign}$$

To determine the general form of the equation, eliminate fractions, remove parentheses, and then collect all terms on the left side of the equation.

$$y - 1 = \frac{1}{2}(x + 3)$$
$$2(y - 1) = 2 \cdot \frac{1}{2}(x + 3) \qquad \text{Multiply both sides by 2}$$
$$2y - 2 = x + 3 \qquad\qquad \text{Remove parentheses}$$
$$2y - 2 - x - 3 = x + 3 - x - 3 \quad \text{Subtract } x \text{ and 3}$$
$$-x + 2y - 5 = 0$$
$$x - 2y + 5 = 0 \qquad\qquad \text{Multiply by } -1 \quad \blacksquare$$

PRACTICE EXERCISE 2

Find the general form of the equation of a line passing through $(3, -5)$ with slope $-\frac{6}{5}$.

The next example solves the applied problem given in the chapter introduction by using a linear equation.

E X A M P L E 4   The total sales of a newly formed manufacturing company amounted to $55,000 in 1981. During the year 1988, the total sales were $111,000. Assuming that the total sales y in year x can be approximated by a linear equation, find this equation and use it to approximate the total sales expected in 1995.

Suppose we identify 1981 as year 1, $x = 1$, then 1988 corresponds to year 8 and $x = 8$. When $x = 1$, $y = 55,000$ and when $x = 8$, $y = 111,000$ giving us the two points $(x_1, y_1) = (1, 55,000)$ and $(x_2, y_2) = (8, 111,000)$ on the line describing total sales. First find the slope of the line.

$$m = \frac{y_2 - y_1}{x_2 - x_1} = \frac{111,000 - 55,000}{8 - 1} = \frac{56,000}{7} = 8000$$

Now use 8000 for m and $(1, 55,000)$ for (x_1, y_1) in the point-slope form.

$$y - 55,000 = 8000(x - 1)$$
$$y - 55,000 = 8000x - 8000$$
$$y = 8000x + 47,000$$

We leave the equation solved for y because we want the value of y when x is 15 (in the year 1995).

$$y = 8000(15) + 47,000 \qquad x = 15$$
$$= 120,000 + 47,000$$
$$= 167,000$$

Assuming conditions do not change, in 1995 the company can expect total sales of about $167,000.   ◪

DEFINITION: SLOPE-INTERCEPT FORM OF THE EQUATION OF A LINE

The equation of a line with slope m and y-intercept $(0, b)$ is determined using the **slope-intercept form**

$$y = mx + b.$$

E X A M P L E 5   Find the slope-intercept form of the equation of the line with slope -2 and y-intercept $(0, 7)$.

Substituting -2 for m and 7 for b we obtain

$$y = mx + b$$
$$y = -2x + 7.   ◪$$

The importance of the slope-intercept form is that once this form is obtained, the slope (the coefficient of x) can be read directly, as can the y-coordinate of the y-intercept (the constant term). Given any form of the equation of a line, if the equation is solved for y, the coefficient of x is the slope and the constant term is the y-coordinate of the y-intercept.

Using this technique and the following postulates, we can determine if two equations represent parallel or perpendicular lines.

POSTULATE 11.2

Two distinct lines with slopes m_1 and m_2 are parallel if and only if $m_1 = m_2$.

POSTULATE 11.3

Two lines with slopes m_1 and m_2 are perpendicular if and only if $m_1 m_2 = -1$.

EXAMPLE 6   Determine whether the lines with equations $3x - 2y + 7 = 0$ and $2x + 3y - 6 = 0$ are parallel or perpendicular.

We solve each equation for y (write each in slope-intercept form) to determine the slope.

$$3x - 2y + 7 = 0 \qquad\qquad 2x + 3y - 6 = 0$$
$$-2y = -3x - 7 \qquad\qquad 3y = -2x + 6$$
$$y = \frac{3}{2}x + \frac{7}{2} \qquad\qquad y = -\frac{2}{3}x + 2$$
$$\downarrow \qquad\qquad\qquad\qquad \downarrow$$
$$m_1 = \frac{3}{2} \qquad\qquad\qquad m_2 = -\frac{2}{3}$$

Because $m_1 = \frac{3}{2}$ and $m_2 = -\frac{2}{3}$ are negative reciprocals or $m_1 m_2 = \left(\frac{3}{2}\right)\left(-\frac{2}{3}\right) = -1$, the two lines are perpendicular.   ▨

EXAMPLE 7   Determine whether the lines with equations $x - 4y + 2 = 0$ and $3x - 12y + 6 = 0$ are parallel.

Solve each equation for y.

$$x - 4y + 2 = 0 \qquad\qquad 3x - 12y + 6 = 0$$
$$-4y = -x - 2 \qquad\qquad -12y = -3x - 6$$
$$y = \frac{1}{4}x + \frac{2}{4} \qquad\qquad y = \frac{3}{12}x + \frac{6}{12}$$
$$y = \frac{1}{4}x + \frac{1}{2} \qquad\qquad y = \frac{1}{4}x + \frac{1}{2}$$
$$\downarrow \qquad\qquad\qquad\qquad \downarrow$$
$$m_1 = \frac{1}{4} \qquad\qquad\qquad m_2 = \frac{1}{4}$$

Because $m_1 = m_2$ we are tempted to conclude that the two lines are parallel. However, the y-intercepts are also equal $\left(\text{both are } \left(0, \frac{1}{2}\right)\right)$ so the equations determine the same line. In situations like this we say that the lines **coincide.**   ▨

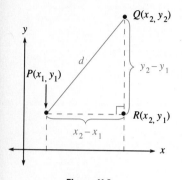

Figure 11.3
Distance Formula

Calculating the distance between two given points is a direct application of the Pythagorean Theorem. Suppose the two points, $P(x_1, y_1)$ and $Q(x_2, y_2)$, are in the first quadrant, as in Figure 11.13. Then the numbers $x_2 - x_1$ and $y_2 - y_1$ are the lengths of the legs of the triangle with vertices P, Q, and R. Applying the Pythagorean Theorem, we obtain

$$d^2 = (x_2 - x_1)^2 + (y_2 - y_1)^2,$$

and because d is positive,

$$d = \sqrt{(x_2 - x_1)^2 + (y_2 - y_1)^2}$$

gives the length of the hypotenuse of the triangle—that is, the distance between the points (x_1, y_1) and (x_2, y_2). Because $(x_2 - x_1)$ and $(y_2 - y_1)$ are both squared, $(x_1 - x_2)^2 = (x_2 - x_1)^2$ and $(y_1 - y_2)^2 = (y_2 - y_1)^2$, and it is immaterial which point we label (x_1, y_1) and which point we label (x_2, y_2). Although our particular points are in the first quadrant, the same results hold regardless of the location of the two points.

We have just given a paragraph-style proof for the next theorem.

THEOREM 11.1   DISTANCE FORMULA

The **distance** between two points with coordinates (x_1, y_1) and (x_2, y_2) is determined with the formula

$$d = \sqrt{(x_2 - x_1)^2 + (y_2 - y_1)^2}.$$

E X A M P L E   8

(a) Find the distance between $(6, 8)$ and $(2, 5)$.

Let $(x_1, y_1) = (6, 8)$ and $(x_2, y_2) = (2, 5)$ and substitute into the distance formula.

$$\begin{aligned} d &= \sqrt{(x_2 - x_1)^2 + (y_2 - y_1)^2} \\ &= \sqrt{(2 - 6)^2 + (5 - 8)^2} \\ &= \sqrt{(-4)^2 + (-3)^2} = \sqrt{16 + 9} = \sqrt{25} = 5 \end{aligned}$$

The distance between the given points is 5 units.

(b) Find the distance between $(2, 2)$ and $(5, 5)$.

Substituting into the distance formula, we have

$$\begin{aligned} d &= \sqrt{(5 - 2)^2 + (5 - 2)^2} \\ &= \sqrt{3^2 + 3^2} = \sqrt{2 \cdot 3^2} = 3\sqrt{2}. \end{aligned}$$ ◪

Another formula, the *midpoint formula,* can be established using similar triangles. It can be shown that the coordinates of a point midway between two given points are found by averaging the corresponding x-coordinates and y-coordinates.

THEOREM 11.2   MIDPOINT FORMULA

The coordinates $(\bar{x}, \bar{y})$ of the midpoint of a line segment joining (x_1, y_1) and (x_2, y_2) are determined with the **midpoint formula**

$$(\bar{x}, \bar{y}) = \left(\frac{x_1 + x_2}{2}, \frac{y_1 + y_2}{2} \right).$$

That is, the x-coordinate of the midpoint is the average of the x-coordinates of the points, and the y-coordinate of the midpoint is the average of the y-coordinates of the points.

NOTE: In the distance formula we subtract x-coordinates and y-coordinates whereas in the midpoint formula we add x-coordinates and y-coordinates. Do not confuse these operations when working with the two formulas.   □

E X A M P L E 9   Find the midpoint between the points $(-2, 1)$ and $(6, -3)$.

Substitute into the midpoint formula.

$$(\bar{x}, \bar{y}) = \left(\frac{x_1 + x_2}{2}, \frac{y_1 + y_2}{2} \right) = \left(\frac{-2 + 6}{2}, \frac{1 + (-3)}{2} \right)$$

$$= \left(\frac{4}{2}, \frac{-2}{2} \right) = (2, -1)$$

The midpoint is $(2, -1)$.   ◪

ANSWERS TO PRACTICE EXERCISES: **1.** $-\dfrac{7}{3}$ **2.** $6x + 5y + 7 = 0$

11.2 *EXERCISES*

In Exercises 1–6, find the slope of the line passing through the given pair of points.

1. $(-5, 2)$ and $(-1, -6)$ **2.** $(7, -1)$ and $(3, 3)$ **3.** $(1, 7)$ and $(3, 3)$

4. $(-4, 1)$ and $(-1, 3)$ **5.** $(-2, 7)$ and $(-5, 7)$ **6.** $(4, 2)$ and $(4, -2)$

Find the general form of the equation of the line satisfying the conditions given in Exercises 7–24.

7. Through $(3, -1)$ with slope -2.

8. Through $(-2, -4)$ with slope -3.

9. Through $(2, -4)$ parallel to a line with slope $\dfrac{1}{3}$.

10. Through $(-3, 5)$ parallel to a line with slope -4.

11. Through $(7, -1)$ and $(5, 3)$.

12. Through $(-7, 1)$ and $(3, -5)$.

13. Through $(2, 3)$ and is a vertical line.

14. Through $(-1, 7)$ and is a vertical line.

15. Through $(2, 3)$ with slope 0.

16. Through $(5, -3)$ with slope 0.

17. With x-intercept $(-5, 0)$ and slope 2.

18. With x-intercept $(3, 0)$ and slope $\dfrac{1}{2}$.

19. With y-intercept $(0, 4)$ and slope -3.

20. With y-intercept $(0, 2)$ and slope -5.

21. Through $(4, -1)$ and $\left(7, \dfrac{2}{3}\right)$.

22. Through $\left(2, \dfrac{1}{2}\right)$ and $(-3, 8)$.

23. With x-intercept $(-2, 0)$ and y-intercept $(0, 5)$.   **24.** With x-intercept $(4, 0)$ and y-intercept $(0, -3)$.

In Exercises 25–30, find the slope and y-intercept of the line by writing the equation in slope-intercept form.

25. $5x - 2y + 4 = 0$

26. $6x + 2y - 10 = 0$

27. $5x + 4 = 0$

28. $x + 7 = 0$

29. $-2y + 4 = 0$

30. $3y + 9 = 0$

31. Find the slope-intercept form of the equation of the line passing through the points $(-2, 3)$ and $(1, -3)$. What is the slope? the y-intercept?

32. Find the slope-intercept form of the equation of the line passing through the points $(4, -5)$ and $(2, 8)$. What is the slope? the y-intercept?

In Exercises 33–38, determine whether the lines with given equations are parallel, perpendicular, or neither.

33. $5x - 3y + 8 = 0$
$3x + 5y - 7 = 0$

34. $2x - y + 3 = 0$
$x + 2y - 5 = 0$

35. $2x - y + 7 = 0$
$-6x + 3y - 1 = 0$

36. $5x - 2y + 3 = 0$
$-10x + 4y + 3 = 0$

37. $2x + 1 = 0$
$-3y - 4 = 0$

38. $4x - y + 7 = 0$
$3x - 1 = 0$

39. Find the general form of the equation of the line passing through $(-1, 4)$ and parallel to the line with equation $-3x - y + 4 = 0$.

40. Find the general form of the equation of the line passing through $(-2, 5)$ and parallel to the line with equation $4x + 2y - 9 = 0$.

41. Find the general form of the equation of the line passing through $(-1, 4)$ and perpendicular to the line with equation $-3x - y + 4 = 0$.

42. Find the general form of the equation of the line passing through $(4, -7)$ and perpendicular to the line with equation $x - 5y + 7 = 0$.

Find the distance between the points given in Exercises 43–48.

43. $(3, 2)$ and $(-2, -10)$

44. $(3, 6)$ and $(-3, -2)$

45. $(-3, 2)$ and $(8, -5)$

46. $(-2, 1)$ and $(5, -3)$

47. $(5, 2)$ and $(5, -3)$

48. $(7, -1)$ and $(-2, -1)$

Find the midpoint of the line segment between the points given in Exercises 49–54.

49. $(-2, -1)$ and $(4, 1)$

50. $(3, -4)$ and $(1, 6)$

51. $(6, 11)$ and $(5, -3)$

52. $(5, -3)$ and $(-2, -1)$

53. $(-2, -1)$ and $(8, -1)$

54. $(5, -3)$ and $(5, 2)$

In Exercises 55–60, find the general form of the equation of the line perpendicular to the line that contains the given points, and that passes through the point midway between them.

55. $(-3, 2)$ and $(3, -6)$ **56.** $(-5, 1)$ and $(3, -1)$ **57.** $(5, -3)$ and $(7, 1)$

58. $(2, -7)$ and $(6, 1)$ **59.** $(2, -1)$ and $(1, -2)$ **60.** $(-4, 3)$ and $(4, -3)$

Some business problems involve information that can be described by a straight-line graph. You can interpret two pieces of given information as points, and then determine the equation that describes the situation as we did in Example 4. Do this for Exercises 61–62.

61. Jan Pierce makes wood-burning stoves. She discovers that 10 stoves can be constructed for $2800, and 25 stoves can be made for $6125. If y is the cost to make x stoves, find a linear equation describing this relationship. Use the equation to calculate the cost of 31 stoves.

62. In 1985 (call this year 1), the Mutter Manufacturing Company had sales of $40,000. In 1988 (year 4), total sales were $90,000. If y represents total sales in year x, find a linear equation that describes this relationship. Use the equation to approximate total sales for the year 1989 (year 5).

11.3 CIRCLES

The circle is another figure that we can describe in a Cartesian coordinate system. Remember from Chapter 7 that a circle is the set of points in a plane a fixed distance r (the radius) from a given point called the center of the circle. Using this definition, we can give a paragraph-style proof of the following theorem.

> **THEOREM 11.3 STANDARD FORM OF THE EQUATION OF A CIRCLE**
> A circle with center (h, k) and radius r can be described by the equation
> $$(x - h)^2 + (y - k)^2 = r^2.$$
> The equation is called the **standard form** of the equation of a circle.

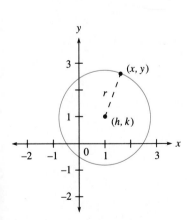

Figure 11.4 Circle

PROOF: To prove this theorem we can use the definition of a circle and the distance formula. Let (x, y) be an arbitrary point on the circle with radius r centered at the point (h, k) as shown in Figure 11.14. By definition, every point (x, y) is r units from (h, k), so in using the distance formula

$$r = \sqrt{(x - h)^2 + (y - k)^2}.$$

Squaring both sides we have

$$r^2 = (x - h)^2 + (y - k)^2,$$

which is the required standard form. ■

Circles centered at the origin have a simplified standard form. For example, consider the equation

$$x^2 + y^2 = 4.$$

We can write this equation in the form

$$(x - 0)^2 + (y - 0)^2 = 2^2$$

to see that the center is $(0, 0)$ and the radius is 2. The graph of this equation is shown in Figure 11.15.

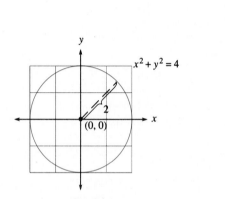

Figure 11.15

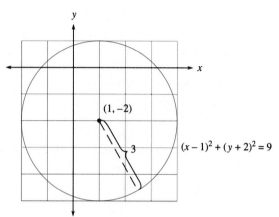

Figure 11.16

 E X A M P L E   1   Graph $(x - 1)^2 + (y + 2)^2 = 9$.

We rewrite the equation in the form $(x - h)^2 + (y - k)^2 = r^2$.

$$(x - 1)^2 + [y - (-2)]^2 = 3^2$$

Thus, the center is $(1, -2)$, and the radius has length 3. The graph is shown in Figure 11.16.   ◪

If we are given the center and radius of a circle, we can determine the standard form of its equation by substituting h, k, and r into $(x - h)^2 + (y - k)^2 = r^2$.

E X A M P L E   2   Find the standard form of the equation of the circle centered at $(2, -1)$ with radius 3.

Substitute $h = 2$, $k = -1$, and $r = 3$ in the standard form.

$$(x - h)^2 + (y - k)^2 = r^2$$
$$(x - 2)^2 + (y - (-1))^2 = 3^2 \qquad \text{Watch the sign on } k$$
$$(x - 2)^2 + (y + 1)^2 = 9 \quad ◪$$

When the standard form of the equation of a circle is expanded, the resulting equation is called the **general form** of the equation of a circle. This is actually a second-degree equation of the form

$$x^2 + y^2 + Dx + Ey + F = 0.$$

E X A M P L E 3   Find the general form of the equation of the circle centered at $(2, -1)$ with radius 3.

From Example 2, the standard form of this equation is

$$(x - 2)^2 + (y + 1)^2 = 9.$$

Expanding, we have

$$x^2 - 4x + 4 + y^2 + 2y + 1 = 9.$$

The general form is

$$x^2 + y^2 - 4x + 2y - 4 = 0. \quad \blacksquare$$

Deriving the general form of the equation of a circle from the standard form is simply a matter of squaring and collecting like terms. The reverse procedure, deriving the standard form from the general form, involves completing the squares in both variables x and y. The next example illustrates.

E X A M P L E 4   Find the standard form of the equation of a circle with general form

$$x^2 + y^2 - 6x + 10y + 29 = 0.$$

Determine the center and radius.

Start by writing the equation with the variable terms on the left side and the constant term on the right, leaving space as shown.

$$x^2 - 6x \qquad + y^2 + 10y \qquad = -29$$

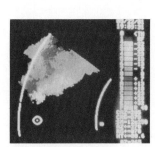

A meteorologist can track the direction of a severe thunderstorm or hurricane using radar. Circles on the screen identify the coordinates of points corresponding to weather disturbances.

To complete the square in x, add $\left[\frac{1}{2}(-6)\right]^2 = 9$, and to complete the square in y, add $\left[\frac{1}{2}(10)\right]^2 = 25$ to the left side. To maintain equality, we also add 9 and 25 to the right side.

$$x^2 - 6x + 9 + y^2 + 10y + 25 = -29 + 9 + 25$$
$$(x^2 - 6x + 9) + (y^2 + 10y + 25) = 5$$
$$(x - 3)^2 + (y + 5)^2 = 5 \qquad \text{Factor}$$

This is the standard form of the equation of the circle with center $(3, -5)$. Notice that because $r^2 = 5$, the radius is $\sqrt{5}$. $\blacksquare$

11.3 EXERCISES

In Exercises 1–12, find the standard and general forms of the equation of a circle centered at the given point and with the given radius.

1. $(-3, 2), r = 1$ **2.** $(2, -1), r = 3$ **3.** $(5, -3), r = 2$

4. $(1, 4), r = 5$ **5.** $(-1, -1), r = 3$ **6.** $(-3, -2), r = 6$

7. $\left(-\frac{1}{2}, 0\right), r = 1$ **8.** $(0, 4), r = 2$ **9.** $(0, 0), r = 1$

10. $\left(2, \frac{1}{2}\right), r = 5$ **11.** $(-2, -4), r = 6$ **12.** $(0, 0), r = 7$

In Exercises 13–18, find the standard form of the equation of each circle and identify the center and the radius.

13. $x^2 + y^2 - 2x + 4y - 20 = 0$ **14.** $x^2 + y^2 - 10x + 4y + 20 = 0$

15. $x^2 + y^2 + 6x = 0$ **16.** $x^2 + y^2 + 8y = 0$

17. $4x^2 + 4y^2 + 40x - 4y + 37 = 0$ **18.** $4x^2 + 4y^2 + 4x - 8y + 1 = 0$

Graph each equation in Exercises 19–30.

19. $x^2 + y^2 = 25$ **20.** $x^2 + y^2 = 9$ **21.** $(x - 2)^2 + (y + 3)^2 = 4$

22. $(x - 1)^2 + (y + 1)^2 = 4$ **23.** $(x - 1)^2 + (y - 2)^2 = 9$ **24.** $(x + 2)^2 + (y + 2)^2 = 4$

25. $x^2 + (y - 3)^2 = 4$ **26.** $(x - 3)^2 + y^2 = 9$ **27.** $2x^2 + 2y^2 = 32$

28. $3x^2 + 3y^2 = 27$ **29.** $x^2 + y^2 - 4x = 0$ **30.** $x^2 + y^2 + 6y = 0$

11.4 PROOFS INVOLVING POLYGONS

Analytic geometry allows us to use algebra to prove certain geometric theorems. The following proofs of earlier theorems show how. Theorems are assigned the original numbers given to them earlier in the text.

> **THEOREM 6.7**
>
> The median from the right angle in a right triangle is one-half the length of the hypotenuse.

PROOF: We can place the right $\triangle ABC$ in a coordinate system with right angle C at the origin, as shown in Figure 11.17. Because B is on the y-axis, we label it $(0, b)$. Similarly A is labeled $(a, 0)$. Because CD is the median from right $\angle C$,

D is the midpoint of the hypotenuse and $AD = BD = \frac{1}{2}AB$. We must show that $DC = AD$. Use the distance formula.

$$AD = \sqrt{\left(a - \frac{a}{2}\right)^2 + \left(0 - \frac{b}{2}\right)^2} = \sqrt{\left(\frac{a}{2}\right)^2 + \left(-\frac{b}{2}\right)^2} = \sqrt{\left(\frac{a}{2}\right)^2 + \left(\frac{b}{2}\right)^2}$$

$$DC = \sqrt{\left(\frac{a}{2} - 0\right)^2 + \left(\frac{b}{2} - 0\right)^2} = \sqrt{\left(\frac{a}{2}\right)^2 + \left(\frac{b}{2}\right)^2}$$

Thus, $DC = AD = \frac{1}{2}AB$, and the theorem is proved. ■

THEOREM 4.26

The diagonals of a parallelogram bisect each other.

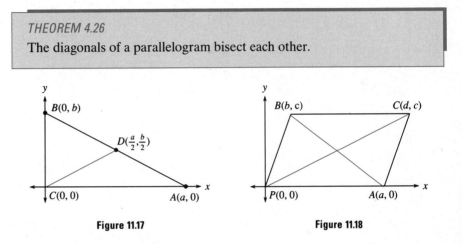

Figure 11.17   **Figure 11.18**

PROOF: Because $\overline{BC} \parallel \overline{PA}$, we label B and C with the same y-coordinate, as shown in Figure 11.18. Also because $BC = PA$,

$$d - b = a - 0$$
$$d = a + b.$$

Thus, $C(d, c)$ becomes $C(a + b, c)$. We now determine the midpoints of the diagonals $\overline{PC}$ and $\overline{AB}$.

$$\text{midpoint of } \overline{PC} = \left(\frac{a + b + 0}{2}, \frac{c + 0}{2}\right) = \left(\frac{a + b}{2}, \frac{c}{2}\right)$$
$$\text{midpoint of } \overline{AB} = \left(\frac{a + b}{2}, \frac{0 + c}{2}\right) = \left(\frac{a + b}{2}, \frac{c}{2}\right)$$

The midpoints of the segments are the same, $\left(\frac{a + b}{2}, \frac{c}{2}\right)$. It is the point of inter-section of $\overline{PC}$ and $\overline{AB}$. Thus, the diagonals of a parallelogram bisect each other. ■

THEOREM 4.29

The diagonals of a rhombus are perpendicular.

PROOF: Because a rhombus is a parallelogram, we label B and C in Figure 11.19 with the same y-coordinate. Also, as in the proof of Theorem 4.26, C can be labeled $C(a + b, c)$. To prove the theorem, we refer to Postulate 11.3 and prove that the product of the slopes of $\overline{PC}$ and $\overline{AB}$ is -1.

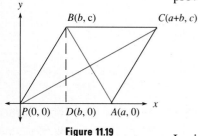

Figure 11.19

$$\text{slope of } \overline{PC} = \frac{c - 0}{a + b - 0} = \frac{c}{a + b}$$

$$\text{slope of } \overline{AB} = \frac{0 - c}{a - b} = \frac{-c}{a - b}$$

$$\text{product of slopes} = \left(\frac{c}{a + b}\right)\left(\frac{-c}{a - b}\right) = \frac{-c^2}{a^2 - b^2}$$

In right $\triangle BDP$ (Figure 11.19), $BD = c$, $PD = b$, and $BP = a$ because $BP = AP$. Thus, by the Pythagorean Theorem

$$a^2 = b^2 + c^2$$
$$a^2 - b^2 = c^2.$$

Therefore,

$$\text{product of slopes} = \frac{-c^2}{a^2 - b^2} = \frac{-c^2}{c^2} = -1.$$

By Postulate 11.3, the diagonals of a rhombus are perpendicular. ■

> ### THEOREM 4.36
> The segment joining the midpoints of two sides of a triangle is parallel to and equal to one-half the third side.

PROOF: Let C be the midpoint of $\overline{BP}$ and D be the midpoint of $\overline{AB}$ in $\triangle ABP$ of Figure 11.20. We must show that $\overline{CD} \parallel \overline{PA}$ and that $CD = \frac{1}{2}PA$. Calculate the coordinates of C and D using the midpoint formula.

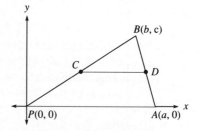

Figure 11.20

$$\text{midpoint of } \overline{BP} = \left(\frac{b+0}{2}, \frac{c+0}{2}\right) = \left(\frac{b}{2}, \frac{c}{2}\right) = C\left(\frac{b}{2}, \frac{c}{2}\right)$$

$$\text{midpoint of } \overline{AB} = \left(\frac{a+b}{2}, \frac{0+c}{2}\right) = \left(\frac{a+b}{2}, \frac{c}{2}\right) = D\left(\frac{a+b}{2}, \frac{c}{2}\right)$$

Now calculate the slopes of $\overline{PA}$ and $\overline{CD}$.

$$\text{slope of } \overline{PA} = \frac{0-0}{a-0} = 0$$

$$\text{slope of } \overline{CD} = \frac{\dfrac{c}{2} - \dfrac{c}{2}}{\dfrac{a+b}{2} - \dfrac{b}{2}} = \frac{0}{\dfrac{a}{2}} = 0$$

Thus, $\overline{CD} \parallel \overline{PA}$ by Postulate 11.2.

Because both $\overline{PA}$ and $\overline{CD}$ are horizontal, the length of each is the difference in the x-coordinates of their endpoints. Because the coordinates of A are $(a, 0)$ and the coordinates of P are $(0, 0)$, then

$$\text{length of } \overline{PA} = PA = a - 0 = a.$$

Also, because D is $\left(\frac{a+b}{2}, \frac{c}{2}\right)$ and C is $\left(\frac{b}{2}, \frac{c}{2}\right)$,

$$\text{length of } \overline{CD} = CD = \frac{a+b}{2} - \frac{b}{2} = \frac{a+b-b}{2} = \frac{a}{2}.$$

Thus, $CD = \frac{1}{2}PA$, and the theorem is proved. ∎

11.4 EXERCISES

In Exercises 1–8, use analytic geometry to prove each theorem.

1. The diagonals of a rectangle are equal.
2. The opposite sides of a parallelogram are equal.
3. The diagonals of an isosceles trapezoid are equal.
4. The medians drawn to the equal sides of an isosceles triangle are equal.
5. The perpendicular bisector of the base of an isosceles triangle passes through the vertex.
6. The median of a trapezoid is parallel to the bases, and its length is one-half the sum of the bases.
7. The triangle formed by joining the midpoints of the sides of an isosceles triangle is isosceles.
8. The medians of any triangle intersect at a point that is two-thirds the distance from the vertex to the midpoint of the opposite side.

KEY TERMS AND SYMBOLS

11.1 Cartesian (rectangular)
  coordinate system,
  p. 351
  coordinate plane, p. 351
  x-axis, p. 351
  y-axis, p. 351
  origin, p. 351
  ordered pair, p. 351
  x-coordinate, p. 351
  y-coordinate, p. 351
  quadrant, p. 352
  plot, p. 352

linear (first-degree)
  equation, p. 352
  solution, p. 352
  graph, p. 352
  table of values, p. 353
  x-intercept, p. 353
  y-intercept, p. 353
11.2 rise of a line, p. 357
  run of a line, p. 357
  slope, p. 357
  general form, p. 359

point-slope form, p. 359
slope-intercept form,
  p. 360
coinciding lines, p. 361
distance formula, p. 362
midpoint formula, p. 363
11.3 standard form of a circle,
  p. 365
  general form of a circle,
  p. 367

PROOF TECHNIQUES

To Prove:

Two Lines are Parallel

1. Show that their slopes are equal. (Postulate 11.2, p. 361)
2. Write the equations in slope-intercept form and compare the slopes.

Two Lines are Perpendicular

1. Show that the product of their slopes is -1. (Postulate 11.3, p. 361)
2. Write equations in slope-intercept form and compare the slopes.

REVIEW EXERCISES

Section 11.1

1. What is the *x*-coordinate of the ordered pair $(-2, 3)$? What is the *y*-coordinate? In which quadrant is the point associated with this ordered pair located?

2. What are the perpendicular lines that form a Cartesian coordinate system called?

3. What is the name of the set of points in a Cartesian coordinate system that corresponds to the solutions of an equation?

4. A linear equation of the form $x = a$ (*a* is a constant) has as its graph a straight line parallel to which axis?

5. Give the coordinates of the points A, B, C, and D in the figure. In which quadrant is each located?

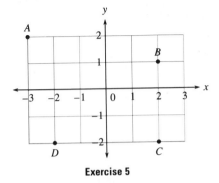

Exercise 5

6. Plot the points $(0, 4)$, $(-1, 2)$, $(2, -3)$, $(-3, 0)$, and $(-1, -4)$.

Determine the intercepts and graph each linear equation in Exercises 7–12.

7. $x + y - 2 = 0$ 8. $3x - 2y - 6 = 0$ 9. $3x - y = 0$

10. $x - 2y = 0$ 11. $3x + 6 = 0$ 12. $4y - 4 = 0$

Section 11.2

Find the slope of the line through each pair of points in Exercises 13–14.

13. $(1, -2)$ and $(3, 7)$ 14. $(-5, -4)$ and $(-6, -2)$

15. If the slope of a line is 5, what is the slope of a line parallel to it? What is the slope of a line perpendicular to it?

16. Find the slope of the line passing through $(3, -1)$ and $(-2, 7)$. What is the slope of a line perpendicular to this line?

17. What is the distance between the points $(4, -1)$ and $(-2, -1)$?

18. What is the midpoint of the line segment joining $(1, -4)$ and $(3, 2)$?

19. If a linear equation is solved for y, the resulting equation is what form of the equation of the line? In this form, what is the numerical coefficient of x called?

20. Find the general form of the equation of the line with slope -4 and passing through $(-1, 3)$.

21. Find the slope-intercept form of the equation of the line with equation $4x + 2y - 10 = 0$. What is the slope? the y-intercept?

22. Find the general form of the equation of a line passing through the points $(-2, 5)$ and $(6, 9)$.

23. Find the general form of the equation of the line that passes through the point $(-3, 1)$ and is perpendicular to the line $2x + y - 3 = 0$.

24. What are the intercepts and the slope of the line with equation $3y - 15 = 0$?

25. Do the following equations represent lines that are parallel, perpendicular, or neither?

$$3x - y + 2 = 0$$
$$x + 3y - 7 = 0$$

26. Virginia Winn can assemble 9 units in 2 hr and 19 units in 5 hr. If y is the number of units she can assemble in x hours, find a linear equation describing this relationship. Use the equation to determine the number of units she can assemble in 8 hours.

Section 11.3

In Exercises 27–30, find the standard and general forms of the equation of a circle centered at the given point with the given radius.

27. $(-5, 2)$, $r = 4$ **28.** $(2, 6)$, $r = 3$ **29.** $\left(\frac{1}{2}, \frac{1}{2}\right)$, $r = 2$ **30.** $(0, 5)$, $r = 1$

In Exercises 31–32, find the standard form of the equation of each circle and give the center and radius.

31. $x^2 + y^2 - 4x + 16y + 19 = 0$ **32.** $4x^2 + 4y^2 + 4x - 24y + 33 = 0$

Graph each equation in Exercises 33–34.

33. $(x - 1)^2 + (y - 3)^2 = 9$ **34.** $(x + 2)^2 + (y - 2)^2 = 16$

Section 11.4

35. Prove that the lines joining the midpoints of adjacent sides of a quadrilateral form a parallelogram.

PRACTICE TEST

Items 1–5 refer to the equation $3x + 2y = 12$.

1. Write the equation in slope-intercept form.

2. What is the x-intercept? **3.** What is the y-intercept?

4. What is the slope? **5.** Graph the equation.

6. Find the missing coordinate so that the ordered pair $(10, \ )$ is a solution to $y = 4x + 30$.

7. Graph $y = x + 3$. **8.** Graph $y - 4 = 0$.

9. Assume that the two points $P(-2, 7)$ and $Q(4, -5)$ are given.
   (a) What is the slope of the line passing through P and Q?
   (b) What is the slope of a line perpendicular to the line passing through P and Q?
   (c) What is the distance between P and Q?
   (d) What are the coordinates of the midpoint of the line segment joining P and Q?

10. Do the following equations represent lines that are parallel, perpendicular, or neither?
$$2x - 5y + 7 = 0$$
$$-2x + 5y + 7 = 0$$

11. Find the general form of the equation of the line with slope -3 and passing through $(-2, 4)$.

12. Find the general form of the equation of the line passing through $(-1, 7)$ and $(4, -3)$.

Find the standard and general forms of the equation of a circle centered at the given point with the given radius.

13. $(-2, -1), r = 5$

14. $(4, -2), r = 3$

15. Find the standard form of the equation of the circle with general form $x^2 + y^2 + 14x - 12y + 60 = 0$. Find the center and radius of the circle.

Graph the equations in Exercises 16–17.

16. $x^2 + y^2 = 16$

17. $(x - 2)^2 + (y + 1)^2 = 9$

18. Prove that the line segments joining the midpoints of opposite sides of a quadrilateral bisect each other.

12

TRIANGLE TRIGONOMETRY

The word *trigonometry* is derived from Greek words that mean *three-angle measurement,* or *triangle measurement.* Historically, trigonometry was developed as a tool for finding the measurements of parts (sides and angles) of a triangle when other parts were known. As a result, it became indispensable in areas such as navigation, surveying, architecture, and astronomy. More recently, the study of trigonometry has expanded even further, but we will restrict our work here to triangle trigonometry. The following application is one we can solve using this basic trigonometry. Its solution appears in Section 12.3, Example 3.

A small forest fire is sighted due south of a fire lookout tower on Woody Mountain. From a second tower, located 11.0 mi due east of the first, a ranger observes that the fire is 51.2° west of due south. How far is the fire from the tower on Woody Mountain?

12.1 THE TRIGONOMETRIC RATIOS

In Chapter 3 we defined a right triangle to be a triangle that contains a right angle (90°); and we said the side opposite the right angle is called the hypotenuse of the triangle. Because the sum of the measures of the angles of a triangle is 180°, we determined that the remaining two angles are complementary acute angles. The sides opposite these angles are called the legs of the triangle.

From now on, we will label the angles of a triangle with capital letters. The sides of the triangle will be indicated by lower-case letters corresponding to the capital letter of the angle opposite the side. For example, the right triangle with angles A, B, and C and sides a, b, and c is shown in Figure 12.1. As in Chapter 6, we will identify the right angle with the letter C. The side **opposite** A is a and the side **adjacent** to A is b.

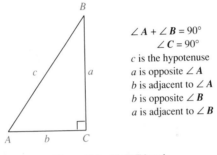

$\angle A + \angle B = 90°$
$\angle C = 90°$
c is the hypotenuse
a is opposite $\angle A$
b is adjacent to $\angle A$
b is opposite $\angle B$
a is adjacent to $\angle B$

Figure 12.1   Right Triangle

In Chapter 5, we learned that similar triangles are triangles that have two equal corresponding angles. As a result, two right triangles are similar if an acute angle of one is equal to an acute angle of the other. Also, we know that corresponding sides of similar triangles are proportional. Consider the two similar right triangles in Figure 12.2. It follows that

$$\frac{a}{c} = \frac{a'}{c'}, \frac{b}{c} = \frac{b'}{c'}, \quad \text{and} \quad \frac{a}{b} = \frac{a'}{b'}.$$

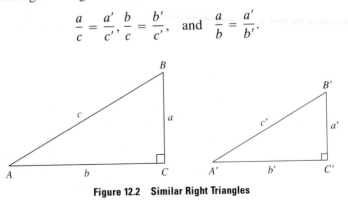

Figure 12.2   Similar Right Triangles

Because these three ratios are independent of the size of the right triangle containing $\angle A$, we can use them to define three trigonometric ratios.

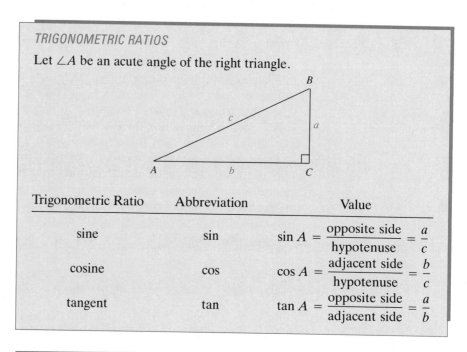

TRIGONOMETRIC RATIOS

Let $\angle A$ be an acute angle of the right triangle.

Trigonometric Ratio	Abbreviation	Value
sine	sin	$\sin A = \dfrac{\text{opposite side}}{\text{hypotenuse}} = \dfrac{a}{c}$
cosine	cos	$\cos A = \dfrac{\text{adjacent side}}{\text{hypotenuse}} = \dfrac{b}{c}$
tangent	tan	$\tan A = \dfrac{\text{opposite side}}{\text{adjacent side}} = \dfrac{a}{b}$

E X A M P L E   1   Determine $\sin A$, $\cos A$, and $\tan A$ for $\angle A$ in Figure 12.3.

We must first find the value of c. We can use the Pythagorean Theorem.

$$c^2 = a^2 + b^2$$

In our example, $a = 3$ and $b = 4$.

$$c^2 = (3)^2 + (4)^2$$
$$= 9 + 16 = 25$$
$$c = 5$$

Figure 12.3

We now know the values of a, b, and c and can write the trigonometric ratios.

$$\sin A = \frac{\text{opposite side}}{\text{hypotenuse}} = \frac{a}{c} = \frac{3}{5}$$

$$\cos A = \frac{\text{adjacent side}}{\text{hypotenuse}} = \frac{b}{c} = \frac{4}{5}$$

$$\tan A = \frac{\text{opposite side}}{\text{adjacent side}} = \frac{a}{b} = \frac{3}{4} \quad \blacksquare$$

We can also determine $\sin B$, $\cos B$, and $\tan B$ for $\angle B$ in Figure 12.3. The basic definitions are the same but now "opposite side" means the side opposite $\angle B$.

$$\sin B = \frac{\text{opposite side}}{\text{hypotenuse}} = \frac{b}{c} = \frac{4}{5}$$

$$\cos B = \frac{\text{adjacent side}}{\text{hypotenuse}} = \frac{a}{c} = \frac{3}{5}$$

$$\tan B = \frac{\text{opposite side}}{\text{adjacent side}} = \frac{b}{a} = \frac{4}{3}$$

Notice that the letters representing the sides change when we consider the trigonometric ratios for $\angle B$. Thus, it is better to remember each ratio in terms of the words "opposite side," "adjacent side," and "hypotenuse."

In Chapter 6 we studied the 45°-45°-right triangle and the 30°-60°-right triangle. We will find the trigonometric ratios of these angles in the next example.

E X A M P L E 2   Find the trigonometric ratios for 45°, 30°, and 60°.

First we construct a 45°-45°-right triangle with $a = 1$ and $b = 1$, as in Figure 12.4. From Theorem 6.13, we know that $c = (\sqrt{2})(1) = \sqrt{2}$.

$$\sin 45° = \frac{a}{c} = \frac{1}{\sqrt{2}} = \frac{\sqrt{2}}{2}, \cos 45° = \frac{b}{c} = \frac{1}{\sqrt{2}} = \frac{\sqrt{2}}{2}, \tan 45° = \frac{a}{b} = \frac{1}{1} = 1$$

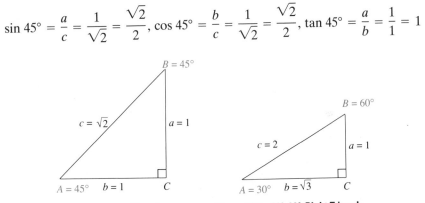

Figure 12.4   45°-45°-Right Triangle   **Figure 12.5   30°-60°-Right Triangle**

Next, we construct a 30°-60°-right triangle, as in Figure 12.5. From Theorem 6.14, if $a = 1$, then $b = \sqrt{3}$ and $c = 2$. We can determine the trigonometric ratios for both 30° and 60° using Figure 12.5.

$$\sin 30° = \frac{a}{c} = \frac{1}{2}, \cos 30° = \frac{b}{c} = \frac{\sqrt{3}}{2}, \tan 30° = \frac{a}{b} = \frac{1}{\sqrt{3}} = \frac{\sqrt{3}}{3}$$

$$\sin 60° = \frac{b}{c} = \frac{\sqrt{3}}{2}, \cos 60° = \frac{a}{c} = \frac{1}{2}, \tan 60° = \frac{b}{a} = \frac{\sqrt{3}}{1} = \sqrt{3}$$

Before calculators were available, tables of values were used to find trigonometric ratios for angles other than 30°, 45°, and 60°. Now we can use calculators to determine the ratios.

To find a trigonometric ratio with a calculator, place it in the mode for degrees by pressing the $\boxed{\text{DGR}}$ or $\boxed{\text{DEG}}$ key. (Read your instruction manual

When a football player kicks a field goal, the angle of the trajectory and the velocity of the ball determine the distance the ball will travel.
Trigonometry can be used to determine the best angle for the ball to be kicked at a constant velocity to achieve the maximum distance.

to determine how your calculator operates.) Note that many calculators compute degrees only in decimal form and cannot accept minutes and seconds.

The $\boxed{\sin}$, $\boxed{\cos}$, and $\boxed{\tan}$ keys are used to find the sine, cosine, and tangent ratios. The next example offers practice in using a calculator.

EXAMPLE 3   Use a calculator to find each trigonometric ratio.
(a) sin 29°

First make sure the calculator is set in the degree mode. Enter 29 and then press the $\boxed{\sin}$ key. The result is 0.4848, correct to four decimal places. The following diagram illustrates this process.

$$29\,\boxed{\sin} \longrightarrow \boxed{0.4848096}$$

Here $\boxed{0.4848096}$ indicates the display of the calculator.

(b) cos 71.5°

The calculator will accept decimal parts of a degree. Thus, we enter 71.5 and press $\boxed{\cos}$.

$$71.5\,\boxed{\cos} \longrightarrow \boxed{0.3173047}$$

Thus, cos 71.5° = 0.3173, correct to four decimal places.

(c) tan 85.6°

Enter 85.6 and press the $\boxed{\tan}$ key.

$$85.6\,\boxed{\tan} \longrightarrow \boxed{12.996160}$$

Thus, tan 85.6° = 12.9962, correct to four decimal places.   ▨

We can now find a trigonometric ratio for a given angle. Suppose we reverse the process and find the angle when the trigonometric ratio is given. Most calculators use the $\boxed{\text{INV}}$ key followed by the $\boxed{\sin}$, $\boxed{\cos}$, or $\boxed{\tan}$ key. Some calculators use one key labeled $\boxed{\sin^{-1}}$ or $\boxed{\arcsin}$.

EXAMPLE 4   Use a calculator to find the angle with the given trigonometric ratio.
(a) sin A = 0.5948

Set the calculator in degree mode. The following sequence yields the desired result.

$$0.5948\,\boxed{\text{INV}}\,\boxed{\sin} \longrightarrow \boxed{36.498376}$$

Thus, $A \approx 36.5°$.

(b) tan A = 0.1263

Follow the given sequence of steps.

$$0.1263\,\boxed{\text{INV}}\,\boxed{\tan} \longrightarrow \boxed{7.1983432}$$

Therefore, $A \approx 7.2°$ ▨

12.1 EXERCISES

In Exercises 1–6, two sides of right $\triangle ABC$ ($\angle C$ is the right angle) are given. Find the three trigonometric ratios for $\angle A$ and $\angle B$.

1. $a = 5$ and $b = 12$ **2.** $a = 3$ and $b = 10$ **3.** $a = 7$ and $c = 15$

4. $b = 7$ and $c = 20$ **5.** $a = 0.6$ and $b = 0.8$ **6.** $a = \sqrt{5}$ and $b = \sqrt{11}$

In Exercises 7–12, $\angle A$ is an acute angle in a right triangle. Use the given trigonometric ratio to find the other two ratios.

7. $\sin A = \dfrac{5}{7}$ **8.** $\cos A = \dfrac{1}{3}$ **9.** $\tan A = \dfrac{\sqrt{5}}{2}$

10. $\sin A = \dfrac{1}{4}$ **11.** $\cos A = \dfrac{\sqrt{33}}{7}$ **12.** $\tan A = 7$

In Exercises 13–18, use your calculator to find each trigonometric ratio, correct to four decimal places.

13. $\sin 65°$ **14.** $\cos 5°$ **15.** $\tan 38°$

16. $\sin 12.4°$ **17.** $\cos 83.8°$ **18.** $\tan 57.1°$

In Exercises 19–24, use your calculator to find $\angle A$, to the nearest tenth of a degree.

19. $\sin A = 0.1258$ **20.** $\cos A = 0.8018$ **21.** $\tan A = 0.9301$

22. $\sin A = 0.9511$ **23.** $\cos A = 0.1022$ **24.** $\tan A = 2.5053$

12.2 SOLVING RIGHT TRIANGLES

The six parts of any triangle (three angles and three sides) can often be found when the measures of individual parts are given. This process is called **solving the triangle.** A right triangle can be solved when two of its sides are known or when one side and one acute angle are known. Simply knowing the angle measures of a triangle is not enough to solve the triangle because any similar triangle will have the same angles but, more than likely, different sides.

EXAMPLE 1   Solve the right triangle with $\angle A = 26°$ and $b = 11$.

First draw the triangle and label the parts as in Figure 12.6. $\angle B$ is easy to find because $\angle A$ and $\angle B$ are complementary.

$$\angle B = 90° - 26° = 64°$$

Next, we notice that because

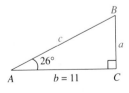

Figure 12.6

$$\tan A = \frac{a}{b}$$

and we know $\angle A$ and b, we can find a.

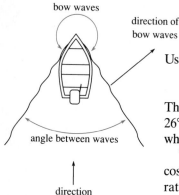

bow waves

direction of
bow waves

angle between waves

direction
of travel

When a boat moves through water at a constant rate of speed, the waves generated extend back from the bow (front) of the boat at a fixed angle. Trigonometry can be used to determine the angle between the bow waves. Knowledge of this angle and the velocity of the bow wave give enough information to calculate the boat's velocity.

$$\tan 26° = \frac{a}{11}$$

$$11(\tan 26°) = a \qquad \text{Multiply both sides by 11}$$

Use the following calculator steps to find a.

$$26 \boxed{\tan} \boxed{\times} 11 \boxed{=} \longrightarrow \boxed{5.3650585}$$

Thus, a is approximately 5. Notice that because b is a whole number and $\angle A = 26°$ is only given to the nearest degree, we round the answer for a to the nearest whole number.

To find c, we can use the Pythagorean Theorem, but instead we will use the cosine ratio. We usually obtain a more accurate answer by using only given parts rather than approximated calculated parts.

$$\cos A = \frac{b}{c}$$

$$\cos 26° = \frac{11}{c}$$

$$c(\cos 26°) = 11 \qquad \text{Multiply both sides by } c$$

$$c = \frac{11}{\cos 26°} \qquad \text{Divide both sides by } \cos 26°$$

$$11 \boxed{\div} 26 \boxed{\cos} \boxed{=} \longrightarrow \boxed{12.238621}$$

The side c is approximately 12. Thus, $\angle B = 64°$, $a = 5$, and $c = 12$. ◪

E X A M P L E   2   Solve the right triangle with $a = 3$ and $b = 7$.

Figure 12.7 gives a reasonably accurate sketch of the triangle. We must find $\angle A$, $\angle B$, and c.

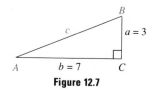

Figure 12.7

To find $\angle A$: $\tan A = \dfrac{3}{7}$

$$3 \boxed{\div} 7 \boxed{=} \boxed{\text{INV}} \boxed{\tan} \longrightarrow \boxed{23.198591}$$

Thus, to the nearest degree, $\angle A = 23°$.
To find $\angle B$: Because $\angle A$ and $\angle B$ are complementary,

$$\angle B = 90° - \angle A = 90° - 23° = 67°.$$

$\angle B$ can also be found by using, for example, $\tan B = \dfrac{7}{3}$, which also gives $\angle B = 67°$.

To find c: We can find c with the Pythagorean Theorem.

$$c^2 = a^2 + b^2$$
$$= 3^2 + 7^2$$
$$= 58$$
$$c = \sqrt{58} \approx 8$$

Thus, $\angle A = 23°$, $\angle B = 67°$, and $c = 8$. ∎

PRACTICE EXERCISE 1

Solve the right triangle in which $b = 10$ and $c = 16$.

E X A M P L E 3   Solve the right triangle in which $c = 8.1$ and $\angle B = 72.5°$.

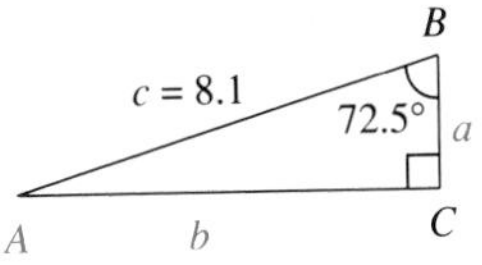

We must find a, b, and $\angle A$. A sketch of the triangle appears in Figure 12.8.

Figure 12.8

To find $\angle A$:

$$\angle A = 90.0° - B$$
$$= 90.0° - 72.5°$$
$$= 17.5°$$

Notice that we give $\angle A$ to the nearest tenth of a degree because $\angle B$ and c are given to the nearest tenth. We will also give a and b to the nearest tenth.

To find a:

$$\cos B = \frac{a}{c}$$

$$\cos 72.5° = \frac{a}{8.1}$$

$$a = (8.1)\cos 72.5° \approx 2.4$$

$$72.5 \boxed{\cos} \boxed{\times} 8.1 \boxed{=} \longrightarrow \boxed{2.435717}$$

To find b:

$$\sin 72.5° = \frac{b}{8.1}$$

$$b = (8.1)\sin 72.5° \approx 7.7$$

$$8.1 \boxed{\times} 72.5° \boxed{\sin} \boxed{=} \longrightarrow \boxed{7.7251073}$$

Thus, $\angle A = 17.5°$, $a = 2.4$, and $b = 7.7$. ∎

ANSWER TO PRACTICE EXERCISE: **1.** $a = 12$, $\angle A = 51°$, $\angle B = 39°$

12.2 EXERCISES

Solve each right triangle in Exercises 1–18. Remember that $\angle C = 90°$.

1. $a = 9$ and $\angle A = 60°$

2. $b = 4$ and $\angle B = 45°$

3. $c = 12$ and $\angle A = 20°$

4. $c = 29$ and $\angle B = 55°$

5. $a = 12$ and $c = 13$

6. $a = 6$ and $b = 8$

7. $b = 7$ and $c = 15$

8. $a = 19$ and $b = 48$

9. $a = 9.2$ and $c = 10.1$

10. $a = 4.3$ and $b = 6.1$ **11.** $\angle A = 26.7°$ and $c = 12.0$ **12.** $\angle A = 62.2°$ and $b = 7.3$

13. $b = 3.2$ and $\angle B = 10.8°$ **14.** $c = 16.6$ and $\angle B = 12.8°$ **15.** $\angle A = 22.5°$ and $c = 28.7$

16. $\angle B = 41.3°$ and $a = 0.8$ **17.** $c = 21.9$ and $\angle B = 81.7°$ **18.** $a = 19.3$ and $\angle A = 26.6°$

12.3 APPLICATIONS INVOLVING RIGHT TRIANGLES

In this section we will investigate a variety of applied problems, all of which depend in some way upon finding unknown parts of a right triangle. For the most part, we will use a process of *indirect measurement:* finding a particular length, distance, or angle without applying a measuring instrument such as a tape measure or protractor.

EXAMPLE 1 A farmer purchased a field that has the shape of an isosceles triangle. If the vertex angle is 90° and the base is 155 m long, what are the perimeter and area of the field?

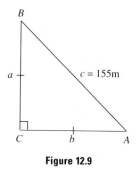

Figure 12.9

First construct the triangle representing the field, as shown in Figure 12.9. Notice that the legs of the right triangle are equal (the triangle is isosceles) and the hypotenuse is 155 m. We know that because $a = b$ then $\angle A = \angle B$. But $\angle A$ and $\angle B$ are complementary and thus, $\angle A = \angle B = 45°$.

From the definition of the sine ratio

$$\sin 45° = \frac{a}{c} = \frac{a}{155}.$$

Thus,

$$a = 155 \sin 45°.$$

The calculator steps are as follows.

$$155 \,\boxed{\times}\, 45 \,\boxed{\sin}\, \boxed{=} \longrightarrow \boxed{109.60155}$$

Thus, $a \approx 110$ m and $b = a \approx 110$ m.

To find the perimeter, we add the measures of the sides of the triangle.

$$155 + 110 + 110 = 375$$

The perimeter is 375 m.

Archimedes (287–212 B.C.)

Archimedes used mathematics to solve many practical problems of his day. He frequently used geometry in his scientific studies as well as in the design of weapons of war. His approximation of π as being between $3\frac{10}{71}$ and $3\frac{1}{7}$ was particularly accurate for his time.

The area of a triangle is one-half the base multiplied by the height. From Figure 12.9, we can use b for the base and a for the height.

$$A = \frac{1}{2}ba$$

$$= \frac{1}{2}(110)(110)$$

$$= 6050$$

The area of the farmer's field is 6050 m². ☑

Next we learn how to determine the height of a building by measuring its shadow.

EXAMPLE 2 A building casts a shadow 200 ft long when the sun makes an angle of 23° with the horizontal. Find the height of the building.

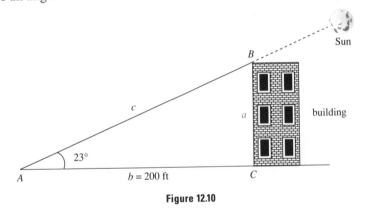

Figure 12.10

Construct the triangle in Figure 12.10. We must determine a in $\triangle ABC$.

$$\tan A = \frac{a}{b} \qquad \text{We know } \angle A \text{ and } b, \text{ so } a \text{ can be found}$$

$$\tan 23° = \frac{a}{200}$$

$$a = (200)\tan 23°$$

$$200 \boxed{\times} 23 \boxed{\tan} \boxed{=} \longrightarrow \boxed{84.894963}$$

The building is approximately 85 ft high. ☑

Forest rangers can use the technique of Example 2 to determine the height of a tree. They can also use right-triangle trigonometry to locate forest fires. The following example solves the applied problem given in the chapter introduction.

E X A M P L E 3   A small forest fire is sighted due south of a fire lookout tower on Woody Mountain. From a second tower, located 11.0 mi due east of the first, a ranger observes that the fire is 51.2° west of due south. How far is the fire from the tower on Woody Mountain?

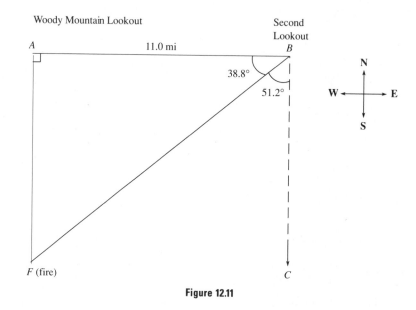

Figure 12.11

In Figure 12.11, $\angle CBF$ is 51.2° and $\angle FAB$ is a right angle. In $\triangle ABF$, $\angle ABF = 38.8°$ because it is the complement of $\angle CBF$. AF is the distance from the fire to the lookout on Woody Mountain.

$$\tan 38.8° = \frac{AF}{11.0}$$
$$AF = (11.0) \tan 38.8°$$

$$11.0 \boxed{\times} 38.8 \boxed{\tan} \boxed{=} \longrightarrow \boxed{8.844227}$$

Thus, the fire is approximately 8.8 mi from the lookout on Woody Mountain. ▨

PRACTICE EXERCISE 1

Find the distance of the fire in Example 3 from the second lookout tower.

E X A M P L E 4   An airplane leaves the East Coast of the United States and flies for two hours at 350 km/hr in a direction that is 32.7° to the east of due north. Assuming that the East Coast is a straight north-south line, how far is the airplane from the coastline?

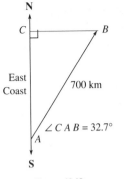

Figure 12.12

The sketch in Figure 12.12 shows the essential elements of the problem. AB is 700 km because 350 km/hr for 2 hr gives

$$\text{distance} = (\text{rate})(\text{time})$$
$$= (350)(2)$$
$$= 700.$$

We must find BC.

$$\sin 32.7° = \frac{BC}{700}$$
$$BC = (700)\sin 32.7°$$
$$700 \boxed{\times} 32.7 \boxed{\sin} \boxed{=} \longrightarrow \boxed{378.16822}$$

The airplane is approximately 378 km from the coast. ◪

ANSWER TO PRACTICE EXERCISE: **1.** 14.1 mi

12.3 EXERCISES

Solve the following problems.

1. A playground is in the shape of an isosceles triangle with vertex angle 90° and equal sides 82 ft. Find (a) its perimeter and (b) its area.

2. A holding area for cattle is in the shape of an isosceles triangle with a vertex angle of 90° and a base of 22 m. What are (a) its perimeter and (b) its area?

3. A large plot of land is in the shape of an equilateral triangle with sides 6.2 mi. What are (a) its perimeter and (b) its area?

4. A machine part is in the shape of an equilateral triangle with an altitude of length 10.8 cm. Find (a) its perimeter and (b) its area.

5. When the sun is at an elevation of 49° above level ground (see the figure below) a pole casts a shadow 40 ft in length. What is the height of the pole?

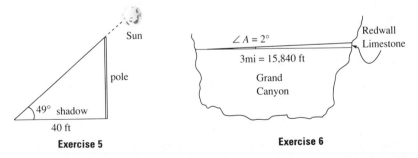

Exercise 5

Exercise 6

6. From a point in the Grand Canyon 3 mi from the base of the Redwall Limestone but on the same horizontal plane, a geologist sights the top of the Redwall at an angle of 2° (see the figure above). How many feet thick is the Redwall Limestone formation at this point?

7. Two buildings are 70.0 ft apart on opposite sides of a street. From the top of one, which is 80.0 ft high, the angle to the top of the other is 37.5°. How tall is the second building?

8. From a point 1200.0 ft from the base of Devil's Tower, the angle to the top of the rock is 35.8°. How high is the tower?

9. The angle down from the level of a 100-ft tower to a small forest fire is 6°. How far from the base of the tower is the forest fire?

10. A balloon is 1375 ft high. The angle down from this level to a house is 25°. How far is the house from a point directly below the balloon?

11. From Fire Lookout B, located 8.0 mi due south of Lookout A, a fire is observed. If the fire is due east of A, and from B it is 31.7° east of due north, how far is the fire from B?

12. Phoenix is 140.0 mi due south of Flagstaff, and Winslow is 60.0 mi due east of Flagstaff. If a pilot were to fly from Phoenix to Winslow, how many degrees east of due north should she head?

13. An airplane leaves an airport and flies 200 km in a direction of 40° east of due south. How far east of the airport is the airplane at this time? How far south?

14. A ship sails 30.0 mi in a direction of 37.7° west of due north. How far north has it sailed? How far west?

15. At an altitude of 2500 ft, the engine on a small plane suddenly fails. What angle of glide with the horizontal is needed to reach an airport runway 3 mi away?

16. A train tunnel through a mountain is 2000 m in length. If the track makes an angle of 2° with the horizontal, what is the change in elevation from one end of the tunnel to the other?

17. If the distance between the two levels in the MetroCenter Shopping Mall is 16.8 ft and the angle that the escalator makes with the horizontal is 32.2°, how far would a person travel while riding from the first to the second level?

18. The chairlift at a ski resort makes an average angle of 17.5° with the horizontal. If the vertical rise is 800 m, what is the approximate length of the ride to the top of the lift?

19. Two points A and B are on opposite sides of a lake as shown in the figure. A surveyor at point P, due west of B and 1500 ft from B, finds ∠APB to measure 43°. How far apart are points A and B?

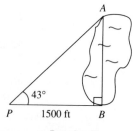

Exercise 19

20. A small boat leaves an island and sails due north for 2 hr at an average rate of 28.0 mph. It then changes course, sailing due east for 1 hr at the same rate before it runs aground on a reef. If a rescue ship sails from the island, what direction should it go to reach the incapacitated boat?

REVIEW EXERCISES

Section 12.1

In Exercises 1–2, two sides of a right $\triangle ABC$ ($\angle C$ is the right angle) are given. Find the three trigonometric ratios for $\angle A$ and $\angle B$.

1. $a = 16$ and $b = 12$

2. $a = 15$ and $c = 20$

3. If $\angle A$ is an acute angle in a right triangle and $\tan A = \dfrac{7}{3}$, find $\sin A$ and $\cos A$.

4. If $\angle B$ is an acute angle in a right triangle and $\cos B = \dfrac{1}{3}$, find $\sin B$ and $\tan B$.

In Exercises 5–7, use your calculator to find each trigonometric ratio correct to four decimal places.

5. $\sin 21°$

6. $\cos 65.8°$

7. $\tan 82.1°$

In Exercises 8–10, use your calculator to find $\angle A$ to the nearest tenth of a degree.

8. $\sin A = 0.2196$

9. $\cos A = 0.9108$

10. $\tan A = 1.6432$

Section 12.2

Solve each right triangle in Exercises 11–16. Remember that $\angle C = 90°$.

11. $b = 6$ and $\angle A = 30°$

12. $c = 12$ and $\angle B = 42°$

13. $a = 2.4$ and $\angle A = 72.3°$

14. $a = 7$ and $b = 11$

15. $b = 12$ and $c = 20$

16. $a = 6.2$ and $c = 9.6$

Section 12.3

Solve the following problems.

17. A garden is in the shape of an isosceles triangle with vertex angle 90° and a base of 14 yd. What are (a) its perimeter and (b) its area?

18. The sun makes an angle of 64.3° with the horizontal and a tree casts a shadow 28.0 m in length. What is the height of the tree?

19. A fire is sighted due west of Lookout A. From Lookout B, 5.2 mi due south of A the angle of the fire is 51.7° west of due north. How far is the fire from B? from A?

20. A boat sails for 2 hr at 25 mph in a direction of 20° west of due south. How far west has it sailed? How far south?

PRACTICE TEST

1. Find the three trigonometric ratios of $\angle A$ and $\angle B$ in the right triangle with $a = 5$ and $c = 9$. ($\angle C = 90°$)

2. If $\angle A$ is an acute angle in a right triangle and $\sin A = \dfrac{2}{5}$, find $\cos A$ and $\tan A$.

3. Find the three trigonometric ratios of 68.9°.

4. If $\cos A = 0.1096$, find $\angle A$.

5. Solve the right triangle with $a = 1.1$ and $c = 4.5$.

6. Solve the right triangle with $b = 15$ and $\angle A = 38°$.

7. A machine part is in the shape of an isosceles triangle with vertex angle 90°. If each of the equal sides is 8 cm, what is (a) the perimeter and (b) the area of the part?

8. An airplane leaves an airport and flies 300 mi in a direction 72° east of due north. How far north of the airport is the airplane? How far east?

Postulates of Geometry

P1.1 Given any two distinct points in space, there is exactly one line that passes through them.

P1.2 Given any three distinct points in space not on the same line, there is exactly one plane that passes through them.

P1.3 The line determined by any two distinct points in a plane is also contained in the plane.

P1.4 No plane contains all points in space.

P1.5 (Ruler Postulate) There is a one-to-one correspondence between the set of all points on a line and the set of all real numbers.

P1.6 (Reflexive Law) Any quantity is equal to itself. $(x = x)$

P1.7 (Symmetric Law) If x and y are any two quantities and $x = y$, then $y = x$.

P1.8 (Transitive Law) If x, y, and z are any three quantities with $x = y$ and $y = z$, then $x = z$.

P1.9 (Addition-Subtraction Law) If w, x, y, and z are any four quantities with $w = x$ and $y = z$, then $w + y = x + z$ and $w - y = x - z$.

P1.10 (Multiplication-Division Law) If w, x, y, and z are any four quantities with $w = x$ and $y = z$, then $wy = xz$ and $\dfrac{w}{y} = \dfrac{x}{z}$ (provided $y \neq 0$ and $z \neq 0$).

P1.11 (Substitution Law) If x and y are any two quantities with $x = y$, then x can be substituted for y in any expression containing y.

P1.12 (Distributive Law) If x, y, and z are any three quantities, then $x(y + z) = xy + xz$.

P1.13 (Segment Addition Postulate) Let A, B, and C be three points on the same line with B between A and C. Then $AC = AB + BC$, $BC = AC - AB$, and $AB = AC - BC$.

P1.14 (Right-Angle Postulate) All right angles are equal.

P1.15   (Angle Addition Postulate) Let A, B, and C be points that determine $\angle ABC$ with P a point in the interior of the angle. Then $\angle ABC = \angle ABP + \angle PBC$, $\angle PBC = \angle ABC - \angle ABP$, and $\angle ABP = \angle ABC - \angle PBC$.

P2.1   (Midpoint Postulate) Each line segment has exactly one midpoint.

P2.2   (Perpendicular Bisector Postulate) Each given line segment has exactly one perpendicular bisector.

P2.3   There is exactly one line perpendicular to a given line passing through a given point on the line.

P2.4   There is exactly one line perpendicular to a given line passing through a given point not on that line.

P2.5   (Angle Bisector Postulate) Each angle has exactly one bisector.

P3.1   (SAS = SAS) If two sides and the included angle of one triangle are equal, respectively, to two sides and the included angle of a second triangle, then the triangles are congruent.

P3.2   (ASA = ASA) If two angles and the included side of one triangle are equal, respectively, to two angles and the included side of a second triangle, then the triangles are congruent.

P3.3   (SSS = SSS) If three sides of one triangle are equal, respectively, to three sides of a second triangle, then the triangles are congruent.

P4.1   (Parallel Postulate) For a given line ℓ and a point P not on ℓ, one and only one line through P is parallel to ℓ.

P4.2   A polygon has the same number of angles as sides.

P4.3   (Area of a Rectangle) The area of a rectangle with length ℓ and width w is determined with the formula $A = \ell w$.

P4.4   (Additive Property of Areas) If lines divide a given area into several smaller nonoverlapping areas, the given area is the sum of the smaller areas.

P4.5   Two congruent polygons have the same area.

P5.1   (AAA $\sim$ AAA) Two triangles are similar if three angles of one are equal, respectively, to three angles of the other.

P7.1   (Circumference of a Circle) The circumference C of any circle with radius r and diameter d is determined with the formula $C = 2\pi r = \pi d$.

P7.2   (Area of a Circle) The area of a circle with radius r is determined with the formula $A = \pi r^2$.

P7.3   (Congruent Circles) If two circles are congruent then their radii and diameters are equal. Conversely, if the radii or diameters are equal, then two circles are congruent.

P7.4   (Arc Addition Postulate) Let A, B, and C be three points on the same circle with B between A and C. Then $\overset{\frown}{AC} = \overset{\frown}{AB} + \overset{\frown}{BC}$, $\overset{\frown}{BC} = \overset{\frown}{AC} - \overset{\frown}{AB}$, and $\overset{\frown}{AB} = \overset{\frown}{AC} - \overset{\frown}{BC}$.

P7.5   If a line is perpendicular to a radius of a circle and passes through the point where the radius intersects the circle, then the line is a tangent.

P7.6   A radius drawn to the point of tangency of a tangent is perpendicular to the tangent.

P7.7 (Area of a Sector) The area of a sector of a circle with radius r whose arc has measure $m°$ is determined with the formula $A = \dfrac{m}{360}\pi r^2$.

P7.8 (Arc Length) The length of an arc measuring $m°$ in a circle with radius r is determined with the formula $L = \dfrac{m}{360}2\pi r = \dfrac{m}{180}\pi r$.

P8.1 (Trichotomy Law) If a and b are real numbers, exactly one of the following is true: $a < b$, $a = b$, or $a > b$.

P8.2 (Transitive Law) If a, b, and c are real numbers with $a < b$ and $b < c$, then $a < c$.

P8.3 (Addition Properties of Inequalities) If a, b, c, and d are real numbers with $a < b$ and $c < d$, then $a + c < b + c$ and $a + c < b + d$.

P8.4 (Subtraction Properties of Inequalities) If a, b, c, and d are real numbers with $a < b$ and $c = d$, then $a - c < b - c$, $a - c < b - d$, and $c - a > d - b$.

P8.5 (Multiplication Properties of Inequalities) If a, b, and c are real numbers with $a < b$, then $ac < bc$ if $c > 0$ and $ac > bc$ if $c < 0$.

P8.6 (Division Properties of Inequalities) If a, b, and c are real numbers with $a < b$ then $\dfrac{a}{c} < \dfrac{b}{c}$ if $c > 0$ and $\dfrac{a}{c} > \dfrac{b}{c}$ if $c < 0$.

P8.7 (The Whole is Greater Than its Parts) If a, b, and c are real numbers with $c = a + b$ and $b > 0$, then $c > a$.

P8.8 Each angle of a triangle is greater than $0°$.

P9.1 The intersection of two distinct planes is a line.

P9.2 (Surface Area and Volume of a Sphere) For a sphere with radius r, the surface area is $SA = 4\pi r^2$ and the volume is $V = \dfrac{4}{3}\pi r^3$.

P11.1 Associated with each point in the plane there is one and only one ordered pair of numbers.

P11.2 Two distinct lines with slopes m_1 and m_2 are parallel if and only if $m_1 = m_2$.

P11.3 Two lines with slopes m_1 and m_2 are perpendicular if and only if $m_1 m_2 = -1$.

THEOREMS AND COROLLARIES OF GEOMETRY

T2.1 (Addition Theorem for Segments) If B is a point between A and C on segment $\overline{AC}$, Q is a point between P and R on segment $\overline{PR}$, $AB = PQ$ and $BC = QR$, then $AC = PR$.

T2.2 (Subtraction Theorem for Segments) If B is a point between A and C on segment $\overline{AC}$, Q is a point between P and R on segment $\overline{PR}$, $AC = PR$, and $AB = PQ$, then $BC = QR$.

T2.3 (Addition Theorem for Angles) If D is a point in the interior of $\angle ABC$, S is a point in the interior of $\angle PQR$, $\angle ABD = \angle PQS$, and $\angle DBC = \angle SQR$, then $\angle ABC = \angle PQR$.

T2.4 (Subtraction Theorem for Angles) If D is a point in the interior of $\angle ABC$, S is a point in the interior of $\angle PQR$, $\angle ABC = \angle PQR$, and $\angle DBC = \angle SQR$, then $\angle ABD = \angle PQS$.

T2.5 Two equal supplementary angles are right angles.

T2.6 Complements of equal angles are equal.

C2.7 Complements of the same angle are equal.

T2.8 Supplements of equal angles are equal.

C2.9 Supplements of the same angle are equal.

T2.10 If A, B, and C are three points on a line, with B between A and C, and $\angle ABD$ and $\angle DBC$ are adjacent angles, then $\angle ABD$ and $\angle DBC$ are supplementary angles.

T2.11 Vertical angles are equal.

T3.1 (Transitive Law for Congruent Triangles) If $\triangle ABC \cong \triangle DEF$ and $\triangle DEF \cong \triangle GHI$, then $\triangle ABC \cong \triangle GHI$.

T3.2 (Segment Bisector Theorem) Construction 2.3 gives the perpendicular bisector of a given line segment.

T3.3 Every point on the perpendicular bisector of a segment $\overline{AB}$ is equidistant from A and B.

T3.4 (Angle Bisector Theorem) Construction 2.6 gives the bisector of a given angle.

T3.5 If two sides of a triangle are equal, then the angles opposite these sides are also equal.

C3.6 If a triangle is equilateral, then it is equiangular.

T3.7 If two angles of a triangle are equal, then the sides opposite these angles are also equal.

C3.8 If a triangle is equiangular, then it is equilateral.

T3.9 The bisector of the vertex angle of an isosceles triangle is the perpendicular bisector of the base of the triangle.

T3.10 The perpendicular bisector of the base of an isosceles triangle passes through the vertex of its vertex angle and bisects the vertex angle.

C3.11 The bisector of the vertex angle of an isosceles triangle coincides with the altitude and median drawn from that vertex.

T4.1 If two lines in a plane are both perpendicular to a third line, then they are parallel.

T4.2 If two lines are cut by a transversal and a pair of alternate interior angles are equal, then the lines are parallel.

T4.3 If two lines are cut by a transversal and a pair of corresponding angles are equal, then the lines are parallel.

T4.4 If two lines are cut by a transversal and a pair of alternate exterior angles are equal, then the lines are parallel.

T4.5 If two lines are cut by a transversal and two interior angles on the same side of the transversal are supplementary, then the lines are parallel.

T4.6 (Converse of Theorem 4.1) If two lines are parallel and a third line is perpendicular to one of them, then it is also perpendicular to the other.

T4.7 (Converse of Theorem 4.2) If two parallel lines are cut by a transversal, then all pairs of alternate interior angles are equal.

T4.8 (Converse of Theorem 4.3) If two parallel lines are cut by a transversal, then all pairs of corresponding angles are equal.

T4.9 (Converse of Theorem 4.4) If two parallel lines are cut by a transversal, then all pairs of alternate exterior angles are equal.

T4.10 (Converse of Theorem 4.5) If two parallel lines are cut by a transversal, then all pairs of interior angles on the same side of the transversal are supplementary.

T4.11 The sum of the measures of the angles of a triangle is 180°.

C4.12 Any triangle can have at most one right angle or at most one obtuse angle.

C4.13 If two angles of one triangle are equal, respectively, to two angles of another triangle, then the third angles are also equal.

C4.14 (AAS = AAS) If two angles and any side of one triangle are equal, respectively, to two angles and the corresponding side of another triangle, then the triangles are congruent.

C4.15 The measure of an exterior angle of a triangle is equal to the sum of the measures of the nonadjacent interior angles.

T4.16 The sum of the measures of the angles of a polygon with n sides is given by the formula $S = (n - 2)180°$.

C4.17   The measure of each angle of a regular polygon with n sides is given by the formula $a = \dfrac{(n-2)180°}{n}$.

T4.18   The sum of the measures of the exterior angles of a polygon, one at each vertex, is 360°.

C4.19   The measure of each exterior angle of a regular polygon with n sides is determined with the formula $e = \dfrac{360°}{n}$.

T4.20   Each diagonal divides a parallelogram into two congruent triangles.

C4.21   The opposite sides and opposite angles of a parallelogram are equal.

T4.22   If both pairs of opposite sides of a quadrilateral are equal, then the quadrilateral is a parallelogarm.

T4.23   If both pairs of opposite angles of a quadrilateral are equal, then the quadrilateral is a parallelogram.

T4.24   Consecutive angles of a parallelogram are supplementary.

T4.25   If two opposite sides of a quadrilateral are equal and parallel, then the quadrilateral is a parallelogram.

T4.26   The diagonals of a parallelogram bisect each other.

T4.27   If the diagonals of a quadrilateral bisect each other, then the quadrilateral is a parallelogram.

T4.28   All four sides of a rhombus are equal.

T4.29   The diagonals of a rhombus are perpendicular.

T4.30   If the diagonals of a parallelogram are perpendicular, then the parallelogram is a rhombus.

T4.31   The diagonals of a rhombus bisect the angles of the rhombus.

T4.32   All angles of a rectangle are right angles.

T4.33   The diagonals of a rectangle are equal.

T4.34   If the diagonals of a parallelogram are equal, then the parallelogram is a rectangle.

T4.35   Two parallel lines are always the same distance apart.

T4.36   The segment joining the midpoints of two sides of a triangle is parallel to the third side and equal to one-half of it.

T4.37   The base angles of an isosceles trapezoid are equal.

T4.38   The diagonals of an isosceles trapezoid are equal.

T4.39   The median of a trapezoid is parallel to the bases and equal to one-half their sum.

T4.40   If three or more parallel lines intercept equal segments on one transversal, then they intercept equal segments on all transversals.

C4.41   The area of a square with sides of length s is determined with the formula $A = s^2$.

T4.42   The area of a parallelogram with length of base b and height h is determined with the formula $A = bh$.

T4.43   The area of a triangle with length of base b and height h is determined with the formula $A = \dfrac{1}{2}bh$.

T4.44 The area of a trapezoid with length of bases b and b' and height h is determined with the formula $A = \frac{1}{2}(b + b')h$.

T4.45 The area of a rhombus with diagonals of length d and d' is determined with the formula $A = \frac{1}{2}dd'$.

T5.1 (Means-Extremes Property) In any proportion, the product of the means is equal to the product of the extremes. That is, if $\frac{a}{b} = \frac{c}{d}$, then $ad = bc$.

T5.2 (Reciprocal Property of Proportions) The reciprocals of both sides of a proportion are also proportional. That is, if $\frac{a}{b} = \frac{c}{d}$, then $\frac{b}{a} = \frac{d}{c}$.

T5.3 (Means Property of Proportions) If the means are interchanged in a proportion, a new proportion is formed. That is, if $\frac{a}{b} = \frac{c}{d}$, then $\frac{a}{c} = \frac{b}{d}$.

T5.4 (Extremes Property of Proportions) If the extremes are interchanged in a proportion, a new proportion is formed. That is, if $\frac{a}{b} = \frac{c}{d}$, then $\frac{d}{b} = \frac{c}{a}$.

T5.5 (Addition Property of Proportions) If the denominators in a proportion are added to their respective numerators, a new proportion is formed. That is, if $\frac{a}{b} = \frac{c}{d}$, then $\frac{a + b}{b} = \frac{c + d}{d}$.

T5.6 (Subtraction Property of Proportions) If the denominators in a proportion are subtracted from their respective numerators, a new proportion is formed. That is, if $\frac{a}{b} = \frac{c}{d}$, then $\frac{a - b}{b} = \frac{c - d}{d}$.

T5.7 If a, b, c, d, e, and f are numbers satisfying $\frac{a}{b} = \frac{c}{d} = \frac{e}{f}$, then $\frac{a + c + e}{b + d + f} = \frac{a}{b}$.

T5.8 If three terms of one proportion are equal, respectively, to three terms of another proportion, then the remaining terms are also equal. That is, if $\frac{a}{b} = \frac{c}{x}$ and $\frac{a}{b} = \frac{c}{y}$, then $x = y$.

T5.9 (AA ~ AA) Two triangles are similar if two angles of one are equal, respectively, to two angles of the other.

T5.10 If $\triangle ABC \cong \triangle DEF$, then $\triangle ABC \sim \triangle DEF$.

T5.11 (Transitive Law for Similar Triangles) If $\triangle ABC \sim \triangle DEF$ and $\triangle DEF \sim \triangle GHI$, then $\triangle ABC \sim \triangle GHI$.

T5.12 (Triangle Proportionality Theorem) A line parallel to one side of a triangle that intersects the other two sides divides the two sides into proportional segments.

T5.13   (Triangle Angle-Bisector Theorem) The bisector of one angle of a triangle divides the opposite side into segments that are proportional to the other two sides.

T5.14   If a line intersects and divides two sides of a triangle into proportional segments, then the line is parallel to the third side.

T5.15   (SAS ~ SAS) If an angle of one triangle is equal to an angle of another triangle and the including sides are proportional, then the triangles are similar.

T5.16   (SSS ~ SSS) If three sides of one triangle are proportional to the three corresponding sides of another, then the triangles are similar.

T5.17   The areas of two similar triangles have the same ratio as the squares of the lengths of any two corresponding sides.

T6.1   (Zero-Product Rule) If a and b are real numbers such that $ab = 0$, then $a = 0$ or $b = 0$.

T6.2   (Quadratic Formula) The solutions to $ax^2 + bx + c = 0$, $a \neq 0$, are given by the formula $x = \dfrac{-b \pm \sqrt{b^2 - 4ac}}{2a}$.

T6.3   (LA = LA) If a leg and acute angle of one right triangle are equal, respectively, to a leg and the corresponding acute angle of another right triangle, then the two right triangles are congruent.

T6.4   (HA = HA) If the hypotenuse and an acute angle of one right triangle are equal, respectively, to the hypotenuse and an acute angle of another triangle, then the two right triangles are congruent.

T6.5   (LL = LL) If the two legs of one right triangle are equal, respectively, to the two legs of another right triangle, then the two right triangles are congruent.

T6.6   (HL = HL) If the hypotenuse and a leg of one right triangle are equal, respectively, to the hypotenuse and a leg of another right triangle, then the two right triangles are congruent.

T6.7   The median from the right angle in a right triangle is one-half the length of the hypotenuse.

T6.8   The altitude from the right angle to the hypotenuse in a right triangle forms two right triangles that are similar to each other and to the original triangle.

C6.9   The altitude from the right angle to the hypotenuse in a right triangle is the mean proportional between the segments of the hypotenuse.

C6.10   If the altitude is drawn from the right angle to the hypotenuse in a right triangle, then each leg is the mean proportional between the hypotenuse and the segment of the hypotenuse adjacent to the leg.

T6.11   (The Pythagorean Theorem) In a right triangle, the square of the length of the hypotenuse is equal to the sum of the squares of the lengths of the legs.

T6.12   (Converse of the Pythagorean Theorem) If the sides of a triangle have lengths a, b, and c, and $a^2 + b^2 = c^2$, then the triangle is a right triangle.

T6.13 (45°-45°-Right Triangle Theorem) In a 45°-45°-right triangle, the hypotenuse is $\sqrt{2}$ times as long as each (equal) leg.

T6.14 (30°-60°-Right Triangle Theorem) In a 30°-60°-right triangle, the leg opposite the 30°-angle is one-half as long as the hypotenuse, and the leg opposite the 60°-angle is $\sqrt{3}$ times as long as the leg opposite the 30°-angle and $\dfrac{\sqrt{3}}{2}$ times as long as the hypotenuse.

T7.1 The diameter d of a circle is twice the radius r of the circle. That is, $d = 2r$.

T7.2 The measure of an inscribed angle is one-half the measure of its intercepted arc.

C7.3 Inscribed angles that intercept the same or equal arcs are equal.

C7.4 Every angle inscribed in a semicircle is a right angle.

T7.5 When two chords of a circle intersect, each angle formed is equal to one-half the sum of its intercepted arc and the arc intercepted by its vertical angle.

T7.6 In the same circle, the arcs formed by equal chords are equal.

T7.7 In the same circle, the chords formed by equal arcs are equal.

T7.8 A line drawn from the center of a circle perpendicular to a chord bisects the chord and the arc formed by the chord.

T7.9 A line drawn from the center of a circle to the midpoint of a chord (not a diameter) or to the midpoint of the arc formed by the chord is perpendicular to the chord.

T7.10 In the same circle, equal chords are equidistant from the center of the circle.

T7.11 In the same circle, chords equidistant from the center of the circle are equal.

T7.12 The perpendicular bisector of a chord passes through the center of the circle.

T7.13 If two chords intersect inside a circle, the product of the lengths of the segments of one chord is equal to the product of the lengths of the segments of the other.

T7.14 If two secants intersect forming an angle outside the circle, then the measure of this angle is one-half the difference of the intercepted arcs.

T7.15 If two secants are drawn to a circle from an external point, the product of the lengths of one secant segment and its external segment is equal to the product of the lengths of the other secant segment and its external segment.

T7.16 The angle formed by a tangent and a chord has a measure one-half its intercepted arc.

T7.17 The angle formed by the intersection of a tangent and a secant has a measure one-half the difference of the intercepted arcs.

T7.18 The angle formed by the intersection of two tangents has a measure one-half the difference of the intercepted arcs.

T7.19 Two tangent segments to a circle from the same point have equal lengths.

T7.20   If a secant and a tangent are drawn to a circle from an external point, the length of the tangent segment is the mean proportional between the length of the secant segment and its external segment.

T7.21   If two circles are tangent internally or externally, the point of tangency is on their line of centers.

T7.22   If two circles intersect in two points, then their line of centers is the perpendicular bisector of their common chord.

T7.23   If a quadrilateral is inscribed in a circle, the opposite angles are supplementary.

T7.24   If a parallelogram is inscribed in a circle, then it is a rectangle.

T7.25   If a circle is divided into n equal arcs, $n > 2$, then the chords formed by these arcs form a regular n-gon.

T7.26   If a circle is divided into n equal arcs, $n > 2$, and tangents are constructed to the circle at the endpoints of each arc, then the figure formed by these tangents is a regular n-gon.

T7.27   All radii of a regular polygon are equal in length.

T7.28   All central angles of a regular polygon have the same measure.

T7.29   The measure a of each central angle in a regular n-gon is determined with the formula $a = \dfrac{360°}{n}$.

T7.30   Every apothem of a regular polygon has the same length.

T7.31   The apothem of a regular polygon bisects its respective side.

T7.32   (Area of a Regular Polygon) The area of a regular polygon with apothem of length a and perimeter p is determined with the formula $A = \dfrac{1}{2}ap$.

T8.1   An exterior angle of any triangle is greater than each remote interior angle.

T8.2   If two sides of a triangle are unequal, then the angles opposite those sides are unequal in the same order.

T8.3   If two angles of a triangle are unequal, then the sides opposite those angles are unequal in the same order.

T8.4   (The Triangle Inequality Theorem) The sum of the lengths of any two sides of a triangle is greater than the length of the third side.

T8.5   (SAS Inequality Theorem) If two sides of one triangle are equal to two sides of another triangle, and the included angle of the first is greater than the included angle of the second, then the third side of the first triangle is greater than the third side of the second triangle.

T8.6   (SSS Inequality Theorem) If two sides of one triangle are equal to two sides of another triangle, and the third side of the first is greater than the third side of the second, then the included angle of the first triangle is greater than the included angle of the second triangle.

T8.7   In the same circle or in congruent circles, the greater of two central angles intercepts the greater of two arcs.

T8.8   In the same circle or in congruent circles, the greater of two arcs is intercepted by the greater of two central angles.

T8.9   In the same circle or in congruent circles, the greater of two chords forms the greater arc.

T8.10   In the same circle or in congruent circles, the greater of two arcs has the greater chord.

T8.11   In the same circle or in congruent circles, the greater of two unequal chords is nearer the center of the circle.

T8.12   In the same circle or in congruent circles, if two chords have unequal distances from the center of the circle, the chord nearer the center is greater.

C8.13   Every diameter of a circle is greater than any other chord that is not a diameter.

T9.1   (Lateral Area of a Right Prism) The lateral area of a right prism is determined with the formula $LA = ph$, where p is the perimeter of a base and h is the height of the prism.

T9.2   (Volume of a Right Prism) The volume of a right prism is determined with the formula $V = Bh$, where B is the area of a base and h is the height.

T9.3   (Lateral Area of a Regular Pyramid) The lateral area of a regular pyramid is determined with the formula $LA = \frac{1}{2}p\ell$, where p is the perimeter of the base and ℓ is the slant height.

T9.4   (Volume of a Pyramid) The volume of a pyramid is determined with the formula $V = \frac{1}{3}Bh$, where B is the area of the base and h is the height.

T9.5   (Surface Area of a Cylinder) The total surface area of a cylinder is determined with the formula $SA = 2\pi rh + 2\pi r^2$, where r is the radius of a base and h is the height, the length of the altitude.

T9.6   (Volume of a Cylinder) The volume of a cylinder is determined with the formula $V = \pi r^2 h$, where r is the radius of a base and h is the height.

T9.7   (Lateral Area and Volume of a Right Cone) The lateral area of a right cone is determined with the formula $LA = \pi r\ell$, where r is the radius of the base, h is the height, and $\ell = \sqrt{r^2 + h^2}$ is the slant height. Also, the volume is $V = \frac{1}{3}\pi r^2 h$.

T10.1   The locus of all points equidistant from two given points A and B is the perpendicular bisector of $\overline{AB}$.

T10.2   The locus of all points equidistant from the sides of an angle is the angle bisector.

T10.3   The bisectors of the angles of a triangle meet at a point equidistant from the sides of the triangle.

T10.4   The perpendicular bisectors of the sides of a triangle intersect at a point that is equidistant from the vertices of the triangle.

T10.5   The altitudes of a triangle are concurrent.

T10.6   The medians of a triangle meet at a point that is two-thirds the distance from the vertex to the midpoint of the opposite side.

T11.1   (Distance Formula) The distance between two points with coordinates $(x_1,\ y_1)$ and $(x_2,\ y_2)$ is determined with the formula $d = \sqrt{(x_2 - x_1)^2 + (y_2 - y_1)^2}$.

T11.2   (Midpoint Formula) The coordinates $(\bar{x}, \bar{y})$ of the midoint of the line segment joining (x_1, y_1) and (x_2, y_2) are determined with the midpoint formula $(\bar{x}, \bar{y}) = \left(\dfrac{x_1 + x_2}{2}, \dfrac{y_1 + y_2}{2} \right)$. That is, the x-coordinate of the midpoint is the average of the x-coordinates of the points, and the y-coordinate of the midpoint is the average of the y-coordinates of the points.

T11.3   (Standard Form of the Equation of a Circle) A circle with center (h, k) and radius r can be described by the equation $(x - h)^2 + (y - k)^2 = r^2$. The equation is called the standard form of the equation of a circle.

CONSTRUCTIONS IN GEOMETRY

C2.1 Construct a line segment with the same length as a given line segment.

C2.2 Construct an angle with the same measure as a given angle.

C2.3 Construct a bisector of a given line segment.

C2.4 Construct a line perpendicular to a given line passing through a given point on the line.

C2.5 Construct a line perpendicular to a given line passing through a given point not on that line.

C2.6 Construct a bisector of a given angle.

C3.1 Construct a triangle that is congruent to a given triangle.

C3.2 Construct a triangle with two sides and the included angle given.

C3.3 Construct a triangle with two angles and the included side given.

C3.4 Construct an equilateral triangle when given a single side.

C3.5 Construct an altitude of a given triangle.

C3.6 Construct a median of a given triangle.

C4.1 Construct the line parallel to a given line that passes through a point not on the given line.

C4.2 Construct a rectangle when two adjacent sides are given.

C4.3 Construct a square when a side is given.

C4.4 Divide a given segment into a given number of equal segments.

C5.1 Construct a segment proportional to three given line segments.

C5.2 Construct a polygon similar to a given polygon.

C7.1 Construct a tangent to a circle at a given point on the circle.

C7.2 Construct a tangent to a circle from a point outside the circle.

C7.3 Construct a common external tangent to two given circles that are not congruent.

C7.4 Construct a common internal tangent to two given circles that are not congruent.

C7.5 Construct a circle that is circumscribed around a given regular polygon.

C7.6 Construct a circle that is inscribed in a given regular polygon.

C10.1 Inscribe a circle in a given triangle.

ANSWERS TO SELECTED EXERCISES

FOR THE STUDENT

If you need further help as you study geometry, you may want to get a copy of the *Student Solutions Manual* that accompanies this textbook from your college bookstore. This manual provides step-by-step solutions to the odd-numbered section exercises and to all practice exercises and review exercises in the text.

CHAPTER 1

1.1

1. 28   **3.** 243   **5.** -32 **7.** 34   **9.** S   **11.** Undefined terms give us a starting point in the development of an axiomatic system.   **13.** The terms "postulate" and "axiom" are actually synonyms; there is no difference, but "postulate" is used in geometry.   **15.** When we look up a word like "happy" in a dictionary of synonyms, eventually we will circle back and arrive at "happy" again.   **17.** Because every SQA is a REC by the postulate, and every REC contains a RAL by the definition, we can conclude that every SQA contains a RAL.   **19.** We would probably obtain different theorems.   **21.** The conclusion follows logically; deductive reasoning   **23.** The conclusion does not follow logically; this is a fallacy in reasoning   **25.** The conclusion does not follow logically; inductive reasoning   **27.** The conclusion follows logically; deductive reasoning   **29.** The conclusion follows logically; deductive reasoning   **31.** We cannot conclude you necessarily are an ogg, only that *if* you are a pon *then* you are an ogg.   **33.** a nickel and a 50¢ piece (We are told *one* is not a nickel, not *both* are not nickels.)   **35.** You have 5 apples. (We were not asked what was left.)   **37.** 12   **39.** A decimal point, because 2.3 is greater than 2 and less than 3.   **41.** The two daughters cross, one returns with the boat and stays while the man crosses. The other daughter returns with the boat and picks up the remaining daughter.   **43.** The defendant told the jury to look at the piece of paper remaining in the hat and let them draw the appropriate conclusion.   **45.** Locate the midpoints of six of the edges as shown in the figure at left and cut the cube along the dashed lines. Each resulting piece will have one surface a regular hexagon as shown in the figure on page 405.   **47.** The fallacy here is that the pieces do not actually fit together as shown in the figure at right in the text. In fact, there is a small strip missing through the diagonal from upper left to lower right that accounts for the "extra" little square. The exaggerated figure illustrates.

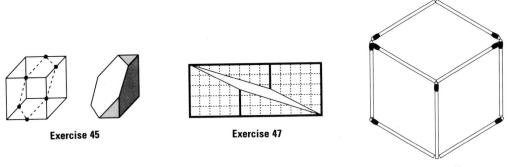

Exercise 45   **Exercise 47**

Exercise 49

49. We have used the nine matches to form a cube (a regular six-sided figure) shown in perspective. (We did not say the figure had to be a plane figure.)

51. Smith is the engineer. Jones is the brakeman. Rodriquez is the fireman.

1.2
1. exactly one   **3.** C is in plane $\mathcal{P}$ **5.** Postulate 1.1   **7.** Postulate 1.4   **9.**

11. A:2; B: -3; C:3.5; D: $-\dfrac{1}{2}$ **13.** 5   **15.** 7   **17.** a **19.** false

21. false   **23.** 2. Postulate 1.9 4. Postulate 1.9 6. Postulate 1.10
8. Postulate 1.7 9. Postulate 1.11   **25.** yes; yes   **27.** yes; yes   **29.** By Postulate 1.2, three distinct points (the ends of the legs) all lie in the same plane (the floor on which the stool stands). However, four points (the ends of a 4-legged stool) need not all lie in the same plane (the floor on which that stool stands).

1.3
1. true   **3.** true   **5.** true   **7.** false   **9.** true   **11.** false   **13.** true   **15.** false   **17.** true   **19.** true   **21.** 65°
23. 90°   **25.** 115°   **27.** 72°   **29.** 53°20′   **31.** 106°   **33.** 122°25′   **35.** 45°   **37.** 40°   **39.** 25°   **41.** 60°
43. acute   **45.** straight   **47.** acute   **49.** acute   **51.** right   **53.** 107°10′   **55.** 139°20′16″   **57.** two   **59.** two

61. It depends on the length of $\overline{AB}$. If $AB < 5$, there is only one. If $AB \geq 5$, there are two.   **63.** $\dfrac{n(n-1)}{2}$ **65.** $x = 9$

67. $y = 20$

Chapter 1 Review Exercises
1. undefined terms, definitions, axioms or postulates, and theorems   **2.** They serve as a starting point when building the system.

3. 17   **4.** K   **5.** $\dfrac{1}{32}$ **6.** The conclusion follows logically; deductive reasoning   **7.** The conclusion does not follow logically; this is a fallacy of reasoning   **8.** The conclusion does not follow logically; inductive reasoning   **9.** The conclusion follows logically; deductive reasoning   **10.** We cannot conclude that tomorrow is a holiday, only that *if* it is Sunday, *then* tomorrow is a holiday.   **11.** 3 minutes   **12.** 3   **13.** true   **14.** false   **15.** true   **16.** true   **17.**
18. Reflexive law   **19.** Symmetric law   **20.** Transitive law
21. Multiplication-division law   **22.** Substitution law   **23.** false   **24.** true
25. true   **26.** true   **27.** false   **28.** true   **29.** false   **30.** false   **31.** true
32. false   **33.** true   **34.** false   **35.** 65°29′15″   **36.** 43°17′9″   **37.** 20°   **38.** no   **39.** exactly one   **40.** two

Chapter 1 Practice Test
1. A postulate is a statement that is assumed true without proof; a theorem is a statement that is proved using deductive reasoning.
2. Inductive reasoning involves reaching a general conclusion based on specific observations. Deductive reasoning involves reaching a specific conclusion based on assumed general conditions.   **3.** deductive reasoning   **4.** No. The specific observations that lead to the conclusion may not be sufficient in number.   **5.** $\dfrac{1}{5}$ **6.** No, this inductive reasoning does not necessarily lead to the same conclusion.   **7.** infinitely many   **8.** $y + 2$ **9.** true   **10.** true   **11.** false   **12.** true   **13.** true   **14.** false
15. true   **16.** false   **17.** 56°20′

CHAPTER 2

2.1

1. 1. Given 2. Premise 1 3. Premise 3 4. Premise 2
3. *Given:* The weather report is accurate.
Prove: School will be canceled.
Proof:

STATEMENTS	REASONS
1. The weather report is accurate.	1. Given
2. We will get 12 inches of snow.	2. Premise 1
3. The streets will be treacherous.	3. Premise 2
4. School will be canceled.	4. Premise 3

∴ If the weather report is accurate, then school will be canceled.

5. No. We proved that *if* the weather report is accurate, *then* we know that school will be canceled.   **7.** No, the theorem could not be proved.   **9.** If a person is a United States citizen, then he/she is the president.   **11.** If a figure is a rectangle, then the figure is a square.   **13.** The tree is not a pine.   **15.** I did not receive an A in the course.   **17.** If it doesn't rain, then I will not stay indoors.   **19.** If I don't have the flu, then I am not running a fever.   **21.** If I am not getting wet, then I am not taking a shower.   **23.** If I am unhealthy, then I do not drink orange juice.   **25.** If a person serves on a jury, then he/she is not one of your sons.

2.2

1. 1. Given 2. Given 3. Given 4. Given 5. Addition-subtraction law (Post. 1.9) 6. Segment addition postulate (Post. 1.13) 7. Substitution law (Post. 1.11)   **3.** 1. B is the midpoint of $\overline{AC}$ 2. Def. of midpoint 3. Segment addition postulate (Post. 1.13) 4. Substitution law (Post. 1.11) 5. $2AB = AC$ 6. Multiplication-division law (Post. 1.10)   **5.** 1. B is the midpoint of $\overline{AC}$ 3. Given 4. $PQ = \dfrac{PR}{2}$ 5. $AC = PR$ 6. Multiplication-division law (Post. 1.10) 7. Substitution law (Post. 1.11)   **7.** 1. $\angle ABC$ is a right angle 2. Def. of right angle 3. Angle addition postulate (Post. 1.15) 4. Transitive law (Post. 1.8) 5. Given 6. Def. of complementary angles 7. Symmetric and transitive laws using Statements 4 and 6   **8.** $\angle ABD = \angle 1$ **9.** $\angle 1$ and $\angle 2$; $\angle 2$ and $\angle 3$; $\angle 3$ and $\angle 4$; $\angle 4$ and $\angle 1$ **11.** 150°   **13.** 150°   **15.** 15°   **17.** yes   **19.** yes   **21.** no   **23.** yes   **25.** yes   **27.** no   **29.** yes
31. *Given:* $\angle A$ and $\angle C$ are supplementary
   $\angle B$ and $\angle D$ are supplementary
   $\angle C = \angle D$
Prove: $\angle A = \angle B$
Proof:

STATEMENTS	REASONS
1. $\angle A$ and $\angle C$ are supp.	1. Given
2. $\angle A + \angle C = 180°$	2. Def. of supp. $\angle$
3. $\angle B$ and $\angle D$ are supp.	3. Given
4. $\angle B + \angle D = 180°$	4. Def. of supp. $\angle$
5. $\angle A + \angle C = \angle B + \angle D$	5. Sym. and trans. laws
6. $\angle C = \angle D$	6. Given
7. $\angle A = \angle B$	7. Add.-subtr. law

33. *Given:* $PR = QS$
Prove: $PQ = RS$
Proof:

STATEMENTS	REASONS
1. $PR = QS$	1. Given
2. $QR = QR$	2. Reflexive law
3. $PR - QR = QS - QR$	3. Add.-subtr. law
4. $PQ = PR - QR$	4. Seg.-add. post.
5. $RS = QS - QR$	5. Seg.-add. post.
6. $PQ = RS$	6. Substitution law

35. *Given:* ∠3 and ∠4 are complementary
        ∠1 and ∠3 are supplementary
        ∠2 and ∠4 are supplementary
Prove: ∠1 + ∠2 = 270°
Proof:

STATEMENTS	REASONS
1. ∠3 and ∠4 are complementary	1. Given
2. ∠3 and ∠4 = 90°	2. Def. of comp. ∠
3. ∠1 and ∠3 are supplementary	3. Given
4. ∠1 + ∠3 = 180°	4. Def. of supp. ∠
5. ∠2 and ∠4 are supplementary	5. Given
6. ∠2 + ∠4 = 180°	6. Def. of supp. ∠
7. ∠1 + ∠2 + ∠3 + ∠4 = 360°	7. Add.-subtr. law using Statements 4 and 6
8. ∠1 + ∠2 + 90° = 360°	8. Substitution law using Statement 2
9. ∠1 + ∠2 = 270°	9. Add.-subtr. law

2.3
1. Use Construction 2.1.   **3.** Use Construction 2.2.   **5.** Use Construction 2.2 twice.   **7.** Use Construction 2.3.   **9.** Use Construction 2.3.   **11.** Use Construction 2.4.   **13.** Use Construction 2.6.   **15.** Use Construction 2.6 three times.   **17.** A ruler is used to measure lengths, but a straightedge is used only to draw a straight line between two points.   **19.** exactly one **21.** Use Construction 2.5.   **23.** Construct the bisector of the angle with vertex at the bridge and sides containing the ranger station and the cabin. Locate the point of intersection of this bisector (ray) and the edge of the forest.

Chapter 2 Review Exercises
1. *Given:* It is Gleep.
      Prove: It is Grob.
      Proof:

STATEMENTS	REASONS
1. It is Gleep.	1. Given
2. It is Glop.	2. Premise 1
3. It is Gunk.	3. Premise 3
4. It is Grob.	4. Premise 2
∴ If it is Gleep, then it is Grob.	

2. *Given:* It is a tree.
      Prove: It can be destroyed by fire.
      Proof:

STATEMENTS	REASONS
1. It is a tree.	1. Given
2. It is made of wood	2. Premise 2
3. It will burn.	3. Premise 3
4. It can be destroyed by fire.	4. Premise 1
∴ If it is a tree, it can be destroyed by fire.	

3. If it is cold, then it is ice.   **4.** If you want the best for someone, then you love that person.   **5.** My car is not red.   **6.** The moon is made of green cheese.   **7.** If the runner does not win the race, then she is not in excellent condition.   **8.** If a painting is not a Picasso, then it is not valuable.   **9.** If I do not climb the mountain, then the weather is not good.   **10.** If you drive, then do not drink.   **11.** 42°   **12.** 138°   **13.** 17°   **14.** 163°   **15.** 1. Given 2. ∠1 + ∠2 = 90° 3. ∠1 and ∠3 are vertical angles 4. Vertical angles are equal 5. ∠2 and ∠4 are vertical angles 6. Vertical angles are equal 8. ∠3 and ∠4 are complementary

16. *Given:* ∠ABC and ∠CBD are supplementary
        $\overrightarrow{BE}$ bisects ∠ABC
        $\overrightarrow{BF}$ bisects ∠CBD
Prove: $\overleftrightarrow{BE} \perp \overleftrightarrow{BF}$
Proof:

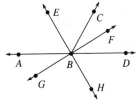

STATEMENTS	REASONS
1. $\overrightarrow{BE}$ bisects ∠ABC	1. Given
2. ∠ABE = ∠EBC	2. Def. of ∠ bisector
3. $\overrightarrow{BF}$ bisects ∠CBD	3. Given
4. ∠CBF = ∠FBD	4. Def. of ∠ bisector
5. ∠ABC = ∠ABE + ∠EBC	5. Angle add. post.
6. ∠ABC = ∠EBC + ∠EBC	6. Substitution law
7. ∠ABC = 2∠EBC	7. Distributive law
8. ∠CBD = ∠CBF + ∠FBD	8. Angle add. post.

9. $\angle CBD = \angle CBF + \angle CBF$	9. Substitution law
10. $\angle CBD = 2\angle CBF$	10. Distributive law
11. $\angle ABC + \angle CBD = 2\angle EBC + 2\angle CBF$	11. Add.-subtr. law
12. $\angle ABC$ and $\angle CBD$ are supplementary	12. Given
13. $\angle ABC + \angle CBD = 180°$	13. Def. of supp. $\angle$'s
14. $2\angle EBC + 2\angle CBF = 180°$	14. Sym. and trans. laws
15. $\angle EBC + \angle CBF = 90°$	15. Mult.-div. law
16. $\angle EBF = \angle EBC + \angle CBF$	16. Angle add. post.
17. $\angle EBF = 90°$	17. Trans. law
18. $\angle EBF$ and $\angle EBG$ are supplementary	18. Def. supp. $\angle$'s
19. $\angle EBG = 90°$	19. From Statements 17 and 18
20. $\angle EBF$ and $\angle EBG$ are adjacent	20. Def. adj. $\angle$'s
21. $\overleftrightarrow{BE} \perp \overleftrightarrow{BF}$	21. Def. of $\perp$ lines

17. First bisect $\overline{PQ}$ determining midpoint T; then use Construction 2.1 to construct segment $\overline{CD}$ on m with length PT.   **18.** Use Construction 2.6.   **19.** Use Construction 2.3.   **20.** infinitely many   **21.** exactly one   **22.** Use Construction 2.2.

Chapter 2 Practice Test

1. *Given:* I rob a bank.
Prove: I will go to jail.
Proof:

STATEMENTS	REASONS
1. I rob a bank.	1. Given
2. I will be arrested.	2. Premise 4
3. I will go on trial.	3. Premise 2
4. I will be convicted.	4. Premise 1
5. I will go to jail.	5. Premise 3
∴ If I rob a bank, then I will go to jail.	

2. If it is white, then it is milk.   **3.** The road to success is not difficult.   **4.** If it is not a collie, then it is not a dog.   **5.** If the fruit is not ripe, then it is not picked.
6. *Given:* $\angle 1 = \angle 2$
Prove: $\angle 3 = \angle 4$
Proof:

STATEMENTS	REASONS
1. $\angle 1 = \angle 2$	1. Given
2. $\angle 3$ and $\angle 1$ are supplementary	2. Adj. $\angle$'s whose noncommon sides are in line are supp.
3. $\angle 4$ and $\angle 2$ are supplementary	3. Same as Reason 2
4. $\angle 3 = \angle 4$	4. Supp's of $=$ $\angle$'s are $=$ (Theorem 2.8)

7. Use Construction 2.3 to find midpoint C; then use Construction 2.2 to construct the desired angle.   **8.** Use Construction 2.6.

CHAPTER 3

3.1

1. D, E, and F **3.** obtuse triangle   **5.** $\overline{DE}$ **7.** $\angle F$ **9.** $\overline{DE}$ **11.** $\overline{AB}$, $\overline{BC}$, and $\overline{AC}$ **13.** equilateral triangle
15. $\overline{BC}$ **17.** $\angle A$ **19.** It has no hypotenuse because it is not a right triangle.   **21.** 90 cm   **23.** 33 in   **25.** 23 ft
27. base: 15 cm; sides: 45 cm   **29.** $\triangle AED$ and $\triangle AEB$ **31.** $\triangle ECD$ and $\triangle ACD$ **33.** true   **35.** false   **37.** false
39. true   **41.** yes   **43.** yes   **45.** no

3.2

1. congruent by SSS $=$ SSS   **3.** Because right angles are equal, the triangles are congruent by SAS $=$ SAS.   **5.** congruent by
ASA $=$ ASA   **7.** 1. Given 2. $\angle 3 = \angle 4$ 3. $BD = BD$ 4. ASA $=$ ASA   **9.** 1. Given 2. $\perp$ lines form rt. $\angle$'s 3. $\angle ABD =$
$\angle BDC$ 4. $\overline{DB} = \overline{DB}$ 5. Given 6. SAS $=$ SAS

11. *Proof:*

STATEMENTS	REASONS
1. $\overline{AD}$ bisects $\overline{BE}$	1. Given
2. C is the midpoint of $\overline{BE}$	2. Def. of bisector
3. $BC = CE$	3. Def. of midpt.
4. $\overline{BE}$ bisects $\overline{AD}$	4. Given
5. C is the midpoint of $\overline{AD}$	5. Def. of bisector
6. $AC = CD$	6. Def. of midpt.
7. $\angle ACB = \angle DCE$	7. Vert. $\angle$'s are $=$
8. $\triangle ABC \cong \triangle CDE$	8. SAS = SAS

13. *Proof:*

STATEMENTS	REASONS
1. $AD = BD$	1. Given
2. $AE = BC$	2. Given
3. $ED = AD - AE$ and $CD = BD - BC$	3. Seg. add. post.
4. $ED = CD$	4. Add.-subtr. law
5. $\angle D = \angle D$	5. Reflexive law
6. $\triangle ACD \cong \triangle BDE$	6. SAS = SAS

15. *Proof:*

STATEMENTS	REASONS
1. $AB = BC$	1. Given
2. $\angle 1 = \angle 2$	2. Given
3. $\angle 4$ is supplementary to $\angle 1$ and $\angle 3$ is supplementary to $\angle 2$	3. Adj. $\angle$'s whose noncommon sides are in a line are supp.
4. $\angle 4 = \angle 3$	4. Supp. of $=$ $\angle$'s are $=$
5. $\angle 5 = \angle 6$	5. Vert. $\angle$'s are $=$
6. $\triangle ABE \cong \triangle BCD$	6. ASA = ASA

17. *Proof:*

STATEMENTS	REASONS
1. $\triangle ACD$ is equilateral	1. Given
2. $AC = AD$	2. Def. of equilateral $\triangle$
3. $\angle 1 = \angle 2$	3. Given
4. $\angle 3$ and $\angle 1$ are supplementary and $\angle 4$ and $\angle 2$ are supplementary	4. Adj. $\angle$'s whose noncommon sides are in a line are supp.
5. $\angle 3 = \angle 4$	5. Supp. of $=$ $\angle$'s are $=$
6. $BC = DE$	6. Given
7. $\triangle ABC \cong \triangle ADE$	7. SAS = SAS

19. 3   **21.** 4   **23.** 120°   **25.** $\overleftrightarrow{RP}$ **27.** $\overline{RQ}$ **29.** $\angle P$ **31.** The proof follows by repeated use of the transitive law.

3.3

1. 9 cm   **3.** 46°   **5.** 1. Given 2. $\angle 1 = \angle 2$ 3. Adj. $\angle$'s whose noncommon sides are in a line are supp. 4. $\angle 2$ and $\angle ECB$ are supplementary 5. Supp. of $=$ $\angle$'s are $=$ 6. Reflexive law 7. SAS = SAS 8. $\angle G = \angle E$ **7.** 1. Given 2. Given 3. $AB = CD$ 4. Seg. add post. 5. $BD = CD + BC$ 6. Add.-subtr. law 7. Substitution law 8. $\triangle AFC \cong \triangle BDE$ 9. cpoctae

9. *Proof:*

STATEMENTS	REASONS
1. $AC = CE$	1. Given
2. $DC = CB$	2. Given
3. $\angle DCE$ and $\angle ACB$ are vertical angles	3. Def. of vert. $\angle$'s
4. $\angle DCE = \angle ACB$	4. Vert. $\angle$'s are $=$
5. $\triangle DCE \cong \triangle ACB$	5. SAS = SAS
6. $\angle A = \angle E$	6. cpoctae

11. *Proof:*

STATEMENTS	REASONS
1. $BC = CD$	1. Given
2. $\angle 1 = \angle 2$	2. Given
3. $AC = AC$	3. Reflexive law
4. $\triangle ABC \cong \triangle ADC$	4. SAS = SAS
5. $AB = AD$	5. cpoctae

13. *Proof:*

STATEMENTS	REASONS
1. $\angle 1 = \angle 2$	1. Given
2. $\angle 3 = \angle 4$	2. Given
3. $DB = DB$	3. Reflexive law
4. $\triangle ABD \cong \triangle CBD$	4. ASA = ASA
5. $\angle A = \angle C$	5. cpoctae

15. *Proof:*

STATEMENTS	REASONS
1. $\overline{GB} \perp \overline{AF}$ and $\overline{FD} \perp \overline{GE}$	1. Given
2. $\angle 1$ and $\angle 2$ are right angles	2. $\perp$ lines form Rt. $\angle$'s
3. $\angle 1 = \angle 2$	3. Rt. $\angle$'s are equal
4. $GD = BF$ and $GB = FD$	4. Given
5. $\triangle BGF \cong \triangle DFG$	5. SAS = SAS
6. $\angle BGF = \angle DFG$	6. cpoctae
7. $\angle 3 = \angle 4$	7. cpoctae
8. $\angle 5 = \angle DFG - \angle 3$	8. Angle add. post.
9. $\angle 6 = \angle BGF - \angle 4$	9. Angle add. post.
10. $\angle 5 = \angle 6$	10. Add.-subtr. law
11. $\triangle BGC \cong \triangle DFC$	11. ASA = ASA
12. $BC = DC$	12. cpoctae

17. Because AB and DE are corresponding parts of congruent triangles, $\triangle ABC$ and $\triangle DCE$ (by SAS = SAS), $AB = 105$ yd (the same as DE).   **19.** The bridge is composed of triangles because the triangle is a rigid figure that cannot be distorted like a four-sided figure. This is a result of SSS = SSS because there is only one triangle possible with three given sides. The bridge cannot change shape without breaking.   **21.** Pattern your proof after the proof given for Theorem 3.2.   **23.** Pattern your proof after the proof given for Theorem 3.2.

3.4

1. true   **3.** false   **5.** false   **7.** 1. Given 2. $\overline{AC}$ bisects $\angle BAD$ 3. Def. of $\angle$ bisector 4. Reflexive law 5. $\triangle BAE \cong \triangle DAE$ 6. cpoctae 7. $\triangle BDE$ is isosceles   **9.** 1. Given 2. $AC = AD$ 3. Given 4. Def. of midpt. 5. E is the midpt. of $\overline{AD}$ 6. $AE = ED$ 7. Seg. add. post. 8. Substitution law 10. Mult.-div. law 13. $\angle 1 = \angle 2$
11. *Proof:*

STATEMENTS	REASONS
1. $\angle 1 = \angle 2$	1. Given
2. $BE = CE$	2. Sides opp. = $\angle$'s are =
3. $\angle 3 = \angle 4$	3. Given
4. $\angle AEB$ and $\angle DEC$ are vertical angles	4. Def. of vert. $\angle$'s
5. $\angle AEB = \angle DEC$	5. Vert. $\angle$'s are =
6. $\triangle ABE \cong \triangle DCE$	6. ASA = ASA
7. $\angle A = \angle D$	7. cpoctae

13. *Proof:*

STATEMENTS	REASONS
1. $AB = AD$	1. Given
2. $\angle 3 = \angle 1$	2. $\angle$'s opp. = sides are =

3. $BC = DC$
4. $\angle 4 = \angle 2$
5. $\angle ABC = \angle 1 + \angle 2$ and
   $\angle ADC = \angle 3 + \angle 4$
6. $\angle ABC = \angle ADC$

3. Given
4. $\angle$'s opp. $=$ sides are $=$
5. Angle add. post.

6. Add.-subtr. law

15. *Proof:*

STATEMENTS	REASONS
1. $\triangle ABC$ is isosceles with base $\overline{BC}$	1. Given
2. $AC = AB$	2. Def. of isos. $\triangle$
3. $\triangle DBC$ is isosceles with base $\overline{BC}$	3. Given
4. $DC = DB$	4. Def. of isos. $\triangle$
5. $\angle ABC = \angle ACB$	5. $\angle$'s opp. $=$ sides are $=$
6. $\angle DBC = \angle DCB$	6. $\angle$'s opp. $=$ sides are $=$
7. $\angle 1 = \angle ABC - \angle DBC$ and $\angle 2 = \angle ACB - \angle DCB$	7. Angle add. post.
8. $\angle 1 = \angle 2$	8. Add.-subtr. law

17. Draw and label two congruent triangles, bisect corresponding angles, and prove the angles are equal as corresponding parts of suitable congruent triangles.   **19.** Make a suitable drawing and show the two segments are equal as corresponding parts of congruent triangles.   **21.** Draw and label two congruent triangles. Use the substitution law together with the fact that corresponding sides of congruent triangles are equal.

3.5
1. Use Construction 3.1.   **3.** Use Construction 3.2.   **5.** Use Construction 3.3.   **7.** Any two obtuse angles will work because their noncommon sides will not intersect.   **9.** Use Construction 3.2   **11.** The sides do not intersect to form a triangle.   **13.** First construct a right angle using Construction 2.3; then use Construction 3.2.   **15.** There are two triangles with these parts.   **17.** Use Construction 3.5 three times.   **19.** Use Construction 3.6 three times.   **21.** Use Construction 2.6 three times.   **23.** They seem to intersect in a common point.   **25.** They seem to intersect in a common point.   **27.** They seem to intersect in a common point.   **29.** Use Construction 3.4

Chapter 3 Review Exercises
1. A, B, and C **2.** $\overline{AB}$, $\overline{BC}$, and $\overline{CA}$ **3.** scalene triangle   **4.** obtuse triangle   **5.** $\overline{AC}$ **6.** $\angle B$ **7.** $\overline{BC}$ **8.** $\angle C$
9. 44 cm   **10.** base: 13 ft; sides: 26 ft   **11.** true   **12.** true   **13.** false   **14.** true   **15.** yes   **16.** SSS $=$ SSS
17. ASA $=$ ASA   **18.** 1. Given 2. $AC = AE$ 3. $\angle A = \angle A$ 4. $\triangle ACF \cong \triangle AEB$ **19.** 1. Given 2. $\overrightarrow{AC}$ bisects $\angle BAD$
3. $\angle BAC = \angle EAC$ 4. Reflexive law 5. $\triangle ABC \cong \triangle AEC$
20. *Proof:*

STATEMENTS	REASONS
1. $AD = BC$	1. Given
2. $AB = DC$	2. Given
3. $DB = DB$	3. Reflexive law
4. $\triangle ADB \cong \triangle BCD$	4. SSS $=$ SSS

21. *Proof:*

STATEMENTS	REASONS
1. $\overline{AC} \perp \overline{BD}$	1. Given
2. $\angle 3 = \angle 4$	2. $\perp$ lines form $=$ adj. $\angle$'s
3. $\angle 1 = \angle 2$	3. Given
4. $EC = EC$	4. Reflexive law
5. $\triangle DEC \cong \triangle BEC$	5. ASA $=$ ASA
6. $DC = BC$	6. cpoctae
7. $AC = AC$	7. Reflexive law
8. $\triangle ABC \cong \triangle ACD$	8. SAS $=$ SAS
9. $AB = AD$	9. cpoctae

22. 2   **23.** 25   **24.** 4   **25.** 35°   **26.** $\overline{BC}$ **27.** $\angle D$ **28.** 1. Given 2. $\overrightarrow{AC}$ bisects $\angle DAB$ 3. Def. of $\angle$ bisector
4. Given 5. $\angle EBA = \angle DBE$ 6. Angle add. post. 7. Substitution law 9. Sym. and trans. laws 10. Mult.-div. law 11. Reflexive law 12. ASA $=$ ASA 13. $AC = BE$

29. *Proof:*

STATEMENTS

1. E is the midpoint of $\overline{AC}$
2. $AE = EC$
3. E is the midpoint of $\overline{BD}$
4. $BE = ED$
5. $\angle AEB$ and $\angle DEC$ are vertical angles
6. $\angle AEB = \angle DEC$
7. $\triangle AEB \cong \triangle DEC$
8. $AB = CD$

REASONS

1. Given
2. Def. of midpt.
3. Given
4. Def. of midpt.
5. Def. of vert. $\angle$'s
6. Vert. $\angle$'s are $=$
7. SAS = SAS
8. cpoctae

30. *Proof:*

STATEMENTS

1. $AB = CD$
2. $BC = DE$
3. $\angle CAE = \angle CEA$
4. $AC = CE$
5. $\triangle ABC \cong \triangle CDE$
6. $\angle B = \angle D$

REASONS

1. Given
2. Given
3. Given
4. Sides opp. $= \angle$'s are $=$
5. SSS = SSS
6. cpoctae

31. *Proof:*

STATEMENTS

1. $\triangle ABD$ is isosceles with base $\overline{BD}$
2. $AB = AD$
3. $\triangle BDE$ is isosceles with base $\overline{BD}$
4. $BE = DE$
5. $AE = AE$
6. $\triangle ABE \cong \triangle ADE$
7. $\angle 3 = \angle 4$
8. $\angle 3$ and $\angle 1$ are supplementary and $\angle 4$ and $\angle 2$ are supplementary
9. $\angle 1 = \angle 2$

REASONS

1. Given
2. Def. of isos. $\triangle$
3. Given
4. Def. of isos. $\triangle$
5. Reflexive law
6. SSS = SSS
7. cpoctae
8. Adj. $\angle$'s whose noncommon sides are in a line are supp.
9. Supp. of $= \angle$'s are $=$

32. Use Construction 3.1.   **33.** First construct a right angle using Construction 2.3; then use the right angle, the segment, and the acute angle with Construction 3.3.   **34.** Use Construction 3.6.   **35.** Use Construction 3.2.

Chapter 3 Practice Test

1. acute triangle   **2.** isosceles triangle   **3.** $\angle A$ **4.** $\angle B = \angle C$ **5.** $\angle B$ **6.** $\overline{BC}$ **7.** yes   **8.** yes   **9.** $\angle ACD$
10. 13 cm   **11.** 33 in   **12.** 28°
13. *Proof:*

STATEMENTS

1. $\angle 1 = \angle 2$
2. D is the midpoint of $\overline{CE}$
3. $CD = DE$
4. $AC = AE$
5. $\angle E = \angle C$
6. $\triangle BCD \cong \triangle FDE$
7. $BD = FD$

REASONS

1. Given
2. Given
3. Def. of midpt.
4. Given
5. Angles opp. $=$ sides are $=$
6. ASA = ASA
7. cpoctae

14. *Proof:*

STATEMENTS

1. $\angle 3 = \angle 4$
2. $AC = AD$
3. $\angle 1$ and $\angle 3$ are supplementary and $\angle 2$ and $\angle 4$ are supplementary

REASONS

1. Given
2. Sides opp. $= \angle$'s are $=$
3. Adj. $\angle$'s whose noncommon sides are in a line are supp.

4. $\angle 1 = \angle 2$
5. $BC = DE$
6. $\triangle ABC \cong \triangle ADE$
7. $\angle 5 = \angle 6$

4. Supp. of $=\angle$'s are $=$
5. Given
6. SAS = SAS
7. cpoctae

15. Use Construction 3.2 followed by Construction 3.5 and Construction 3.6.

CHAPTER 4

4.1

1. 2. Assumption   3. Premise 1   4. Premise 2   5. Premise 3
3. *Given:* The weather is nice.
   Prove: I am healthy.
   Proof:

STATEMENTS	REASONS
1. The weather is nice.	1. Given
2. Assume I am unhealthy.	2. Assumption
3. I don't exercise.	3. Premise 2
4. I'm not playing tennis regularly.	4. Premise 1
5. The weather is bad.	5. Premise 3

This is a contradiction of the given Statement 1 "The weather is nice." Thus, we must conclude
that our assumption in Statement 1 was incorrect. $\therefore$ If the weather is nice, then I am healthy.
5. yes   **7.** yes   **9.** no   **11.** no   **13.** one; only line ℓ **15.** Use the fact that *if* two lines are parallel to a third line and
they intersect, *then* there are two lines through a point that are parallel to a line not containing the point, a contractiion.   **17.** yes
19. no   **21.** no   **23.** Many examples can be given.   **25.** The most obvious answer is the yard lines and the side lines.
27. No, if a triangle has two right angles, the noncommon sides of the angles could not intersect to form the triangle.

4.2

1. $\angle 3$ and $\angle 5$; $\angle 4$ and $\angle 6$ **3.** $\angle 1$ and $\angle 5$; $\angle 2$ and $\angle 6$; $\angle 3$ and $\angle 7$; $\angle 4$ and $\angle 8$ **5.** $\angle 1, \angle 3, \angle 5, \angle 7$ **7.** $\angle 4, \angle 6, \angle 8$
9. $\angle 2 = \angle 4 = \angle 8 = 135°$, $\angle 1 = \angle 3 = \angle 5 = \angle 7 = 45°$ **11.** $\angle 1 = \angle 3 = \angle 5 = \angle 7 = 55°$, $\angle 2 = \angle 4 = \angle 6 = \angle 8 = 125°$
13. no   **15.** yes   **17.** no   **19.** yes
21. *Proof:*

STATEMENTS	REASONS
1. $\angle 1$ is supplementary to $\angle 2$	1. Given
2. $\angle 1$ is supplementary to $\angle CED$	2. Adj. $\angle$'s whose noncommon sides are in a line are supp.
3. $\angle 2 = \angle CED$	3. Supp. of $=\angle$'s are $=$
4. $m \parallel n$	4. Corr. $\angle$'s are $=$

23. *Proof:*

STATEMENTS	REASONS
1. $m \parallel n$	1. Given
2. $\angle BAC = \angle CED$ and $\angle ABC = \angle CDE$	2. Alt. int. $\angle$'s are $=$
3. $AB = DE$	3. Given
4. $\triangle ABC \cong \triangle CDE$	4. ASA = ASA

25. *Proof:*

STATEMENTS	REASONS
1. C is midpoint of $\overline{AE}$ and $\overline{BD}$	1. Given
2. $AC = CE$ and $BC = CD$	2. Def. of midpt.
3. $\angle ACB = \angle DCE$	3. Vert. $\angle$'s are $=$
4. $\triangle ACB \cong \triangle DCE$	4. SAS = SAS
5. $\angle ABC = \angle CDE$	5. cpoctae
6. $m \parallel n$	6. Alt. int. $\angle$'s are $=$

27. *Proof:*

STATEMENTS	REASONS
1. $AC = CE$	1. Given
2. $m \parallel n$	2. Given
3. $\angle BAC = \angle CED$	3. Alt. int. $\angle$'s are $=$
4. $\angle ACB = \angle DCE$	4. Vert. $\angle$'s are $=$
5. $\triangle ABC \cong \triangle CDE$	5. ASA = ASA
6. $DC = CB$	6. cpoctae

29. The proof is similar to the proof of Theorem 4.3 **31.** Use alt. int. $\angle$'s are $=$ with adj. $\angle$'s whose noncommon sides are in a line are supp.   **33.** $x = 90$, $y = 30$ **35.** $x = 10$

4.3

1. (a) triangle (b) 180° (c) 60° (d) 360° (e) 120° (f) 15 cm   **3.** (a) pentagon (b) 540° (c) 108° (d) 360° (e) 72° (f) 25 cm   **5.** (a) heptagon (b) 900° (c) about 128.6° (d) 360° (e) about 51.4° (f) 35 cm   **7.** (a) nonagon (b) 1260° (c) 140° (d) 360° (e) 40° (f) 45 cm   **9.** 11 sides   **11.** There is no polygon.   **13.** 16 sides   **15.** There is no polygon.   **17.** 160°   **19.** decrease   **21.** 60°   **23.** 6 sides   **25.** 4   **27.** Use the fact that if a triangle has two right angles or two obtuse angles then we contradict the fact that the sum of the angles is 180°.   **29.** Use Theorem 4.11, the $\angle$ add. post., and the add.-subtr. law.   **31.** This follows directly from Exercise 30 and Theorem 4.18.   **33.** The desired point is the intersection of the bisectors of $\angle DAB$ and $\angle ABC$.

4.4

1. true   **3.** true   **5.** false   **7.** false   **9.** true   **11.** true   **13.** true   **15.** false   **17.** true   **19.** true   **21.** 1. Sum of $\angle$'s of quadrilateral $= 360°$ 2. Given 3. Substitution law 5. Mult.-div. law 6. $\angle A$ and $\angle B$ are supplementary 7. $\overline{AD} \parallel \overline{BC}$ 8. Substitution law 9. $\overline{AB} \parallel \overline{DC}$ 10. $ABCD$ is a parallelogram   **23.** Use Theorem 4.20.   **25.** Use Corollary 4.21.   **27.** 60° and 120°

29. *Proof:*

STATEMENTS	REASONS
1. $\overline{BE} \perp \overline{AD}$ and $\overline{DF} \perp \overline{BC}$	1. Given
2. $\angle BEA = 90°$ and $\angle DFC = 90°$	2. $\perp$ lines form 90° angles
3. $\angle BEA = \angle DFC$	3. Trans. and sym. laws
4. $\angle A = \angle C$	4. Opp. $\angle$'s of $\square$ are $=$
5. $AB = DC$	5. Sides of rhombus are $=$
6. $\triangle ABE \cong \triangle CDF$	6. AAS = AAS
7. $BE = DF$	7. cpoctae

31. *Proof:*

STATEMENTS	REASONS
1. $AP = QC$	1. Given
2. $ABCD$ is a parallelogram	2. Given
3. $\angle 1 = \angle 2$	3. Alt. int. $\angle$'s are $=$
4. $AB = DC$	4. Opp. sides of $\square$ are $=$
5. $\triangle ABP \cong \triangle DCQ$	5. SAS = SAS
6. $PB = DQ$	6. cpoctae
7. $\angle 3 = \angle 4$	7. Alt. int. $\angle$'s are $=$
8. $AD = BC$	8. Opp. sides of $\square$ are $=$
9. $\triangle APD \cong \triangle BCQ$	9. SAS = SAS
10. $PD = QB$	10. cpoctae
11. $PBQD$ is a parallelogram	11. Opp. sides are $=$

33. Opposite sides of a parallelogram remain parallel even though the parallelogram is distorted.   **35.** Approximately 72 lb

4.5

1. true   **3.** true   **5.** true   **7.** true   **9.** true   **11.** false   **13.** true   **15.** true   **17.** true   **19.** false   **21.** 4.5 cm   **23.** 81°   **25.** 6 cm   **27.** 99°   **29.** 1. By construction 2. $ABCD$ is a trapezoid with median $\overline{EF}$ 3. Def. of median of trapezoid 4. $AE = ED$ and $BF = FC$ 5. Vert. $\angle$'s are $=$ 6. $\overline{AB} \parallel \overline{CD}$ 7. Alt. int. $\angle$'s are $=$ 8. AAS = AAS 9. cpoctae

11. Seg. add. post.   12. Substitution law   15. Line ∥ to one of two ∥ lines is ∥ to other.   **31.** Use Theorems 4.21 and 4.24
33. Construct a diagonal of the quadrilateral; use it as the base of two overlapping triangles, and use Theorem 4.36.   **35.** Use
Theorem 4.36   **37.** Use Construction 4.2   **39.** Yes, by Theorem 4.34.

4.6

1. $A = 12$ cm²; $P = 14$ cm   **3.** $A = 50$ yd²; $P = 32$ yd   **5.** $A = 121$ cm²; $P = 44$ cm   **7.** $A = 36$ yd²; $P = 30$ yd   **9.** A
$= 32$ cm²; $P = 34$ cm   **11.** $A = 337.5$ yd²; $P = 79$ yd   **13.** $A = 15.075$ in²; $P = 16.0$ in   **15.** $A = 29.25$ cm²; $P = 27.5$ cm
17. $A = 80$ yd²; $P = 36$ yd   **19.** $A = 1780$ ft²; $P = 180$ ft   **21.** 1. Given 2. $\overline{AC} \perp \overline{BD}$ 3. Add. prop. of area 4. Form. for
area of △ 5. Area of $\triangle ABC = \frac{1}{2}(d)(BE)$ 6. Substitution law 7. Distributive law 8. $DE + BE = BD = d'$ 9. $A = \frac{1}{2}dd'$

23. Use the definition of median, Theorem 4.39, and Theorem 4.44.   **25.** 216 ft²   **27.** 9.36 ft²   **29.** 64.08 in²   **31.** 7 cm
33. Approximately 2.86 gal   **35.** $1949.44   **37.** $40.83   **39.** Use the fact that the diagonal of a rectangle divides the
rectangle into triangles with the same area.   **41.** 39 ft²   **43.** 12 ft, 10 ft

Chapter 4 Review Exercises

1. The first statement in a direct proof is P, and we form successive statements arriving at Q. The first statement in an indirect proof
is $\sim Q$, and we form successive statements arriving at a contradiction (usually $\sim P$). Thus, our assumption of $\sim Q$ is wrong so we have
Q, and therefore $P \rightarrow Q$.
2.   *Given:* I have the money.
   Prove: I will buy my wife a present.
   Proof:

STATEMENTS	REASONS
1. I have the money	1. Given
2. Assume I don't buy my wife a present.	2. Assumption
3. My wife will be unhappy.	3. Premise 3
4. She won't wash my shirts.	4. Premise 4
5. I can't go to work.	5. Premise 1
6. I won't have any money.	6. Premise 2

This is a contradiction of Statement 1 "I have money." Thus, the assumption that "I don't buy my
wife a present" is incorrect, so we must conclude that I bought her a present.
∴ If I have the money, then I will buy my wife a present.
3. For a given line ℓ and a point P not on ℓ, one and only one line through P is parallel to ℓ.   **4.** yes   **5.** no   **6.** yes
7. no   **8.** yes   **9.** yes   **10.** $\angle 1, \angle 2$, and $\angle 3$ **11.** 30   **12.** 80
13. *Proof:*

STATEMENTS	REASONS
1. $\ell \parallel m$ and $\angle 1 = \angle 2$	1. Given
2. $\angle 3 = \angle 1$ and $\angle 4 = \angle 2$	2. Alt. int. ∠'s are =
3. $\angle 3 = \angle 4$	3. Trans. and sym. laws
4. $AB = AC$	4. Sides opp. = ∠'s are =
5. $\triangle ABC$ is isosceles	5. Def. of isos. △

14. $y = 140, x = -60$ **15.** 720°   **16.** 120°   **17.** 360°   **18.** 60°   **19.** 15 sides   **20.** 22 sides   **21.** 35°   **22.** no
23. no   **24.** They are equal.   **25.** yes   **26.** true   **27.** true   **28.** false   **29.** true   **30.** true   **31.** false
32. false   **33.** true   **34.** true   **35.** true
36. *Proof:*

STATEMENTS	REASONS
1. $\triangle ABF \cong \triangle EFD$	1. Given
2. B is the midpoint of $\overline{AC}$	2. Given
3. $AB = ED$	3. cpoctae
4. $AB = BC$	4. Def. of midpt.
5. $ED = BC$	5. Trans. and sym. laws
6. F is the midpoint of $\overline{AD}$	6. Given
7. $BF = \frac{1}{2}CD$	7. Seg. joining midpts. of sides is = $\frac{1}{2}$ other side

8. $BE = BF + FE$ 8. Seg. Add. post.
9. $BF = FE$ 9. cpoctae
10. $BE = BF + BF$ 10. Substitution law
11. $BE = 2BF$ 11. Distributive law

12. $BF = \dfrac{1}{2}BE$ 12. Mult.-div. post.

13. $\dfrac{1}{2}BE = \dfrac{1}{2}CD$ 13. Substitution law

14. $BE = CD$ 14. Mult.-div. post.
15. $BCDE$ is a parallelogram   15. Opp. sides are $=$

37. yes   **38.** yes   **39.** no   **40.** no   **41.** They are equal.   **42.** no   **43.** Use Construction 4.4.   **44.** Use the fact that the diagonals are perpendicular bisectors of each other and Construction 2.3.
45. *Proof:*

STATEMENTS	REASONS
1. $ABCE$ is a rectangle	1. Given
2. $BCDE$ is a parallelogram	2. Given
3. $AC = BE$	3. Diag. of rect. are $=$
4. $BE = CD$	4. Opp. sides of $\square$ are $=$
5. $AC = CD$	5. Transitive law
6. $\triangle ACD$ is isosceles	6. Def. of isos. $\triangle$

46. 36 ft   **47.** It is multiplied by 4.   **48.** Approximately 2.8 gal of paint.   **49.** 126 cm²   **50.** $A = 2535$ cm²; $P = 253$ cm
51. Show that the diagonals of the resulting parallelogram are equal.   **52.** Use Exercise 22 in Section 4.6.

Chapter 4 Practice Test

1. Exactly one   **2.** true   **3.** false   **4.** true   **5.** true   **6.** 1080°   **7.** 360°   **8.** 135°   **9.** true   **10.** true
11. false   **12.** false   **13.** Use corresponding parts of congruent triangles $\triangle AED$ and $\triangle BFC$.   **14.** Show opposite sides are equal.   **15.** Use Construction 2.2 and a construction similar to Construction 4.3.   **16.** 340 ft²   **17.** 8 yd   **18.** $879.10
19. 88 cm   **20.** Draw a perpendicular segment from P to DC and use the fact that it gives the height of both the parallelogram and the triangle.

CHAPTER 5

5.1

1. $\dfrac{4}{7}$ **3.** $\dfrac{1}{4}$ **5.** 50 mph   **7.** $\dfrac{1}{6}$ **9.** $\dfrac{4}{3}$ **11.** 2   **13.** $\dfrac{7}{4}$ **15.** 5 or -5 **17.** 5   **19.** 4   **21.** 12 or -12 **23.** 24 or -24 **25.** 6500 votes   **27.** 130 mi   **29.** 805 lb   **31.** Take the appropriate measurements and approximate the various ratios.   **33.** Pattern the proof after the ones given in text.   **35.** Pattern the proof after the ones given in text.

5.2

1. 15 cm   **3.** 12 cm   **5.** 103°   **7.** always   **9.** sometimes   **11.** always   **13.** sometimes   **15.** never
17. sometimes   **19.** $EC = 10$ ft   **21.** $AD = 16$ yd, $BD = 6$ yd   **23.** $AD = 10$ ft, $BD = 5$ ft   **25.** $DC = 3$ ft
27. $BD = 18$ in, $DC = 27$ in   **29.** 120 ft   **31.** First show $\triangle AED \sim \triangle EBF$ and use the fact that opposite sides of a rectangle are equal.   **33.** Use the fact that $\triangle ABF \sim \triangle ADP$, $\triangle AFC \sim \triangle APE$, and Theorem 5.12.   **35.** Use Theorem 5.7.   **37.** Use Construction 5.1.   **39.** 3 ft

5.3

1. yes   **3.** no   **5.** yes   **7.** no   **9.** yes   **11.** 25 to 1   **13.** 100 yd²   **15.** Use SAS $\sim$ SAS.   **17.** Use SAS $\sim$ SAS.
19. Use AA $\sim$ AA.

Chapter 5 Review Exercises

1. $\dfrac{4}{5}$ **2.** 60 mph   **3.** 3   **4.** 40   **5.** 2   **6.** 5   **7.** 20   **8.** 20 or -20 **9.** 180 mi   **10.** false   **11.** true

12. true   **13.** 3.5 cm   **14.** 2 cm   **15.** 8 cm   **16.** 95°   **17.** yes   **18.** no   **19.** $EC = 21$ ft   **20.** $AE = 37.5$ ft, $EC = 12.5$ ft   **21.** $FC = 20$ cm   **22.** $BF = 21$ ft, $FC = 56$ ft   **23.** Show $\overline{DE} \parallel \overline{BC}$ and use Theorem 5.12.   **24.** Use Theorem 5.12 and Theorem 5.13.   **25.** Use Construction 5.1.   **26.** They are similar.   **27.** They are not similar.   **28.** 5 yd² **29.** Use SAS ~ SAS.   **30.** Use SSS ~ SSS and alternate interior angles are equal.   **31.** 40 ft

Chapter 5 Practice Test
1. 40 mph   **2.** 7   **3.** 12 or -12 **4.** 91 lb   **5.** true   **6.** 2 ft   **7.** yes   **8.** 7.5   **9.** 8.75   **10.** yes   **11.** no **12.** 80 ft   **13.** Use SSS ~ SSS, then alt. int. ∠'s are $=$.

CHAPTER 6

6.1
1. 2   **3.** -9 **5.** 11   **7.** -15 **9.** 6.24   **11.** 20.76   **13.** $2\sqrt{3}$ **15.** $3\sqrt{5}$ **17.** $5\sqrt{2}$ **19.** $10\sqrt{3}$ **21.** $\dfrac{\sqrt{5}}{5}$ **23.** $\dfrac{\sqrt{5}}{5}$ **25.** $\dfrac{3\sqrt{10}}{5}$ **27.** $\dfrac{\sqrt{10}}{5}$ **29.** 9, -9 **31.** $2\sqrt{2}, -2\sqrt{2}$ **33.** 2, -3 **35.** $\dfrac{1}{2}, -3$ **37.** $-1 \pm \sqrt{2}$ **39.** $\dfrac{1 \pm \sqrt{37}}{6}$ **41.** 9, -9 **43.** $2\sqrt{11}, -2\sqrt{11}$ **45.** 4, -5 **47.** $3\sqrt{70}, -3\sqrt{70}$ **49.** 1. Given   2. Add.-subtr. law 3. Factor and use substitution law   4. Zero-product rule   5. Add.-subtr. law   6. Substitution Law   **51.** 42.8 cm²

6.2
1. LA = LA   **3.** LL = LL   **5.** HL = HL   **7.** 10 cm   **9.** 6 cm   **11.** 10 ft   **13.** 27 cm   **15.** 8 yd   **17.** $\sqrt{30}$ ft, approximately 5.48 ft   **19.** $(-4 + \sqrt{65})$ cm, approximately 4.06 cm   **21.** $4\sqrt{5}$ yd, approximately 8.94 yd   **23.** Use HL = HL.   **25.** Use HL = HL followed by the property stating that sides opposite equal angles in a triangle are equal.   **27.** 80 cm² **29.** 16 cm²   **31.** 30 cm²

6.3
1. 5 cm   **3.** 60 ft   **5.** $\sqrt{85}$ yd   **7.** 18 cm   **9.** yes   **11.** yes   **13.** no   **15.** $10\sqrt{2}$ ft   **17.** $3\sqrt{2}$ yd   **19.** 6 cm **21.** $3\sqrt{6}$ ft   **23.** $3\sqrt{2}$ yd   **25.** $\dfrac{1}{2}$ cm   **27.** 20 ft   **29.** 8 yd   **31.** $7\sqrt{3}$ cm   **33.** 2 ft   **35.** $\dfrac{3}{2}$ yd   **37.** 2 cm **39.** $4\sqrt{2}$ in   **41.** 17.0 ft   **43.** 230.9 ft   **45.** $5\sqrt{3}$ ft   **47.** Let each equal leg have length a with the hypotenuse of length c. Show that $c = \sqrt{2}a$ using the Pythagorean Theorem.   **49.** $d = \sqrt{3}x$ **51.** $\sqrt{3}$ times the length of the original segment. The process can be continued to find a segment $\sqrt{n}$ times the length of the original segment for $n = 2, 3, 4, 5, \ldots$.   **53.** The quadrilateral is clearly a rhombus; prove that it contains a right angle by showing the angle is the supplement of the angle formed by adding the two acute angles of the right triangle.   **55.** (a) Show $\angle DAB$ is a right angle by using an argument similar to that given in Exercise 53.   (b) Use the formula for the area of a triangle three times.   (c) Show that $\overline{ED} \parallel \overline{CB}$ because each is $\perp$ to the same line.   (d) Use the formula for the area of a trapezoid.   (e) When the two expressions are equated, the result simplifies to $a^2 + b^2 = c^2$.

Chapter 6 Review Exercises
1. 5   **2.** -5 **3.** 8   **4.** 9.11   **5.** 12.61   **6.** 97.68   **7.** $2\sqrt{7}$ **8.** $5\sqrt{3}$ **9.** $7\sqrt{2}$ **10.** $\dfrac{\sqrt{7}}{7}$ **11.** $\dfrac{\sqrt{6}}{6}$ **12.** $\dfrac{\sqrt{10}}{2}$ **13.** 8, -8 **14.** 1, -7 **15.** 2, $-\dfrac{1}{5}$ **16.** $\dfrac{-7 \pm \sqrt{17}}{4}$ **17.** $\pm 7\sqrt{2}$ **18.** 8, -10 **19.** $\pm 10\sqrt{30}$ **20.** HA = HA   **21.** HL = HL   **22.** LL = LL   **23.** 22 cm   **24.** 20 ft   **25.** $8\sqrt{7}$ cm   **26.** 4 yd   **27.** 915 ft² **28.** Use the fact that the diagonals of a rhombus are perpendicular, and Corollary 6.9.   **29.** Use the fact that the diagonals of a rhombus are perpendicular, and Corollary 6.10.   **30.** 2400 yd²   **31.** $\sqrt{185}$ cm   **32.** 22 ft   **33.** no   **34.** 60 cm **35.** $20\sqrt{3}$ yd   **36.** 4 ft   **37.** $6\sqrt{3}$ cm   **38.** 3 yd   **39.** $\dfrac{14\sqrt{3}}{3}$ ft   **40.** $\dfrac{1}{5}$ yd   **41.** 10 cm   **42.** $3\sqrt{2}$ ft   **43.** 19.3 ft² **44.** 1 mi   **45.** Use the Pythagorean Theorem to find an altitude and then use the formula for the area of a triangle.

Chapter 6 Practice Test

1. 7   **2.** 20.76   **3.** $4\sqrt{5}$ **4.** $\dfrac{\sqrt{33}}{11}$ **5.** $1, -\dfrac{3}{2}$ **6.** $\pm 10\sqrt{3}$ **7.** Show the triangles are congruent by HL = HL and use corresponding parts of congruent triangles.   **8.** $2\sqrt{15}$ cm   **9.** 13.5 yd   **10.** 4 ft or 16 ft   **11.** $3\sqrt{2}$ cm   **12.** 7 yd   **13.** 17.0 yd   **14.** 5 cm   **15.** $\dfrac{\sqrt{3}}{4}x^2$

CHAPTER 7

7.1

1. 22 in   **3.** $\dfrac{3}{2}$ ft   **5.** 8 in   **7.** $\dfrac{1}{3}$ ft   **9.** $C = 14\pi$ yd; $A = 49\pi$ yd^2 **11.** $C = \dfrac{4}{3}\pi$ cm; $A = \dfrac{4}{9}\pi$ cm^2 **13.** $C = 25.75$ ft; $A = 52.78$ ft^2 **15.** $C = 37.68$ mi; $A = 113.04$ mi^2 **17.** 0.25 in^2 **19.** 12-inch: $0.0398 per in^2; 16-inch: $0.0373 per in^2; the 16-inch pizza costs less   **21.** 235.74 mm^2 **23.** 142.25 cm^2 **25.** (a) central angle (b) inscribed angle (c) 120° (d) 240° (e) 60° (f) 120°   **27.** (a) 64° (b) 296° (c) 64° (d) 32°   **29.** Use the facts that $\angle ABC = 90°$ and that $\angle ABC$ is an inscribed angle.   **31.** Draw auxiliary line segment $\overline{OC}$ and show $\angle COD = \angle DOB$.   **33.** 5.8404×10^8 mi   **35.** 1111.19 mi

7.2

1. 94°   **3.** 60°   **5.** 12 in   **7.** 90°   **9.** 10 cm   **11.** 11 ft   **13.** 4 cm   **15.** 15 cm   **17.** 40°   **19.** 30°   **21.** 4 ft   **23.** 9 cm   **25.** $\widehat{AC} = 20°$; $\angle ABC = 10°$ **27.** Prove that $\angle AED = 90°$ using Theorem 7.5.   **29.** Prove two triangles are congruent; then use the fact that an isosceles triangle has also been formed.   **31.** The proof is similar to that of Theorem 7.6.   **33.** The proof is similar to that of Theorem 7.10.   **35.** 13 in

7.3

1. 35°   **3.** 36°   **5.** 40°   **7.** 35°   **9.** 34°   **11.** 29°   **13.** 50°   **15.** 50°   **17.** 17 cm   **19.** 8 cm   **21.** 12.5 yd   **23.** two   **25.** Apply Theorem 7.18.   **27.** Extend $\overleftrightarrow{AB}$ and $\overleftrightarrow{CD}$ to meet at point P and apply Theorem 7.19 twice.   **29.** The proof is similar to that of Theorem 7.17.   **31.** Use Construction 7.1.   **33.** Use Construction 7.3.   **35.** Use Construction 7.4.   **37.** 89.4 mi   **39.** 6.7 ft

7.4

1. 94°   **3.** 96°   **5.** 90°   **7.** 90°   **9.** Refer to Theorem 7.25.   **11.** Refer to Theorem 7.25.   **13.** Refer to Theorem 7.25.   **15.** Use Construction 4.3 and Construction 7.6.   **17.** Use Construction 3.4 and Construction 7.5.   **19.** 120°   **21.** 72°   **23.** 40°   **25.** 8 sides   **27.** 18 sides   **29.** Use Theorem 7.23 together with the fact that the opposite angles of a parallelogram are equal.   **31.** Use the fact that each side is an equal chord of the circumscribed circle so that all radii of the inscribed circle must be equal.   **33.** Circumscribe a circle around the regular polygon and use the fact that equal chords (sides of the polygon) have equal central angles.   **35.** $4\sqrt{3}$

7.5

1. $A = \dfrac{25}{4}\pi$ cm^2; $L = \dfrac{5}{2}\pi$ cm   **3.** $A = 2\pi$ cm^2; $L = \pi$ cm   **5.** $A = 4.6$ yd^2 **7.** $A = 13.0$ ft^2 **9.** $A = 45.1$ cm^2 **11.** 4.4 cm^2 **13.** 24 yd   **15.** 400 yd^2 **17.** $384\sqrt{3}$ ft^2 **19.** 30 cm   **21.** 452.2 ft^2 **23.** 0.6 cm^2 **25.** Use Theorem 7.8.   **27.** 3612.3 yd^2

Chapter 7 Review Exercises

1. 5.2 cm   **2.** $\dfrac{2}{5}$ yd   **3.** $C = 28.9$ ft; $A = 66.4$ ft^2 **4.** 36.2 in^2 **5.** (a) central angle (b) inscribed angle (c) 50° (d) 310° (e) 25°   **6.** Use the fact that vertical angles are equal.   **7.** 94°   **8.** 135°   **9.** 17 cm   **10.** 10 ft   **11.** They are equal.   **12.** 20°   **13.** 78°   **14.** 9 ft   **15.** Use the fact that equal chords form equal arcs.   **16.** Prove two triangles congruent and use the fact that equal chords form equal arcs.   **17.** 33°   **18.** 35°   **19.** 33°   **20.** 50°   **21.** 38 cm   **22.** $13.\overline{3}$ ft   **23.** Prove $\angle P = \angle B$, making $\triangle PBC$ isosceles.   **24.** Use the fact that tangents to a circle from the same point determine equal tangent segments.   **25.** Use the facts that a radius drawn to a tangent is perpendicular to the tangent and that vertical angles are equal.   **26.** Use the fact that tangents to a circle from the same point determine equal tangent segments.   **27.** Use Construction 7.1.   **28.** 200.1 mi   **29.** 92°   **30.** 90°   **31.** Divide the circle into four equal arcs to form the square.

32. Construct the regular hexagon by inscribing it in a circle; then use Construction 7.6.   **33.** 20°   **34.** 36 sides   **35.** 14 cm
36. Prove two triangles congruent and use cpoctae.   **37.** 20.4 in²   **28.** 9.0 cm²   **39.** 2338.3 yd²   **40.** 754.4 ft²
41. $(8\pi - 16)$ yd²   **42.** 3.6 ft

Chapter 7 Practice Text
1. $C = 39.6$ cm; $A = 124.6$ cm²   **2.** 65°   **3.** 35°   **4.** 3 cm   **5.** 95°   **6.** 4 cm   **7.** 8 cm   **8.** 15 ft   **9.** Use
Construction 7.2.   **10.** 80°   **11.** Use Construction 3.4 and Construction 7.6.   **12.** 45°   **13.** 196 yd²   **14.** 5.8 cm²
15. Use the fact that lines perpendicular to the same line are parallel.

CHAPTER 8

8.1
1. Transitive law   **3.** Division property of inequalities   **5.** Subtraction property of inequalities   **7.** The whole is greater than
its parts   **9.** Transitive law   **11.** $x < 4$ **13.** $x > -\dfrac{7}{3}$ **15.** $x < 8$ **17.** $x < -18$ **19.** false   **21.** true   **23.** true
25. false   **27.** true   **29.** true   **31.** $AC > DE$ **33.** $\angle C > \angle D$ **35.** The triangle inequality theorem prevents its
construction.   **37.** Prove $\triangle QBC \cong \triangle DEF$, which gives $QC = DF$. Use $AC > QC$ and substitute.   **39.** Prove $\triangle ABR \cong$
$\triangle RBQ$, making $AR = RQ$. Use $AC = AR + RC = RQ + RC > QC = DF$. The last equality follows because $\triangle BQC \cong \triangle EDF$.
41. Use the fact that $BC > AC$ and $CE > CD$ with the addition property of inequalities.   **43.** This follows from the triangle
inequality theorem.   **45.** Use the triangle inequality theorem three times.   **47.** Use the hint and the triangle inequality
theorem.   **49.** (a) 62 mi and 1240 mi (b) 217 mi and 1519 mi (c) 279 mi and 1457 mi

8.2
1. true   **3.** true   **5.** false   **7.** true   **9.** $\overset{\frown}{AB} > \overset{\frown}{CD}$ **11.** $\overset{\frown}{AB} > \overset{\frown}{CD}$ **13.** $AB > CD$ **15.** $\overline{CD}$ **17.** Use the
technique illustrated in the proof of Theorem 8.11 and refer to the proof given for the same circle in text.   **19.** Use the technique
illustrated in the proof of Theorem 8.11 and Exercise 18.   **21.** Proof is similar to the one in Practice Exercise 1.   **23.** First
prove the result for the same circle; then use the information in the proof for congruent circles.   **25.** Show that the side of a
square is shorter than a side of the triangle and apply Theorem 8.11 and the definition of the apothem.   **27.** The proof follows
from Theorem 8.3 followed by Theorem 8.9.

Chapter 8 Review Exercises
1. Transitive law   **2.** Subtraction property of inequalities   **3.** Division property of inequalities   **4.** Trichotomy law
5. The whole is greater than its parts   **6.** $x < -17$ **7.** $y > -7$ **8.** false   **9.** true   **10.** true   **11.** false   **12.** true
13. $AC < DE$ **14.** no   **15.** Use SAS inequality theorem.   **16.** true   **17.** false   **18.** false   **19.** false   **20.** true
21. no   **22.** Use the fact that $\triangle ABC$ is a right triangle.

Chapter 8 Practice Test
1. Subtraction property of inequalities   **2.** $x < -2$ **3.** $\angle E > \angle B$ **4.** $\angle C < \angle F$ **5.** 18 ft; 2 ft   **6.** $\angle 1 > \angle 2$
7. $\overline{UV}$ **8.** $XY > UV$ **9.** Use the SAS inequality theorem.   **10.** Show that $AB > BC$.

CHAPTER 9

9.1
1. true   **3.** false   **5.** true   **7.** false   **9.** true   **11.** true   **13.** $f = 7, v = 6, e = 11$ **15.** $f = 6, v = 6, e = 10$
17. $f = 8, v = 6, e = 12$ **19.** 20   **21.** 90 m   **23.** 4   **25.** No, consider the angles.

9.2
Some answers are determined using a calculator with no rounding until the final step.   **1.** true   **3.** true   **5.** 150 ft²
7. 104 ft²   **9.** 19.4 yd²   **11.** 5697 in²   **13.** 512 m³   **15.** 12.4 cm³   **17.** 1124.6 ft³   **19.** 2248 m³   **21.** 924 m²
23. 26,293 cm²   **25.** 2420 yd³   **27.** 13,033 cm³   **29.** 46.8 yd³   **31.** 2420 cm²   **33.** approximately 51 gal   **35.** $d =$
$\sqrt{\ell^2 + w^2 + h^2}$

9.3
Some answers are determined using a calculator and 3.14 for π with no rounding until the final step.   **1.** true   **3.** 414 cm²
5. 3233.2 in²   **7.** 361.5 m²   **9.** 1231 cm³   **11.** 23,436.6 in³   **13.** 2019.0 m³   **15.** 75 yd²   **17.** 6331.5 in²   **19.** 2211 ft²

21. 38 in³   **23.** 707.6 cm³   **25.** 14,503 m³   **27.** 53.6 cm³; 107.2 cm³   **29.** 3514 in²; 14,469 in³   **31.** 1.6 cans   **33.** 2.2 gal   **35.** 1.8 gal

9.4
Some answers are determined using a calculator and 3.14 for π with no rounding until the final step.   **1.** 1809 m²   **3.** 7.07 ft²   **5.** 393.9 yd²   **7.** 1436 ft³   **9.** 3.05 m³   **11.** 11,002.9 yd³   **13.** 72 cm²   **15.** 244 cm³   **17.** $V_{cube} = 125$ in³; $V_{sph} = 173$ in³; V_{sph} is larger   **19.** $r = 3$ units   **21.** 137,189 ft³   **23.** $1393.41   **25.** 490.7 ft³   **27.** 1008.1 m³   **29.** 93.0 g

Chapter 9 Review Exercises
Some answers are determined using a calculator and 3.14 for π with no rounding until the final step.   **1.** true   **2.** true   **3.** true   **4.** true   **5.** false   **6.** false   **7.** $f = 6, v = 8, e = 12$ **8.** $f = 12, v = 8, e = 18$ **9.** 24   **10.** $49.50   **11.** 942.6 cm²   **12.** 191.4 ft²   **13.** 6434.9 in³   **14.** 519.9 m³   **15.** 118,221 cm²   **16.** 327.2 ft²   **17.** 1038.8 in³   **18.** 34,835.6 m³   **19.** 810 cm²   **20.** 2.9 in³   **21.** 127.1 ft²   **22.** 5418.6 cm²   **23.** 18,294,111 in³   **24.** 89,464.0 m²   **25.** 305.6 cm²   **26.** 3067.8 ft²   **27.** 3292.5 m³   **28.** 20,561.6 in³   **29.** $SA = 397.2$ yd²; $V = 525.0$ yd³   **30.** 457.1 m³   **31.** 191.0 ft²   **32.** 9021.1 m²   **33.** 3.26 cm³   **34.** 8987.5 in³   **35.** 208.8 cm³   **36.** 94 m³   **37.** 14,862.2 g   **38.** 3468.1 ft²   **39.** 88,510.3 kg   **40.** $321.52

Chapter 9 Practice Test
1. true   **2.** true   **3.** true   **4.** true   **5.** 20   **6.** 447.5 cm²   **7.** 26,914.3 in³   **8.** 1569.7 m²   **9.** 59.5 ft²   **10.** 4390.0 yd³   **11.** 52 in²   **12.** 16,264.4 m³

CHAPTER 10

10.1
1. circle with center P and radius 5   **3.** two circles concentric with the given circle and with radii 1 unit and 7 units   **5.** a line parallel to the given lines and 1.5 units from each of them   **7.** no points if m does not intersect the circle; one point if m is tangent to the circle; two points if m passes through the circle   **9.** a diameter (excluding the endpoints) that is perpendicular to the given chords   **11.** one point that is the intersection of the line equidistant from the given parallel lines and the given intersecting line   **13.** Given a center of a circle tangent to both lines, then it is a radius distance from each of the lines. Given a point on the line equidistant between the parallel lines, there is a circle tangent to both lines with the given point as center.   **15.** Given right $\triangle ABC$, show C must be on $\overset{\frown}{AEB}$ because $\angle ACB = 90°$ and $\overset{\frown}{ADB} = 180°$. Given C on $\overset{\frown}{AEB}$, use $\overset{\frown}{ADB} = 180°$ to show $\angle ACB = 90°$ **17.** The set of all points on and interior to the circle centered at the station with radius 70 mi.

10.2
1. EH **3.** 6 units   **5.** Construct $\perp$ bisectors of $\overline{AB}$ and $\overline{BC}$ and use the intersection of the bisectors as center of the circle.   **7.** 8 units   **9.** $2\sqrt{3}$ units   **11.** $10\sqrt{3}$ m   **13.** Supply details to outline in text. (See Figure 10.11.)   **15.** The location of the distribution center should be at the circumcenter of the triangle with vertices at the three stores.

Chapter 10 Review Exercises
1. two circles concentric with given circle and with radii 8 units and 12 units   **2.** one circle concentric with the given circle and with radius 10 units   **3.** one point at the intersection of the bisector of the angle and the circle   **4.** no points if m does not intersect the square; one point if m passes through a vertex only; two points if m intersects two sides; and a side of the square if m is colinear with the side   **5.** Construct two chords of the given length and use congruent triangles to show that midpoints are all the same distance from center. If a point is that distance from the center, show that a chord of the fixed length can be constructed with the point as its midpoint.   **6.** Construct $\perp$ bisectors of $\overline{AB}$ and $\overline{BC}$ that intersect at the center point of the desired circle.   **7.** $5\sqrt{3}$ cm, $10\sqrt{3}$ cm   **8.** 8 units   **9.** 5 units   **10.** Show that the angle bisectors and $\perp$ bisectors of the sides coincide.

Chapter 10 Practice Test
1. line parallel to the given lines and 3 units from each   **2.** no points if the circles do not intersect; one point if the centers are 2 units or 10 units apart; two points if the circles intersect at two points   **3.** Show that midpoints lie on a diameter and that given a diameter the points are midpoints of parallel chords.   **4.** 7 cm, 14 cm   **5.** 3 cm   **6.** Show that the triangles formed by the medians are congruent triangles.   **7.** The locus is all points on and interior to a semicircle with radius 8 ft centered at the midpoint of the fence.

CHAPTER 11

11.1

1.

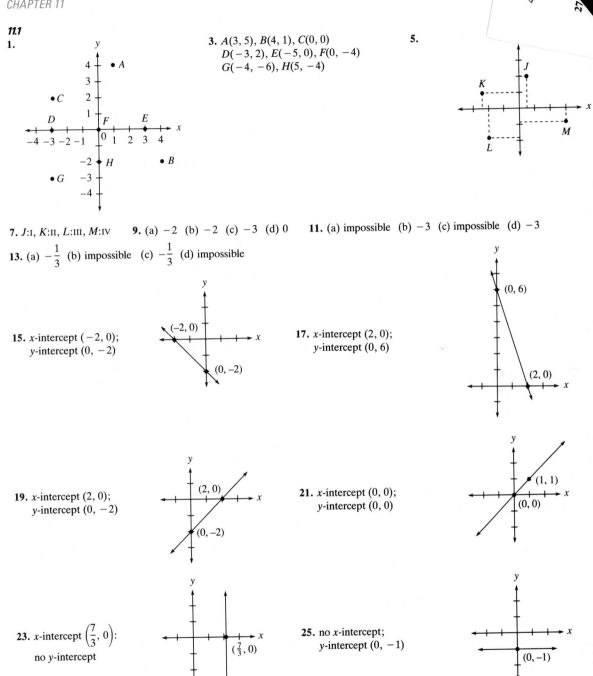

3. $A(3, 5)$, $B(4, 1)$, $C(0, 0)$
$D(-3, 2)$, $E(-5, 0)$, $F(0, -4)$
$G(-4, -6)$, $H(5, -4)$

5.

7. J:I, K:II, L:III, M:IV   **9.** (a) -2 (b) -2 (c) -3 (d) 0 **11.** (a) impossible (b) -3 (c) impossible (d) -3

13. (a) $-\dfrac{1}{3}$ (b) impossible   (c) $-\dfrac{1}{3}$ (d) impossible

15. x-intercept $(-2, 0)$;
y-intercept $(0, -2)$

17. x-intercept $(2, 0)$;
y-intercept $(0, 6)$

19. x-intercept $(2, 0)$;
y-intercept $(0, -2)$

21. x-intercept $(0, 0)$;
y-intercept $(0, 0)$

23. x-intercept $\left(\dfrac{7}{3}, 0\right)$:
no y-intercept

25. no x-intercept;
y-intercept $(0, -1)$

. *x*-intercept $(0, 0)$;
   y-intercept $(0, 0)$

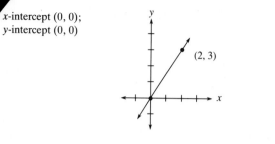

(2, 3)

29. *x*-intercept $(2, 0)$;
   y-intercept $(0, 3)$

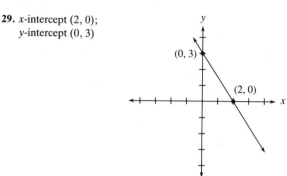

(0, 3)

(2, 0)

31. *x*-intercept $(0, 0)$; every point on the *y*-axis is a *y*-intercept; the graph is the *y*-axis.   **33.** All have the same *y*-intercept, $(0, 1)$.
35. \$475; 200;

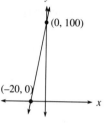

(0, 100)

(−20, 0)

11.2

1. -2 **3.** -2 **5.** 0   **7.** $2x + y - 5 = 0$ **9.** $x - 3y - 14 = 0$ **11.** $2x + y - 13 = 0$ **13.** $x - 2 = 0$
15. $y - 3 = 0$ **17.** $2x - y + 10 = 0$ **19.** $3x + y - 4 = 0$ **21.** $5x - 9y - 29 = 0$ **23.** $5x - 2y + 10 = 0$
25. slope $\frac{5}{2}$; *y*-intercept $(0, 2)$ **27.** slope undefined; no *y*-intercept   **29.** slope 0; *y*-intercept $(0, 2)$ **31.** $y = -2x - 1$;
slope -2; *y*-intercept $(0, -1)$ **33.** perpendicular   **35.** parallel   **37.** perpendicular   **39.** $3x + y - 1 = 0$
41. $x - 3y + 13 = 0$ **43.** 13   **45.** $\sqrt{170}$ **47.** 5   **49.** $(1, 0)$ **51.** $\left(\frac{11}{2}, 4\right)$ **53.** $(3, -1)$ **55.** $3x - 4y - 8 = 0$
57. $x + 2y - 4 = 0$ **59.** $x + y = 0$ **61.** $665x - 3y + 1750 = 0$; \$7455

11.3

1. $(x + 3)^2 + (y - 2)^2 = 1^2$; $x^2 + y^2 + 6x - 4y + 12 = 0$ **3.** $(x - 5)^2 + (y + 3)^2 = 2^2$; $x^2 + y^2 - 10x + 6y + 30 = 0$
5. $(x + 1)^2 + (y + 1)^2 = 3^2$; $x^2 + y^2 + 2x + 2y - 7 = 0$ **7.** $\left(x + \frac{1}{2}\right)^2 + y^2 = 1^2$; $x^2 + y^2 + x - \frac{3}{4} = 0$ or $4x^2 + 4y^2 + 4x -$
$3 = 0$ **9.** $x^2 + y^2 = 1^2$; $x^2 + y^2 - 1 = 0$ **11.** $(x + 2)^2 + (y + 4)^2 = 6^2$; $x^2 + y^2 + 4x + 8y - 16 = 0$ **13.** $(x - 1)^2 +$
$(y + 2)^2 = 5^2$; $(1, -2)$; $r = 5$ **15.** $(x + 3)^2 + y^2 = 3^2$; $(-3, 0)$; $r = 3$ **17.** $(x + 5)^2 + \left(y - \frac{1}{2}\right)^2 = 4^2$; $\left(-5, \frac{1}{2}\right)$; $r = 4$

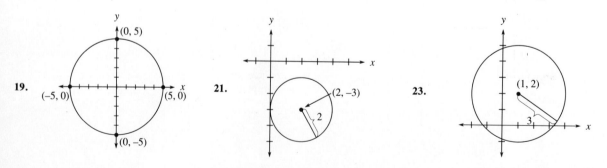

19.
(0, 5)
(−5, 0)
(5, 0)
(0, −5)

21.
(2, −3)
2

23.
(1, 2)
3

25. **27.** **29.**

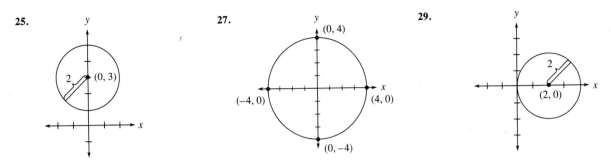

11.4

1. Construct a rectangle with vertices $(0, 0)$, $(a, 0)$, (a, b), and $(0, b)$. Use the distance formula to show the diagonals equal.
3. Label the vertices $(0, 0)$, $(a, 0)$, (b, c), and (d, c). Show that $d = a - b$ and use the distance formula on diagonals.   **5.** Label the vertices $P(0, 0)$, $A(a, 0)$, and $B(b, c)$, where $\angle B$ is the vertex angle. Show that $b = \dfrac{a}{2}$ and that the perpendicular bisector of the base $\overline{PA}$ passes through all points with x-coordinate $\dfrac{a}{2}$.   **7.** Use Theorem 4.36.

Chapter 11 Review Exercises

1. x-coordinate: -2; y-coordinate: 3; II   **2.** axes   **3.** graph of the equation   **4.** y-axis   **5.** $A(-3, 2)$; $B(2, 1)$; $C(2, -2)$; $D(-2, -2)$

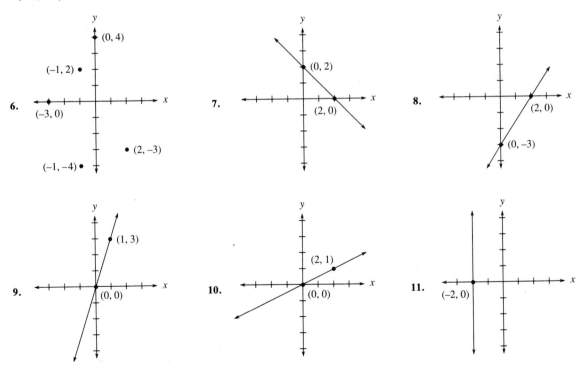

12.

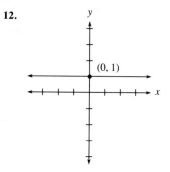

(0, 1)

13. $\dfrac{9}{2}$ **14.** -2 **15.** 5; $-\dfrac{1}{5}$ **16.** $-\dfrac{8}{5}$; $\dfrac{5}{8}$ **17.** 6 **18.** $(2, -1)$ **19.** slope-intercept; slope   **20.** $4x + y + 1 = 0$

21. $y = -2x + 5$; -2; $(0, 5)$ **22.** $x - 2y + 12 = 0$ **23.** $x - 2y + 5 = 0$ **24.** no x-intercept; y-intercept $(0, 5)$; slope 0

25. perpendicular   **26.** $10x - 3y + 7 = 0$; 29 units   **27.** $(x + 5)^2 + (y - 2)^2 = 4^2$, $x^2 + y^2 + 10x - 4y + 13 = 0$

28. $(x - 2)^2 + (y - 6)^2 = 3^2$, $x^2 + y^2 - 4x - 12y + 31 = 0$ **29.** $\left(x - \dfrac{1}{2}\right)^2 + \left(y - \dfrac{1}{2}\right)^2 = 2^2$, $2x^2 + 2y^2 - 2x - 2y - 7 = 0$

30. $x^2 + (y - 5)^2 = 1^2$, $x^2 + y^2 - 10y + 24 = 0$ **31.** $(x - 2)^2 + (y + 8)^2 = 7^2$; $(2, -8)$; $r = 7$ **32.** $\left(x + \dfrac{1}{2}\right)^2 + (y - 3)^2$

$= 1^2$; $\left(-\dfrac{1}{2}, 3\right)$; $r = 1$

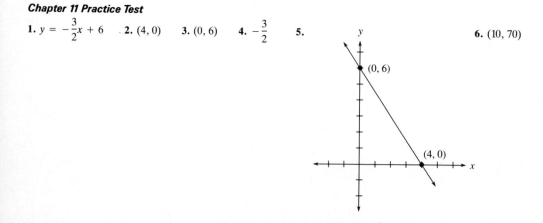

33.

3

(1, 3)

34.

4

(−2, 2)

35. Construct a quadrilateral with one vertex $(0, 0)$ and one side on the x-axis. Label the other vertices and determine the coordinates of the midpoints. Show that the slopes of opposite sides of the quadrilateral with the midpoints as vertices are equal.

Chapter 11 Practice Test

1. $y = -\dfrac{3}{2}x + 6$ **2.** $(4, 0)$ **3.** $(0, 6)$ **4.** $-\dfrac{3}{2}$ **5.**   **6.** $(10, 70)$

(0, 6)

(4, 0)

7.

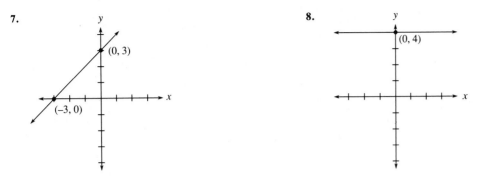

8.

9. (a) -2 (b) $\dfrac{1}{2}$ (c) $6\sqrt{5}$ (d) $(1, 1)$ **10.** parallel   **11.** $3x + y + 2 = 0$ **12.** $2x + y - 5 = 0$ **13.** $(x + 2)^2 +$

$(y + 1)^2 = 5^2; x^2 + y^2 + 4x + 2y - 20 = 0$ **14.** $(x - 4)^2 + (y + 2)^2 = 3^2; x^2 + y^2 - 8x + 4y + 11 = 0$ **15.** $(x + 7)^2 +$

$(y - 6)^2 = 5^2; (-7, 6); r = 5$

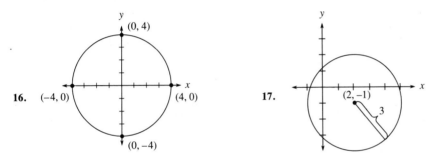

16. $(-4, 0)$ **17.**

18. Construct a quadrilateral with one vertex $(0, 0)$ and one side on the x-axis. Label the vertices and determine the midpoints of the sides. Find the midpoint of each of the two lines. Because the midpoints are the same, the point is the intersection of the two lines.

CHAPTER 12

12.1

1. $\sin A = \dfrac{5}{13}; \cos A = \dfrac{12}{13}; \tan A = \dfrac{5}{12}; \sin B = \dfrac{12}{13}; \cos B = \dfrac{5}{13}; \tan B = \dfrac{12}{5}$ **3.** $\sin A = \dfrac{7}{15}; \cos A = \dfrac{4\sqrt{11}}{15}; \tan A = \dfrac{7\sqrt{11}}{44};$

$\sin B = \dfrac{4\sqrt{11}}{15}; \cos B = \dfrac{7}{15}; \tan B = \dfrac{4\sqrt{11}}{7}$ **5.** $\sin A = 0.6; \cos A = 0.8; \tan A = 0.75; \sin B = 0.8; \cos B = 0.6; \tan B = 1.\overline{3}$

7. $\cos A = \dfrac{2\sqrt{6}}{7}; \tan A = \dfrac{5\sqrt{6}}{12}$ **9.** $\sin A = \dfrac{\sqrt{5}}{3}; \cos A = \dfrac{2}{3}$ **11.** $\sin A = \dfrac{4}{7}; \tan A = \dfrac{4\sqrt{33}}{33}$ **13.** 0.9063 **15.** 0.7813

17. 0.1080 **19.** $7.2°$ **21.** $42.9°$ **23.** $84.1°$

12.2

1. $\angle B = 30°, b = 5, c = 10$ **3.** $\angle B = 70°, a = 4, b = 11$ **5.** $b = 5, \angle A = 67°, \angle B = 23°$ **7.** $a = 13, \angle A = 62°, \angle B$

$= 28°$ **9.** $b = 4.2, \angle A = 65.6°, \angle B = 24.4°$ **11.** $\angle B = 63.3°, a = 5.4, b = 10.7$ **13.** $\angle A = 79.2°, a = 16.8, c = 17.1$

15. $\angle B = 67.5°, a = 11.0, b = 26.5$ **17.** $\angle A = 8.3°, a = 3.2, b = 21.7$

12.3

1. (a) 280 ft (b) 3362 ft²   **3.** (a) 18.6 mi (b) 16.6 mi²   **5.** 46 ft   **7.** 133.7 ft   **9.** 951 ft   **11.** 9.4 mi   **13.** 129 km east, 153 km south   **15.** 9°   **17.** 31.5 ft   **19.** 1399 ft

Chapter 12 Review Exercises

1. $\sin A = \dfrac{4}{5}$, $\cos A = \dfrac{3}{5}$, $\tan A = \dfrac{4}{3}$, $\sin B = \dfrac{3}{5}$, $\cos B = \dfrac{4}{5}$, $\tan B = \dfrac{3}{4}$ **2.** $\sin A = \dfrac{3}{4}$, $\cos A = \dfrac{\sqrt{7}}{4}$, $\tan A = \dfrac{3\sqrt{7}}{7}$, $\sin B =$

$\dfrac{\sqrt{7}}{4}$, $\cos B = \dfrac{3}{4}$, $\tan B = \dfrac{\sqrt{7}}{3}$ **3.** $\sin A = \dfrac{7\sqrt{58}}{58}$, $\cos A = \dfrac{3\sqrt{58}}{58}$ **4.** $\sin B = \dfrac{2\sqrt{2}}{3}$, $\tan B = 2\sqrt{2}$ **5.** 0.3584   **6.** 0.4099

7. 7.2066   **8.** 12.7°   **9.** 24.4°   **10.** 58.7°   **11.** $\angle B = 60°$, $a = 3$, $c = 7$ **12.** $\angle A = 48°$, $a = 9$, $b = 8$ **13.** $\angle B = 17.7°$, $b = 0.8$, $c = 2.5$ **14.** $c = 13$, $\angle A = 32°$, $\angle B = 58°$ **15.** $a = 16$, $\angle A = 53°$, $\angle B = 37°$ **16.** $b = 7.3$, $\angle A = 40.2°$, $\angle B = 49.8°$ **17.** (a) 34 yd (b) 50 yd²   **18.** 58.2 m   **19.** 8.4 mi from B, 6.6 mi from A **20.** 17 mi west, 47 mi south

Chapter 12 Practice Test

1. $\sin A = \dfrac{5}{9}$, $\cos A = \dfrac{2\sqrt{14}}{9}$, $\tan A = \dfrac{5\sqrt{14}}{28}$, $\sin B = \dfrac{2\sqrt{14}}{9}$, $\cos B = \dfrac{5}{9}$, $\tan B = \dfrac{2\sqrt{14}}{5}$ **2.** $\cos A = \dfrac{\sqrt{21}}{5}$, $\tan A = \dfrac{2\sqrt{21}}{21}$ **3.** $\sin 68.9° = 0.9330$, $\cos 68.9° = 0.3600$, $\tan 68.9° = 2.5916$ **4.** $\angle A = 83.7°$ **5.** $b = 4.4$, $\angle A = 14.1°$, $\angle B = 75.9°$ **6.** $\angle B = 52°$, $a = 12$, $c = 19$ **7.** (a) 27 cm (b) 32 cm²   **8.** 93 mi north, 285 mi east

INDEX

Table of Squares and Square Roots

n	n^2	$\sqrt{n}$	$\sqrt{10n}$	n	n^2	$\sqrt{n}$	$\sqrt{10n}$
1	1	1.000	3.162	51	2601	7.141	22.583
2	4	1.414	4.472	52	2704	7.211	22.804
3	9	1.732	5.477	53	2809	7.280	23.022
4	16	2.000	6.325	54	2916	7.348	23.238
5	25	2.236	7.071	55	3025	7.416	23.452
6	36	2.449	7.746	56	3136	7.483	23.664
7	49	2.646	8.367	57	3249	7.550	23.875
8	64	2.828	8.944	58	3364	7.616	24.083
9	81	3.000	9.487	59	3481	7.681	24.290
10	100	3.162	10.000	60	3600	7.746	24.495
11	121	3.317	10.488	61	3721	7.810	24.698
12	144	3.464	10.954	62	3844	7.874	24.900
13	169	3.606	11.402	63	3969	7.937	25.100
14	196	3.742	11.832	64	4096	8.000	25.298
15	225	3.873	12.247	65	4225	8.062	25.495
16	256	4.000	12.649	66	4356	8.124	25.690
17	289	4.123	13.038	67	4489	8.185	25.884
18	324	4.243	13.416	68	4624	8.246	26.077
19	361	4.359	13.784	69	4761	8.307	26.268
20	400	4.472	14.142	70	4900	8.367	26.458
21	441	4.583	14.491	71	5041	8.426	26.646
22	484	4.690	14.832	72	5184	8.485	26.833
23	529	4.796	15.166	73	5329	8.544	27.019
24	576	4.899	15.492	74	5476	8.602	27.203
25	625	5.000	15.811	75	5625	8.660	27.386
26	676	5.099	16.125	76	5776	8.718	27.568
27	729	5.196	16.432	77	5929	8.775	27.749
28	784	5.292	16.733	78	6084	8.832	27.928
29	841	5.385	17.029	79	6241	8.888	28.107
30	900	5.477	17.321	80	6400	8.944	28.284
31	961	5.568	17.607	81	6561	9.000	28.460
32	1024	5.657	17.889	82	6724	9.055	28.636
33	1089	5.745	18.166	83	6889	9.110	28.810
34	1156	5.831	18.439	84	7056	9.165	28.983
35	1225	5.916	18.708	85	7225	9.220	29.155
36	1296	6.000	18.974	86	7396	9.274	29.326
37	1369	6.083	19.235	87	7569	9.327	29.496
38	1444	6.164	19.494	88	7744	9.381	29.665
39	1521	6.245	19.748	89	7921	9.434	29.833
40	1600	6.325	20.000	90	8100	9.487	30.000
41	1681	6.403	20.248	91	8281	9.539	30.166
42	1764	6.481	20.494	92	8464	9.592	30.332
43	1849	6.557	20.736	93	8649	9.644	30.496
44	1936	6.633	20.976	94	8836	9.695	30.659
45	2025	6.708	21.213	95	9025	9.747	30.822
46	2116	6.782	21.448	96	9216	9.798	30.984
47	2209	6.856	21.679	97	9409	9.849	31.145
48	2304	6.928	21.909	98	9604	9.899	31.305
49	2401	7.000	22.136	99	9801	9.950	31.464
50	2500	7.071	22.361	100	10000	10.000	31.623